INTERNATIONAL ENVIRONMENTAL AGREEMENTS AND DOMESTIC POLITICS

International Environmental Agreements and Domestic Politics

The case of acid rain

Edited by
ARILD UNDERDAL
University of Oslo

KENNETH HANF
University of Barcelona

Routledge
Taylor & Francis Group

LONDON AND NEW YORK

First published 2000 by Ashgate Publishing

Reissued 2019 by Routledge
2 Park Square, Milton Park, Abingdon, Oxon, OX14 4RN
52 Vanderbilt Avenue, New York, NY 10017

Routledge is an imprint of the Taylor & Francis Group, an informa business

Notice:
Product or corporate names may be trademarks or registered trademarks, and are used only for identification and explanation without intent to infringe.

Publisher's Note
The publisher has gone to great lengths to ensure the quality of this reprint but points out that some imperfections in the original copies may be apparent.

Disclaimer
The publisher has made every effort to trace copyright holders and welcomes correspondence from those they have been unable to contact.

A Library of Congress record exists under LC control number:

ISBN 13: 978-1-138-71308-6 (hbk)
ISBN 13: 978-1-138-71306-2 (pbk)
ISBN 13: 978-1-315-19956-6 (ebk)

Contents

List of Figures and Tables

Contributors

Karin Bäckstrand is a doctoral student at the Department of Political Science, Lund University, working on a dissertation analysing the use of scientific inputs in the efforts to regulate transboundary air pollution.

Sonja Boehmer-Christiansen is Reader of Geography at the University of Hull.

Caroline du Monteuil was during this work a doctoral student at Science and Technology Policy Research Unit (SPRU), University of Sussex.

Kenneth Hanf was until recently Senior Lecturer in Public Administration at the Erasmus University in Rotterdam. He is presently teaching at the Pompeu Fabra University of Barcelona. He is also engaged in research at the Institute of Advanced Social Studies of the Spanish Superior Council of Scientific Research (IESA-CSIC), also in Barcelona.

Petteri Hiienkoski is Lecturer in Communication and Political Studies at the Helsinki Evangelical College. During most of this project he worked as a Research Fellow at the Finnish Institute of International Affairs.

Stephan Kux is a senior lecturer in Political Science at the University of Zürich.

Torunn Laugen started her work on this project as a student at the Fridtjof Nansen Institute, in Oslo. She is now working in the Norwegian Ministry of Defense.

Rodolfo Lewanski is Senior Lecturer in Public Administration at the Faculty of Political Science, University of Bologna.

Joaquín López-Novo is Lecturer in Sociology at Universidad Complutense de Madrid.

Francesc Morata is Professor of Political Science and Public Administration at the Autonomous University of Barcelona.

Walter Schenkel has a Ph.D. in political science from the University of Zürich.

Henrik Selin is a doctoral student at the Department of Water and Environmental Studies, Linköping University, working on a dissertation analysing the politics of regulating emissions of persistent organic pollutants within the LRTAP regime.

Jim Skea is Director of the Policy Studies Institute, London. Until recently, he was Director of the Economic and Social Research Council's (ESRC's) Global Environmental Change Programme and a Professorial Fellow in the Environment Programme at SPRU, University of Sussex.

Detlef F. Sprinz is a Senior Research Fellow at the Potsdam Institute for Climate Impact Research (PIK), and teaches international relations at the University of Potsdam.

Arild Underdal is Professor of Political Science at the University of Oslo and Senior Research Fellow at the Center for International Climate and Environmental Research (CICERO).

Andreas Wahl was during his work on this project a research assistant at the Potsdam Institute for Climate Impact Research (PIK). He has subsequently joined the civil service of the City of Berlin.

Preface

This book represents the last step along a long road that began with the first of a series of meetings organised within the European Science Foundation's Scientific Programme on Environment, Science and Society. Building on the ideas and experiences exchanged in the course of these meetings, a proposal was submitted to the European Commission for research into the domestic underpinnings of international co-operation. Research on the domestic bases of international environmental agreements and the modelling of national-international linkages was subsequently carried out under the grant EV5V-CT92-0185 from the European Union's Environ-mental Research Programme (Economic and Social Aspects of the Environment). Supplementary grants from national research councils enabled us to include also countries that were not EU members. We would like to thank all these funding agencies, and in particular DG XII of the European Commission for its help it getting the project off the ground and keeping in the right track during its lifetime. Our special debt of gratitude is due to Angela Liberatore of DG XII, who had the thankless task of prodding us to meet long past deadlines and to complete missing reports. Angela acquitted herself of this job with forbearance and patience. She was able to combine effectively firmness and a sense of humour. But her contribution was more than that of an able administrator; she also made important intellectual contributions to its development.

Of course, no project can succeed without the individual and joint efforts of the researcher teams involved. Ultimately, it is to them that the credit is due for the completion of the research tasks we perhaps all too light-heartedly had assumed. Their willingness to undertake the revisions of the original country studies made this book possible. Their constant queries regarding the status of the manuscript provided the extra amount of pressure to keep us on course.

The final research report submitted to the Commission also contained a case study on the Netherlands. Although it was not possible to revise that study in time for this book, we would nevertheless like to thank its authors, Jos Lippman and Julie Raadschelders, for their contributions to the project. Their material is used in the comparative analysis. The initial comparative analysis of the national experiences also drew on data and information from a case study on Hungary carried out by Larry O'Toole. The results of this study are contained in a separate book, entitled *Institutions, Policy and Outputs for Acidification: The Case of Hungary,* published by Ashgate in 1998.

A word of special thanks is due to Tarje Johnsen, who juggled his political science studies with the erratic arrival of still new chapters and requests for revisions, to produce the camera-ready copy for the book.

Kenneth Hanf   Arild Underdal
Barcelona   Oslo

1 The Domestic Basis of International Environmental Agreements

KENNETH HANF

There can be no doubt that governments of industrial countries, by and large, intervene more actively in order to preserve environmental quality now than they did even a short while ago. Public concern over environmental degradation has grown significantly (although by no means steadily) during the last couple of decades, and most governments have responded by introducing new and more ambitious legislation and by strengthening their institutional capacity for managing environmental resources. Most recently, countries throughout the world are confronted with a growing volume of transboundary pollution.

The main concern of the research reported on here is not so much the causes of this proliferation of international policy problems, but rather the kinds of political processes and management problems that such issues generate at the domestic level in the area of environmental policy. In particular, we are interested in the way in which the international dimensions of environmental management *penetrate* domestic society and politics, and how, in turn, these processes within the nation state *feed back* upon the operation and further development of international agreements and institutions. A good deal of promising work on conceptualising institutional arrangements for international co-operation (including in the area of environmental policy) has been done by scholars using the concept of international regimes. We, too, believe that the international regime approach offers a fruitful avenue for obtaining insights into the processes through which institutional arrangements for collaboration among nation states emerge and operate. Initially, much of the work in this field tended to down play the domestic underpinnings of such regimes by stressing interactions occurring at the international level. Moreover, little systematic attention was paid to the implementation of agreements once arrived at. In recent years, several major studies have begun to fill in some of the gaps of earlier work (see e.g. Chayes and Chayes, 1995; Victor, Raustiala and Skolnikoff, 1998; Weiss and Jacobson, 1998). We think that the research recently completed on the experiences with the acidification regime makes a contribution to a more complete understanding of the interaction between international and domestic politics and institutions throughout all phases of the formation and operation of such international regimes.

The research reported on in this book focuses primarily on the conditions for creating such institutional arrangements for developing and implementing effective co-operative solutions to international environmental problems. An analytical framework has been developed and applied to the case of the transboundary air pollution regime, commonly referred to as the 'acid rain regime'. As we will see in the following chapter, this regime combines activities, and the resulting agreements and commitments, within and under the aegis of both the United Nations' Economic Commission of Europe (ECE) and the European Community (EC). In this introductory chapter we examine briefly the general problem of international co-operation and sketch the main lines of a perspective for analysing the interaction between international and domestic politics in the formation and operation of international environmental regimes. These ideas are developed further in the model presented in Chapter 3, which serves as the analytical framework for the national case studies in the book.

Ecological Interdependence and the Problem of International Co-operation

It certainly comes as no surprise to hear that environmental problems do not respect national borders. To the extent that this is so, it is equally clear that the state of the environment in any given country will not be the result of the national efforts of its government alone. This means that the quality of life we enjoy is determined to a significant extent by the activities of citizens of other countries. Consequently, dealing effectively with transboundary environmental problems requires that we find some way of gaining the co-operation of the governments of these countries so as to affect the behaviour of those members of their societies who are the cause of our discomfort.

There is a number of different kinds of international environmental problems which share the characteristic of creating situations in which 'groups of interdependent actors are likely to suffer mutual losses or fail to reap joint gains in the absence of an effective governance system' (Young, 1994, p.20). A defining characteristic of such problems is that they enmesh the different countries, some more than others (depending on the nature of the particular problem at issue), in an ever tighter web of ecological inter-dependence, which binds together citizens of different states and regions of the world. In such a situation, securing the natural basis of life in any one country will depend to large extent on what other states do.

However, although we may 'need' co-operation in order to deal with these types of problems, it is often difficult to achieve it. Indeed, in some cases it may not happen at all. For although international ecological interdependence has grown, sovereign states continue to be the primary actors on the stage of world politics. As a result, the factors that determine the willingness to co-operate and that ultimately shape whatever collective measures are decided upon tend to be defined by national rather the international or global concerns. Despite the pressures from different directions for some form of collaboration to serve the mutual interest in environmental protection, international co-operation in dealing with these problems is not a foregone conclusion.

Given the increasing importance of transboundary or even global environmental problems, what alternatives are there for generating suffi-cient amounts of international co-operation, which will adequately *internalise* the relevant dimensions of the international environmental problems the world currently faces? How are we, that is, to 'match' our institutions for the making and implementation of international environmental policy to the 'biospheric unity of the earth' – if the 'political unification of the world' is unlikely and, if possible, perhaps not even desirable?

The Institutional Preconditions of International Co-operation

Even though nation states continue to be the most important actors in world affairs in a highly decentralised international system, the absence of a central government at the international level does not necessarily rule out the possibility of creating international environmental regimes and the organisations needed to implement and administer them. The term 'anarchy', as this has been used traditionally to characterise international society, does not mean that the international community is entirely without institutions and orderly procedures (Keohane, 1989). There are more than a few signs that world politics is less Hobbesian than we have been led to believe. There is, that is, a number of institutional factors that make co-operation more likely. In fact, there are many signs of a wide range of different forms and arrangements for international co-operation with regard to environmental problems (Caldwell, 1991).

The evidence for what Caldwell refers to as a well-developed structure of international co-operation for environmental protection should remind us that the acknowledged anarchic character of international society need not be equated with a lack of organisation and structure in international relations. On the contrary, there exists a variety of different kinds of institutions that influence and guide the behaviour of individual governments and the interactions between them.

This array of international organisations, procedures and practices is available for organising multilateral negotiations and concluding international treaties, for forming working groups and expert bodies to prepare these negotiations, and for providing important tools and processes to explore possibilities for co-operative action. When the international community confronts a problem demanding the pooling of resources and capabilities, there already are institutions to organise and channel the activities necessary for developing problem-specific institutional arrangements to define and carry the necessary co-operation between nation states. Such institutions shape the incentives states have to engage in co-operative efforts and otherwise make it possible for them to take actions which they otherwise would not find attractive. They affect the costs of alternative courses of actions as well as the way states view the roles they are to play and the assumptions they make regarding the behaviour of others. In this way, international institutions provide both a context within and a set of procedures by means of which representatives of national governments can arrive at common understandings of problems, define shared meanings regarding the situation they face, develop procedures to facilitate negotiations and decision making as well as monitor compliance and guard against defection. Over time, familiarity and mutual trust can emerge out of the separate steps along the path of incremental conditional co-operation, thereby facilitating mutually beneficial policy co-ordination among governments.

If this is so, then the crucial question is not what is going to replace the state but rather how states can collectively manage global and regional interdependence (List and Rittberger, 1992, p.87). One institutional response for coping with interdependence and problems and conflicts that arise from it is the international regime – understood as a form of collective action by states, based on shared principles, norms, rules and decision making procedures with regard to specific issue areas. Insofar as regime theory can undercover the factors affecting the possibility, conditions and consequences of international governance 'beyond anarchy and short of supranational governments', the analysis of such regimes can make an important contribution to the study of the management of the international environment affairs.

Analysts of international regimes as a vehicle for international co-operation recognise that the international community is not confronted with the choice between anarchic, competitive international politics and hierarchically-ordered international policy making. A further, and increasingly prominent, alternative is some form of international collective self-regulation, i.e. the voluntary participation of states and other international actors in collective action to achieve joint gains or to avoid joint losses. International regimes are one manifestation of these efforts at collective self-regulation by states. Like domestic society, international society in many areas is characterised by non-hierarchical politics through which joint regulatory actions can be taken to deal with common problems. In such cases obligations do not emanate from a hierarchical norm- and rule-setting process but from voluntary agreements to play by a set of rules which are binding in the sense that the create convergent expectations regarding acceptable govern behaviour (Rittberger, et al., 1993, p.393). These situations differ from government as this is normally understood in that the compulsion exerted by the rules is not backed up by the threat or use of physical force but is, rather, based on non-hierarchical normative institutions creating a stabilised pattern of behaviour.

Regime Formation: The Interaction of International and Domestic Factors

The encouraging message from these observations is that 'co-operation is possible but depends in part on institutional arrangements' (Keohane, 1989, p.3). While ecological interdependence may make international co-operation imperative, this 'objective need' does not necessarily translate directly into a national preference for such joint action. The apparent discrepancy between 'needs' and 'deeds' underscores the importance of research into the causes of failures and the conditions for success in designing institutional arrangements for effective international co-operation in dealing with transboundary, increasingly global, threats to environmental quality.

The research presented in this book builds consciously on the existing theory and research with respect to international regimes, as well as studies by those applying structural analysis to predict the positions likely to be taken by various national actors participating in such regimes. The phenomenon of international co-operation has been the concern of a good number of scholars for some time, with numerous studies on the topic already published or under way at the moment. However, until recently, these studies have mainly focused on the negotiation process and interactions among states at the international level, paying only limited attention to linkages between domestic and international politics. Together with such authors as Evans, Jacobson and Putnam (1993); Keohane and Milner (1996); Milner (1997); and Schreurs and Economy (1997), our research assumes that the important factors that determine success and failure in attempts to find co-operative solutions to international environmental problems are to be found in the domestic setting of decision making and not only in the interstate game.

Instead of viewing nations in the international arena as unitary rational actors, we assume that national positions are formulated through internal processes of conflict resolution and bargaining. As noted above, effective responses to increasingly serious environmental problems require the collaboration, even joint action, among nation states at the international level. At the same time, however, the nature of these problems ensures that international action to deal with these issues will have serious consequences for and therefore directly involve important elements of domestic society in the different phases of the policy process. Consequently, if we wish to understand what is likely to happen at the international level, it is necessary to examine the processes, structures and values at the national unit level which determine the manner in which national positions on negotiating international agreements are arrived at and the ultimate agreements are then carried out.

If we wish to determine whether or not regimes 'matter' – i.e. whether they exert independent influence on state behaviour – we need to trace carefully the processes, structures and values at the national level which determine the manner in which such agreements are carried out and responded to. We cannot focus exclusively on the structure of the interstate game, looking for the external constraints upon state behaviour that would encourage compliance with international contracts. Nor will we be able to understand how the positions nations take and defend in international negotiations leading to the agreement come about without considering the way in which international negotiations both mobilise and penetrate domestic society. As Moravcsik has observed: 'State priorities and policies are determined by politicians at the head of national government, who are "embedded in domestic and transnational civil society, which decisively constrains their identities and purposes"'. Events and decisions at the international level, he continues, '...create patterns of societal interests that influence governments via the "transmission belt" of domestic politics...' (Moravcsik, 1993, p.483). To the extent that an analysis of international regimes ignores domestic political processes, it will not be able to show the way in which regimes actually come about and can influence policy choices at the operational level.

Each phase of the process of international co-operation can have both an international and domestic dimension. For example, while the pre-negotiation stage will involve international meetings – of national scientists, representatives of concerned governments, working parties and more general multilateral meeting convened by already existing international governmental organisations – there will also be parallel activities and processes of problem definition and position determination at the national level. Not only will the decision by the national governments (the 'states') to enter into negotiations on some kind of formal agreement regulating the scope and content of co-operative action on a particular problem be the product of the interaction of domestic and international decision making processes. Indeed, whether or not a problem requiring international agreement even exists may be viewed differently by countries, in part as a function of their economic structure and the political power of actors representing these different economic interests. Likewise, the national position that the representatives of a given country take to these international negotiations will also be shaped by this same set of processes.

Already in deciding whether or not to join or conclude an agreement, decision-makers will have made their calculations with an eye to the domestic interests that support or oppose the agreement. Even though an agreement is concluded in the belief that adequate political support will be forthcoming (and that the agreement will, therefore, be faithfully carried out), it may still prove difficult, if not impossible, to deliver on the commitments made; or the measures designed to implement the accord may look quite different than what was intended and needed – before the agreement was negotiated. An appreciation of the difficulties involved in moving from international agreement to national action and, ultimately, on to the required behavioural changes on the part of society's members, suggests that instances of *involuntary* defection may be at least as frequent and interesting as 'cheating' by deliberate choice.

International agreements are not formulated, and certainly not implemented, in a vacuum. They enter a 'regulatory space' already occupied by a set of problem definitions and policy strategies, as well as with constellations of supporting and opposing societal and bureaucratic forces. It is through domestic institutions and practices of political representation that the sets of national interests or goals emerge that states then bring to international negotiations. It follows then that if international organisations are to be effective in mobilising 'pressures for international regulation that enhances environmental protection' they will need to promote sufficient political concern for such measures within societies (Keohane, 1994, p.28). It is the *combination* of international and domestic factors that is crucial for putting pressure on national governments to participate in international environmental regimes.

A better understanding of the linkages between national and international decision making would add to the store of knowledge on which the search for more effective and feasible institutional arrangements for collective action among nation states could be based. In particular, explicit attention to problems of implementing international environmental agreements will fill a crucial gap in our understanding of how international agreements are carried out at the national level.

The Phases of International Co-operation

In order to understand the dynamics of the processes through which international co-operation develops with regard to environmental issues, it is necessary to note some features of these problems that set them apart from more traditional questions of foreign and international politics. International environmental problems represent a type of policy concern that differs in important respects from those that have traditionally been considered quintessential 'foreign policy' issues. Most of these 'new' problems and their potential solutions affect *societies* or segments of societies in ways that defy categorisation as solely 'domestic' *or* 'foreign' policy; they typically have domestic as well as external ramifications. Moreover, while the quintessential foreign policy issue is one referring to *national* interests and *state* behaviour, many of the environmental problems added to the agenda in recent decades are of particular salience to some *sub*group or segment of society, and their solutions often involve regulation of corporate behaviour or even the behaviour of individual *citizens* (see e.g. Hanf and Underdal, 1998).

The environmental problems, which are the object of international action, are generated as side effects of otherwise socially beneficial activities. Damage to the environment typically occurs as a *side effect* of perfectly legitimate activities undertaken for other purposes, such as the production and consumption of energy and goods. A policy designed to protect the environment cannot therefore be neatly compartmentalised and just *added* to other policy commitments; in order to succeed it will have to *penetrate* and modify those activities and policies that cause environmental damage in the first place. Efforts to change patterns of production and consumption behaviour with nation states by means of international regulation require measures aimed at these primary activities, thereby mobilising societal and governmental actors in reaction to the actions taken. In this sense, international action on environmental problems penetrates more directly and deeply domestic society than is the case with more traditional issues of foreign policy. As we have noted, an analysis of co-operation at the international level must also examine institutions and processes through which these domestic actors and interests take part in the negotiation and execution of international environmental agreements.

From this perspective it is necessary to identify the actors involved in this process as well as their interactions and to relate them to the national positions adopted for the negotiating table. In order to guide the analysis of the interactions of these different governmental and societal actors during this preliminary phase of international co-operation, we have developed a model of the relationships between government and society in terms of the demand and supply of governmental regulation with regard to the acidification problem. This model is presented in Chapter 3.

Along with most studies on international regimes, we assume that international co-operation is not something that is achieved (or missed) at a single given point in time. On the contrary, it tends to emerge from a long and often arduous process through which individual states come to realise, against the background of the demands and priorities articulated through their national political processes, the need and advantages of joining forces with others. Viewed in terms of a process through which states explore and act upon possibilities to establish a set of rules intended to guide the behaviour of those involved in some issue area, international co-operation consists of a series of separate but interrelated activities extending over time.

In order to examine more closely the interrelationships among these different activities the research analysed the institutional structures and processes through which national preferences on international environmental issues are formulated; the national positions are subsequently promoted and modified during the course of international negotiations; and the agreements reached are then carried out (at both the international and national levels). The distinctions we make between different phases of international co-operation look a lot like the approach taken recently by Oran Young (1998, 1994). He, too, argues that '(a) satisfactory account of regime formation...will require separate but interconnected propositions concerning the several stages of the overall process' (Young, 1998, p.3). For this reason, he finds if useful too 'divide the overall cycle of regime formation into a least three stages: the agenda formation stage, the negotiation stage, and the operationalization stage' (Young, 1998, p.4).

Our project focused on two crucial dimensions of international co-operation: the formulation of national positions and preferences; and the implementation of international environmental agreements. National case studies provided an in-depth empirical analysis of these two phases, and the relationships between them. The phase of institutional bargaining was dealt with only in as far as this was necessary in order to indicate its impact upon the other two stages. In a certain sense, the formation of national preferences falls within the agenda formation phase. An analysis of this broader process through which issues emerge and gain prominence on the international political agenda would itself have to be broken down into sub-phases. More importantly, from the perspective of our project (as we will see below), it would be necessary to examine the way in which events and decisions at the domestic and international levels impact upon and influence each other during this phase. Another distinction between our project and the scheme proposed by Young is that we include both operationalisation and daily operations of a regime in place as part of the post-negotiation implementation phase.

The Formation of National Preferences

When talking about international environmental co-operation, the focus has quite often been on the international negotiations and the success or failure of these efforts to come up with an accord to capture the level and substance of agreement among the concerned parties. Of course, there is activity that precedes and is supposed to follow the formulation of these agreements. Equally important to an understanding of the processes of international co-operation, however, is an examination of the developments that precede any decision by a particular country to join in searching for a mutually acceptable way of dealing with an international problem; and then, once the agreement has been signed, in carrying out the obligations or commitments entered into. Young's agenda formation encompasses those activities through which a particular issue arrives on 'active political agenda at the international level' (Young, 1998, p.6). Only when it has achieved sufficient prominence among both national and international political actors, will they be willing to invest the 'time and political capital needed to embark on explicit negotiations' (Young, 1998, p.5).

The dependent variable for research on the first phase in the process of international co-operation (or regime formation) is the *substantive position* taken by a government on the issue at hand (assuming that this can be ascertained apart from its strategic representation in the negotiations). Differences in national positions are to be accounted for by reference to national interest structures and policy processes through which domestic demands and demands from the international community are aggregated to the negotiating posture assumed by a given country.

Using the hypotheses derived from the assumptions underlying our model of domestic politics, the formation of national positions on the issues on the international agenda is examined to ascertain the way in which they come about as a result of the interactions of societal and governmental actors within the institutional framework of the country in question. A complete analysis of these interactions would involve (i) an inventory of the actors involved; (ii) the mapping of their interactions during this formulation phase; and (iii) an examination of the different resources of their disposal, including their institutional position within the formal process of decision making. The interactions also need to be traced through time, as the individual positions are aggregated through bargaining and coalition politics. For example, it may be the case that the environmental agency is able to control the agenda-setting and diagnosis phases. However, it may have to face other domestic institutional 'competitors' as the negotiations proceed from discussions of general principles (framework conventions) to more specific 'deeds' (protocols) and on to the national implementation of these commitments. It is important to know to what extent patterns of participation and influence in the policy making process change over time as a particular issue is treated both in different phases of the policy process and in different stages in the further elaboration of the regime (Hanf and Underdal, 1998).

From National Preferences to Collective Action

The international regime itself – at least in its initial formal terms – emerges from a process of explicit negotiation among a group of concerned countries 'seeking to reach agreement on the provisions of a document spelling out the terms of a constitutive contract' (Young, 1998, p.11). This is the phase of 'institutional bargaining', which '...is dominated by the initiation of direct and focused negotiation and ends with the signing of an agreement' (Young, 1998, pp.12-15). For Young it is the core of the process of international co-operation (Young, 1994, p.83). Such

international negotiations, through which different sets of national preferences are aggregated, represent a traditional area of interest in studying
international environmental agreements (IEAs). In part because a good deal
of theoretical and empirical work has been done in this area, and given the
primary focus of this project in considering the linkages between domestic
decision making and international collaboration, international environmental
negotiations as such were not studied in depth. We were, however, interested
in the reciprocal influence between the domestic and the international levels
of action during the negotiations. The work of Robert Putnam (1988) on two
level games is useful for focusing on these interactions.

The Implementation of International Agreements

The post-negotiation phase covers all those steps needed to transform an
international agreement signed by the parties who have agreed to its terms
into an actual institutional arrangement in place. It encompasses those
activities through which international commitments are translated into the
programmatic measures and, ultimately, administrative actions at the national level which are intended to bring about the behavioural changes on
the part of the relevant target groups and, thereby, the quality objectives of
the agreements. Young distinguishes between this phase of
'operationalisation' of the regime (which he considers to be part of the
formation of an international regime) and the day-to-day operations of the
regime, which follow. For him, the process of regime formation is over,
'once the provisions of an international regime have passed into the
domestic practice of the members...' (Young, 1998, p.19).

As Haas et al. (1993, p.16) remind us, it is ultimately national
decisions that affect environmental quality, even though international measures may be necessary to reach harmonised national measures. And yet,
the 'implementation game' has more often than not been neglected in
studies of international co-operation, or analysed only in terms of factors
encouraging or curbing defections from the agreements entered into.
However, even if agreement can be reached regarding the nature of the
problem and the type of actions that should be taken to deal with it, there
still remains the challenge of translating this agreement into the actions
required to 'solve' the problem. International environmental agreements
are designed to have an effect on different categories of (societal) activi-

ties, which are perceived as having negative effects on the quality of the environment. Therefore, the effectiveness of these accords depends, in the last analysis, on the actions taken by – and within – the states participating in an IEA to effect behavioural changes *within* their societies. In this sense, the international policy process does not end with the signing of an international agreement or convention. If there is no implementation, there is no 'real' policy and thus no effective co-operation.

In part, this relative neglect can be attributed to the fact that traditional foreign policy and security issues did not raise the same kinds of questions regarding the execution of international agreements. The behavioural prescriptions or proscriptions contained in such treaties were addressed in the first instance to the *state* itself (i.e. to the national government), and, in terms of the domestic forces involved, to a limited set of actors (even though the issues themselves could be of great importance to the society as a whole). International environmental problems, on the other hand, and the agreements drawn to deal with them tend, as we have already noted, to penetrate societies in a more pervasive and direct way. They carry potentially high costs for important interests, such as producers and/or consumers of particular goods and services.

As the number of international environmental agreements establishing some form of international regime, through which processes of international governance take place, has grown in recent years, attention has shifted from issues of regime formation to questions of regime effectiveness (Young and von Moltke, 1994). With this shift in interest toward the *consequences* of international co-operation, questions of implementation have come to the fore. It is the implementation of agreements once arrived at, that is the area *par excellence* for examining the impact of domestic processes and structures on the bottom line of international co-operation. What happens when international agreements 'enter' the national administrative systems responsible for carrying out policy in general will determine how successful efforts to institutionalise international co-operation ultimately will be.

Implementation of international environmental agreements is a complex process. It encompasses those activities through which international commitments are translated into the programmatic measures and, ultimately, administrative actions at the national level which are intended to bring about the behavioural changes on the part of the relevant target groups and, thereby, the quality objectives of the agreements.

Implementation involves multiple channels of interaction between agents of public authorities and those subject to the regime rules and the regulations promulgated to enforce them. 'Running' regimes, once the treaty is concluded, and adjusting them to changes within national and international society, is as difficult and daunting a task as reaching the initial agreement itself. Each of the different sets of decisions through which implementation is effected can be the focus of political interest and pressure from those sectors in society for which the agreement entails costs or benefits. Each phase can involve a different pattern of interaction among governmental and societal actors. It is also obvious that in none of these sub-phases does it make much sense to talk of a sovereign actor rationally toting up the costs and benefits of compliance.

When implementation is seen as a distinct policy 'game', which leaves its own imprint on the actual thrust of environmental policies, we get a different perspective on the problems faced in meeting one's international obligations. Viewed in this way, implementation 'failure' and 'success' are not only a matter of 'will' (deliberate choice) but also as a matter of *ability* and capacity to govern. National decision-makers have to depend on other, sub-national public and private actors to supply important informational inputs or services needed to carry out the implementation strategy chosen. What emerges from such an analysis of the actions through which international environmental regimes operate is a picture of a *multilevel complex* of norm setting and rule implementing activities performed by different sets of actors at both the national and international levels. A number of separate yet linked networks join public and private actors in the performance of these different functions. Together they determine to what extent a regime will effectively penetrate the societal processes within the member countries through which the intended behavioural changes are to be effected.

Dynamics of Regime Development

While each of these phases can be analysed in terms of its own dynamics, it is important to analyse the factors that link decision processes in one phase to what has gone on before or occurs afterwards. It is these links between the different phases of the model, especially between the first and third phases, which are at the heart of our research proposal. In particular, we assume that there are also important links between what happens during implementation (at the national level) and what has occurred in the preceding phases, i.e. both the formulation of national preferences and the negotiations in this connection in the international arena. For example, we assume that a consensual national position in the first phase will enhance the prospects for successful implementation. Also important is the relation between the substance of the international agreement, i.e. the extent to which the terms of the agreement coincide with a given country's own preferences, and success in implementing it.

The principles, rules, norms and decision procedures of a regime continue to change and evolve, with obvious consequences for the signatory states. The initial convention will need to be further specified in protocols before concrete steps can be taken or indeed 'shares' of responsibility – in the form of national quotas – can be apportioned. Over time, as a result of continuing research and debate, there comes a better understanding of the nature of the problem at hand and the range of possibilities for dealing with it. As consensus grows, initially general and non-specific commitments can be filled in and measures to achieve quality objectives sharpened. Consequently, national policies and programs will also have to be adjusted to changes in the normative structure of the regime as it evolves. Actually, the activities leading up to the subsequent rounds of institutional bargaining are themselves parts of the implementation phase of the initial round of international co-operation and the agenda-setting or preliminary round of problem definition and mobilisation leading in the subsequent iteration of co-operation. The interactions now take place within the institutional framework – with both its international and national aspects – through which the regime at that point of time has taken shape. It is in this sense that international conventions usually define a general normative framework within which the further programming – in the form of follow-up protocols or similar instruments – is carried out, spelling out the more specific regulations and other measures, which are to be taken. These initial agreements also lay out a set of arrangements for the subsequent collective decision-making, i.e. regular meetings of national representatives, working groups or some kind of executive office or secretariat.

Organisation of the Book

The organisation of the remaining chapters of the book is as follows. Chapter 2 presents a short overview of the development and structure of what we have called the 'acidification regime'. It sketches the main lines of the international debate and negotiations – the agenda-setting and institutional bargaining phases of international co-operation – leading up to the Geneva Convention of 1979. The different protocols are then examined through which the Convention on Long-range Transboundary Air Pollution was given concrete operational form. The important role of the European Community as a participant in this regime is briefly examined by looking at the regulatory framework its legislative measures created for the core group of EC member countries who are signatories of the Geneva Convention and its protocols. The interactions between the EC and UNECE components of the acidification regime have been crucial for this international effort to deal with the problem of transboundary air pollution.

In Chapter 3 a model of the domestic politics underlying the formation and operation of an international environmental regime is developed, and contrasted with a model viewing the state as a unitary rational actor. In the domestic politics model, government positions in international negotiations and national implementation records are seen as a function of 'societal demand' for and 'governmental supply' of environmental policy.

There then follows a set of national case studies that summarise the relevant experience of the countries covered in the original research project. The countries examined were selected because they have played different roles or taken different positions during the negotiations on the international agreements around which the 'acid rain regime' was formed. Thus, we wanted to include 'pushers' (such as Sweden and Norway), 'draggers' (such as the UK in the early phases), countries with intermediate positions (such as France), and countries who changed their initial positions substantially in the process (Germany being the most important case). Moreover, since the EC plays an increasingly important role in the development of regional policy in this issue-area, we wanted to include EC members from the 'North' as well as from the 'South' of Europe.

For each of these countries the processes of decision making were analysed that led to the decision to participate (as well as the substantive positions taken at a particular time in the negotiations) and the processes through which the agreements arrived at have (or have not) been carried out. We also looked at the way post-agreement experiences impact upon the further development of national preferences and positions taken in subsequent international negotiations. To the extent this has occurred, the case studies also examine how changes in the agreement are incorporated into the implementation process.

The book concludes with a comparative analysis exploring how well the domestic politics model developed in Chapter 3 competes with the unitary rational actor model in accounting for the pattern of variance found among the ten European countries included in this study.

References

Caldwell, L. (1991), *International Environmental Policy*, Duke University Press, Durham.

Chayes, A. and Chayes, A. C. (1995), *The New Sovereignty*, Harvard University Press, Cambridge, MA.

Evans, P., Jacobson, H. and Putnam, R. (eds) (1993), *Double-Edged Diplomacy*, University of California Press, Berkeley, CA.

Haas, P., Keohane, R. O., and Levy, M. A. (eds) (1993), *Institutions for the Earth. Sources of Effective International Environmental Protection*, The MIT Press, Cambridge, MA and London.

Hanf, K. and Underdal, A. (1998), 'Domesticating International Commitments: Linking National and International Decision-Making', in A. Underdal (ed), *The Politics of International Environmental Management*, Kluwer, Dordrecht.

Keohane, R. O. (1989), *International Institutions and State Power. Essays in International Relations*, Westview, Boulder, CO.

Keohane, R. O. and Milner, H. V. (eds) (1996), *Internationalization and Domestic Politics*, Cambridge University Press. Cambridge and New York.

List, M. and Rittberger, V. (1992), 'Regime Theory and International Environmental Management', in A. Hurrell and B. Kingsbury (eds.), *The International Politics of the Environment*, Oxford University Press, Oxford.

Moravcsik, A. (1993), 'Introduction: Integrating International and Domestic Theories of International Bargaining', in P. B. Evans, H. K. Jacobson and R. D. Putnam (eds), *Double-Edged Diplomacy*. University of California Press, Berkeley, CA.

Putnam, R.D. (1998), 'Diplomacy and Domestic Politics: The Logic of Two-Level Games', *International Organization*, vol. 42, pp. 427-460.

Rittberger, V. (ed) (1993), *Regime Theory and International Relations*. Clarendon Press, Oxford.

Schreurs, M.A. and Economy, E.C. (eds) (1997), *The Internationalization of Environmental Protection*, Cambridge University Press, Cambridge.

Victor, D., Raustiala, K. and Skolnikoff, E. B. (eds) (1998), *The Implementation and Effectiveness of International Environmental Agreements*. MIT Press, Cambridge, MA.

Young, O. R. (1994), *International Governance. Protecting the Environment in a Stateless Society,* Cornell University Press, Ithaca and London.

Young, O. R. (1998), *Creating Regimes. Arctic Accords and International Governance.* Cornell University Press, Ithaca and London.

Young, O. R. and von Moltke, K. (1994), 'The Consequences of International Environmental Regimes: Report from the Barcelona Workshop', *International Environmental Affairs,* 6, no.4.

Weiss, E. B. and Jacobson, H. K. (eds) (1998), *Engaging Countries,* MIT Press, Cambridge, MA.

2 The Problem of Long-Range Transport of Air Pollution and the Acidification Regime

KENNETH HANF

Problem Focus: The 'Acidification Regime'

The three phase model described in the first chapter, together with the analytical model developed in chapter three, serve as the point of departure for bringing into focus the domestic bases of the development of the international regime for dealing with the problem of 'acid rain' or, in broader terms, the problem of acidification.

The core of this regime is the Convention on Long-Range Transboundary Air Pollution (1979) and the subsequent substantive protocols dealing with specific pollutants. These include the 1985 Helsinki Protocol on the reduction of sulphur emissions or their transboundary fluxes; the 1988 Sofia Protocol on the freeze of nitrogen oxides; and the 1991 Geneva Protocol on the reduction of emissions of volatile organic compounds. In 1994 work was completed on the new revised SO_2 protocol and preliminary work has begun on the renegotiation of the protocol on nitrogen oxides. At present several new protocols are underway regarding new and combined NO_X and VOC requirements; heavy metals; and persistent organic pollutants (POPs). While the latter two processes were concluded in 1998, the so-called 'super' NO_X protocol, dealing not only with acidification and ground-level ozone but also eutrophication (all caused, to one degree or another, by nitrogen oxides) is scheduled for completion some time in 1999.

A second element of the 'acid rain regime' consists of a number of directives of the EC, initially with the Large Combustion Plants Directive

(1988) as well as the directives dealing with automobile emissions (1988). By explicitly including EC activities in the definition of the overall regime, we are able to examine the manner in which the LRTAP and the measures taken by the member states of the European Union are inter-related and coordinated with one another. In addition, the dynamics of decision making on these issues within the Community have affected the positions taken by these governments within the acidification regime in the narrower sense. At the moment, the European Commission is busy preparing a number of directives that will implement the recently developed acidification and ozone strategies of the European Union.

The main lines of the development of the international treatment of the problem of acid rain are well known. Nevertheless, a brief overview of the evolution of the regime and the relations among its different components is useful for setting the scene for the theoretical discussion and the presentation of the individual country studies in the following chapters. To this end this chapter sketches the general nature of the problem of acid rain as one variant of the broader phenomenon of transboundary air pollution; the emergence of the problem as an issue of international concern; and the gradual development of the international regime established to deal with this problem. Since the primary purpose of this chapter is to provide a common point of reference for the analysis of the material of the later chapters, no attempt will be made go into these matters in great detail. For this reason we have unabashedly made use of the overviews developed by other scholars of the acidification problem and the history of the development of the international co-operative efforts to deal with it. In particular we have drawn heavily on the studies by Levy (1993, 1994), Boehmer-Christiansen and Skea (1991), Sprinz (1993), and Sprinz and Vaahtoranta (1994), Gehring (1994), and Wettestad (1999). These authors have, for their part, synthesised in similar ways (but with different accents) data and information taken from a wide range of original sources.

The Acidification Problem

Sulphur dioxide and oxides of nitrogen released into the air when coal and oil are burned can be carried many hundreds of miles by wind currents before being deposited, either as precipitation or as dry deposition. Since such deposition often occurs through mixing with precipitation, and the damage is often caused through the mechanism of acidification, this phenomenon came to be known as 'acid rain'. In this way oxide of sulphur and nitrogen form acids, which can damage materials and aquatic and terrestrial ecosystems. However, this term now is used to refer to both wet and dry depositions.

In Europe, about half of the air pollution crosses borders, and much of that kills fish and trees, and corrodes buildings and monuments. Acid rain becomes a problem when its steady deposition over time eventually exhausts the ability of a receptor to counteract the acid's effect. Although sulphur dioxide and nitrogen oxides are both also potent causes of respiratory diseases in humans, they are usually safe to humans in the diffuse concentration that persist over long-range distances. Nevertheless, these long-term effects can generate negative consequences for human health, as when heavy metals are leached into drinking water supplies (Levy, 1993, p.78).

Both acid precipitation and the long-range transport of air pollution were familiar long before acid rain became the subject of international controversy. It had been known for some time that emissions from combustion of coal and oil could generate acidic precipitation. Swedish researchers had already started monitoring freshwater acidity levels in the 1940s. In the 1950s and 1960s they began to notice significant increases in these levels. However, at that time the much more serious and immediate problems of local, especially urban, air pollution dominated national regulatory politics. The contemporary understanding of acid rain is, therefore, of more recent origin. In 1968 the Swedish scientist Svante Oden published a paper in which he argued that precipitation over Scandinavia was becoming increasingly acidic, thus inflicting damage on fish and lakes. Most importantly, he asserted that acidic precipitation was to a large extent caused by sulphur compounds from British and Central European industrial emissions.

In addition to sulphur dioxide (SO_2), the most important compounds in this connection are nitrogen oxides (NO_x) and volatile organic compounds (VOCs). While nitrogen oxides are also acidifying compounds, they came on to the political agenda later than sulphur because of the early Scandinavian preoccupation with sulphur emissions and freshwater acidification. By the mid-1980s, however, there was sufficient agreement among the activist governments that NO_x was a major contributor to ecosystem damage (in particular, in connection with the evidence of extensive forest dieback in the forests of Central Europe). Nitrogen oxides come from both power plants and automobile emissions. They are responsible for some acid deposition, and also contribute to the formation of ground-level ozone that can be toxic to humans and harmful to vegetation. The scientific consensus now is that ozone is just as harmful to forest and agricultural crops as sulphur deposition. It also is a public health problem in many urban areas. While all this came to be accepted, the magnitude of the damage caused continues to be disputed (Levy, 1993, p.94).

Volatile organic compounds entered the political agenda in the latter half of the 1980s as scientific appreciation for the role of ground-level ozone in ecosystem damage increased. In the effort to understand the role of acid rain (meaning acidic deposition of sulphate and nitrogen compounds) in terrestrial damage, especially forest die-back, researchers came to realise that most terrestrial damage is not traceable to a single cause, but is rather a host of interacting stresses. These stresses include climatic variation, disease and other natural causes. Ground-level ozone is the chief anthropogenic stress that combines with traditional acid rain to harm plants, materials and human health. Ozone is a potent oxidant, causing corrosion of materials and cellular damage in living organisms. It is a principal ingredient in urban smog. It is created by the interaction of air pollutants and sunlight. In addition to nitrogen, VOCs are the primary additional ozone precursors. These chemicals, including fuels, solvents, cleaners and other industrial compounds, evaporate into the air where they have the potential to create ozone. Ozone itself does not travel because of its high reactivity, but ozone precursors are capable of travelling across borders.

Of the compounds primarily causing the acidification (and ground-level ozone), sulphur is the easiest (relatively speaking) to deal with. Its effects are well understood and most of its emissions can be controlled by regulating combustion of fuels in power generation and large industrial plants. Nitrogen is tougher, both because its effects are more complicated and therefore less well understood, and because its sources include power generation plus transportation. That is one reason why activist countries were forced to abandon their initial hopes for a 30 per cent reduction protocol for NO_x and why those countries that had, nevertheless, voluntarily accepted a reduction target of 30 per cent are experiencing difficulties in meeting their commitments.

In turn, VOCs are easily an order of magnitude more complicated, on both chemical and political grounds, than NO_X. As Sprinz observes (1992, p.63): 'More than 80 substances need to be regulated and the natural science is less well understood as compared to sulphur and nitrogen oxides'.

The Formation of the 'Acidification Regime'

The issue of acid rain entered various international agendas as a result of a sustained campaign by the Scandinavian countries, especially Sweden followed by Norway. Governments in both Norway and Sweden became concerned about the acidification of their lakes and forests due to transboundary air pollution because valuable fish stocks and forest resources were threatened. Their concern led to a spate of diplomatic activity aimed at getting the issue of acid rain and long-range transport of air pollutants placed on the political agenda of the international community. Through various forums they tried to focus international attention on the problem. Drawing on the results of this acidification research, they sought to mobilise international co-operation to reduce sulphur emissions. Other European countries, however, remained sceptical of the claim that their emissions were harming ecosystems in Scandinavia. Today, there is general agreement on the core of the initial Scandinavian claims.

Between the time of Oden's 'discovery' and the successful conclusion of international negotiations on the institutional framework of the acidification regime at the beginnings of the 1980s, the primary lines of conflict were fixed. On the one side were Sweden and Norway, concerned with the serious domestic acid rain damage in lakes and streams, much of it caused, they argued, by imported sulphur dioxide. On the other side was the entire rest of Europe, which was being blamed for exporting sulphur dioxide to Scandinavia. During this period, Sweden and Norway attempted, without success, to persuade their neighbours to reduce emissions of SO_2. These countries, however, had little incentive to comply with these requests since they were, as far as they knew, free of acid damage. Consequently, reducing emissions was bound to be expensive for them with little immediate benefit in terms of better environmental quality for the accused 'polluters'.

Although alerted by Swedish research and the diplomatic efforts of Scandinavian countries to the threat of acid deposition, polluting countries continued to wait until the end of the 1970s for more convincing proof of the damage being caused. Sweden and Norway did not, however, remain idle. In the meantime, they made considerable use of different international organisations to advance their case for stricter air-pollution controls. Since neither was a member of the European Community (EC), these efforts were initially focused on the Organisation of European Co-operation and Development (OECD) and, subsequently, the United Nations Economic Commission for Europe (ECE).

The risk of acid precipitation and long range transport of sulphur compounds first reached the international agenda in 1969, when it was presented as 'a problem which required continued study' at an ad hoc meeting organised by the OECD's Air Management Sector Group (AMSG) to examine a report on the acidity and concentration of sulphate in rain. Much of the impetus and pressure for further research in the OECD originated from a joint Scandinavian research institute, NORDFORSK. In 1970 this institute presented a project proposal to AMSG for a survey to study the long-range transport of air pollution.

While the OECD was considering the Swedish generated request, Sweden pushed ahead within the United Nations framework, with a proposal to organise a conference on the human environment. At the 1972 meeting in Stockholm the Swedish government presented its evidence on the extent of long-range transport of pollutants and the damage that was being caused to soils and lakes. At that time the scientific community received this information with a great deal of scepticism. Nevertheless, all the nations present – including the United Kingdom – accepted Principle 21 of the Conference Declaration which reiterated a traditional principle of international law that states have an obligation to ensure that activities carried out in one country do not cause environmental damage in other countries, or to the global commons. It was the strength of Scandinavian (and Canadian) concern about both transboundary air pollution and marine pollution that was instrumental in gaining acceptance by other countries of this principle (Boehmer-Christiansen and Skea, 1991, p.25).

In the years immediately following 1972 little concrete political progress was made at the international level with regard to air pollution. The world was too busy trying to deal with the problems caused by the energy crisis and a major economic recession. What international environmental diplomacy there was focused on problems of marine pollution (Boehmer-Christiansen and Skea, 1991). On a regional basis, however, the OECD, on behalf of the developed nations, took action on the Norwegian and Swedish proposal for a project on the problems of trans-frontier air pollution.

Early in the 1970s, the OECD set up its Environment Directorate, which acted as a source of information for industrialised countries. The principal OECD activities on this issue during the 1970s were research, publicity and the development of policy principles and recommendations issued by the ministerial level of the OECD Council. Although the OECD lacked any legal mechanism for implementing its policies and principles, the information it generated on the phenomenon of transboundary air pollution played an important role in sensitising countries to the nature and seriousness of the problem. In 1972 the OECD started the first internationally co-ordinated large-scale research program on the long-range transport of air pollutants. Eleven European nations participated in the 'Co-operative Technical Programme to Measure the Long-Range Transport of Air Pollutants'. The results of this study, published in 1977 with a revised and expanded version appearing in 1979, provided the first independent international verification of Scandinavian charges that in southern Norway and Sweden imported sulphur was primarily responsible for the acidification of lakes. Five of the eleven European countries participating in the study received more pollution from abroad than from domestic sources.

Even though the OECD Council had adopted a set of 'Principles Concerning Trans-frontier Pollution' in 1974, this problem continued to have a low priority among the member countries of the organisation. It did, however, become clear from the OECD's work that the problem of transboundary air pollution in Europe could not be dealt with effectively without the active involvement of the Soviet Union and the countries of Eastern Europe. In turn, the importance of the participation of Eastern European, meant that the OECD, with a membership centred on Western industrialised countries, was not the most appropriate forum for tackling this problem. Not only did this organisation not include many of the largest polluters, located in the East bloc countries; it also did not have the power to enforce its policy recommendations. At the same time, the membership of the European Community was also problematic: on the one hand, it did not include the pushers Sweden and Norway, while, on the other hand, many of its members were major emitters of SO_2 who were reluctant to control emissions.

There was another organisation, the United Nations' Economic Commission for Europe, that did include as members the relevant Western and Eastern European states. But the fact that the regime formation activity shifted to this organisational arena had more to do with the role it could play as a forum for East-West co-operation in connection with the broader process of international détente. It was this issue of high politics that became a second, initially unintended, driving force, along side the emerging environmental concern, that pushed the issue of acid rain in the direction of international co-operation.

In the mid-1960s, the former Soviet Union, which sought recognition of the status quo of European borders, particularly that of East Germany, started to promote the idea of closer international co-operation, an initiative that ultimately led to the Conference on Security and Co-operation in Europe (CSCE). In 1969, in the Budapest Declaration of the members of the Warsaw Treaty Organisation, air pollution had been mentioned for the first time as a possible field for such East-West co-operation. Responsibility for investigating the possibilities in this area was assigned in the Final Act of the CSCE in Helsinki to the ECE. Even though transboundary air pollution was, at that time, perceived as primarily of Western European problem with the Nordic countries as the main victims and some Central European countries as the main polluters, it now had entered the sphere of high politics. An important issue linkage between the environmental concern of the Nordic countries and the political interests of the Soviet Union had been created. Consequently, in the second half of the seventies, the main focus for broadly based international pressures to reduce emissions had shifted to the ECE. Sweden and Norway again led efforts to use this new organisational arena for mobilising international action on the issue.

These initiatives received important political boost when, in 1975, Soviet premier Brezhnev called for co-operation between East and West European countries on issues of environment, energy and transport. The main motivation behind the making of this proposal was to further the détente process in the wake of the 1975 Helsinki accords. The different options for such co-operation were considered in the ECE and it was decided that the issues of the long-range transport of air pollution best fit the multiple national agendas in play at that time. After two years of nego-tiations, the ECE Convention on Long-range Transboundary Air Pollution (LRTAP) was signed by 35 Contracting Parties (34 countries and the EC Commission) in Geneva in November 1979.

During the negotiations leading to the agreement, three major blocks of actors emerged. The Nordic countries were the most concerned about the problem, while the Soviet Union and its allies were primarily interested in the procedural issues of a high-level meeting. Although for different reasons, these two groups acted as pushers for some kind of international agreement. The main actors in the third group, the UK and West Germany, either disputed the scientific evident linking sulphur emissions to damage in Nordic countries, or were reluctant to give authority to the ECE in this matter.

The European Community (EC) could not arrive at a unified position. This led to severe tensions inside the Community. An important role in finally unifying the lines of the EC was played by France which, while holding the Presidency of the EC, did not want the EC to be blamed for a failure to reach an agreement. Another dispute that involved the EC was the refusal of the Soviet Union to accept the EC's signature to the Convention. A solution was found that did not mention the EC as such in the Convention but granted a right for signature to certain 'regional organisations' (Article 14). Even though only the EC fit this provision, at least in principle, access was also possible for similar organisations in socialist countries.

The final result of the negotiations on transboundary air pollution was a classical compromise, in the words of Marc Levy, a 'lowest common denominator' compromise (Levy, 1991, p.83). Although the Scandinavians had tried hard to have binding cuts in sulphur emissions included in this agreement, this framework convention was limited to laying down the principle that states should reduce transborder air pollution as much as is economically feasible, and to creating mechanisms for collecting and disseminating information on pollution flows and their effects. No country accepted any explicit commitment, which it considered to be significantly costly. With regard to the principles embodied in LRTAP, they are quite similar to those contained in earlier OECD guidelines and in the 1972 Stockholm Declaration. The signatory countries also agree that they would regularly inform one another of national policy developments in this area. Provisions were also made to promoting collaborative research on the issue. The Convention did not specify any pollutants, but stated that measures should start with sulphur dioxide (SO_2).

It is clear that countries signed the Convention for different reasons. Led by the Soviet Union, the east-bloc countries were primarily interested in establishing east-west co-operation based on a formal international treaty. The Scandinavian countries, led by Sweden, continued to be interested in reducing emissions of SO_2 in neighbouring countries. The rest of Europe was relatively indifferent. The momentum from the Helsinki process, on the one hand, and the OECD's acid rain program, on the other, was sufficient to keep the negotiations going and, ultimately, to bring them to a successful conclusion.

The LRTAP Protocols: The Evolution of the Regime

After the signing of the Convention, the next step in the co-operation was the 1984 EMEP Protocol which basic objective has been to allocate the costs of the monitoring program. As we have seen, the OECD, in response to the Swedish initiatives began the monitoring of SO_2 emissions in 1972. The OECD's monitoring and transport modelling activities were consolidated in 1978 under the auspices of the Co-operative Programme for the Monitoring and Evaluation of the Long-Range Transmission of Air Pollutants in Europe (EMEP) within the UNECE. When the CLRTAP was signed in 1979, EMEP was incorporated into its organisational structure.

The SO₂ Protocol

Although LRTAP was an important victory for the Scandinavian countries, it did not bring the kinds of benefits they had sought. The Convention committed parties to broad principles and joint research activities, but not to any concrete measures to reduce acid rain. Not surprisingly, therefore, these countries were satisfied with the Convention.

Shortly after signing of the CLRTAP the former Soviet Union invaded Afghanistan and détente came to a halt. With this development, one of the driving forces that had led to the agreement thus disappeared. However, Sweden continued to be the primary political force behind the LRTAP agenda, being especially concerned with reducing emissions of SO_2, the cause of Scandinavian freshwater acidification. When in 1982 the

CLRTAP had still not received the required 24 ratification's to enter into force, the Swedish government launched a new initiative by organising a major conference 'to remind nations of the acidification of her lakes and streams' (Boehmer-Christiansen and Skea, 1991, p.27).

This conference on the Acidification of the Environment was intended to renew pressure on countries, which still opposed international acid rain controls. Invitations were issued to attend a conference in Stockholm on the occasion of the tenth anniversary of the 1972 UN Conference on the Human Environment. In contrast to the first Stockholm meeting, participation in 1982 was limited to the signatories of the CLRTAP. The strategic purpose of the meeting was also reflected in the strong emphasis on the issue of environmental acidification. Even though some important East European signatories to the Convention did not attend, the conference marked a starting point for increased efforts to reach a binding international agreement to control and reduce transboundary air pollution.

Conference communiqués from both experts and ministers acknowledged the seriousness of the problem. It was during this conference that Bernard Ulrich, a German forestry researcher, released his findings that the Black Forest was dying and that this forest dieback (*Waldsterben*) could in large measure be attributed to acid rain. This was the first time that terrestrial ecosystem damage on a large scale had entered the international agenda. It was also at this conference that West Germany did a public about face with regard to its position on the acidification issue and subsequently became one of the most forceful pushers for effective international regulation. A number of expert meetings were conducted during the course of the conference summarising the contemporary state of scientific knowledge about acidification. These conclusions pointed towards the need for major emissions reductions.

In light of what they considered to be convincing evidence of the need for decisive international action, the Scandinavian countries proposed that countries should undertake to reduce SO_2 emissions 30 per cent below 1980 levels by the year 1993. The proposal failed to win the support of the UK, the US, France or the East Bloc countries. However, to the surprise of the international community, the Federal Republic of Germany (FRG) did support the proposal. The discovery of 'Waldsterben' set in motion of complex set of domestic political pressures within Germany. These, in turn, led to its conversion to the Scandinavian position in the international negotiations on acid rain calling for reductions in SO_2 emissions.

Although not accepted at the 1982 Stockholm conference, the FRG and the Scandinavian countries continued to push the idea of a 30 per cent SO_2 reduction. A corresponding proposal was presented to the Executive Board of the LRTAP in June 1983, where it received additional support from Canada, Austria and Switzerland. At this meeting of the signatories of the CLRTAP, Norway, Sweden and Finland submitted a 'Proposal for a Concerted Programme for the Reduction of Sulphur Emissions', which called for a 30 per cent reduction of national SO_2 emissions or their transboundary fluxes by 1992, using 1980 emissions as the baseline. Denmark had originally cosponsored the proposal but had backed down due to pressure from other EC countries. West Germany, on the other hand, officially supported the Nordic proposal. The German government also suggested, together with Switzerland and Austria, to add measures to control NO_X emissions. These countries, joined by the Canada and the Soviet Union, took the lead and committed themselves to the unilateral reduction of SO_2 emissions by 30 per cent.

The about-face of Germany was a key turning point for a number of reasons. German support guaranteed not only serious SO_2 reductions by a large polluter; it also added a powerful political ally to the Scandinavian cause. Furthermore, the discovery that terrestrial ecosystems might also be threatened by acid rain drastically altered the geo-political map. What had once been seen as a Scandinavian problem had suddenly become a potentially continent-wide problem.

The following developments were marked by two further conferences ('outside' but still related to the LRTAP) designed as a means for increasing pressure on laggard countries. At the Ottawa Conference, the countries that had committed themselves to a unilateral reduction of SO_2 emissions were joined by France and the Netherlands. As the Soviet Union did not attend the Conference, these ten countries formed the so-called '30 per cent Club', which proceeded to act as an international pressure group in favour of concrete reductions of SO_2 emissions. Support for this initiative did, however, come from the East Bloc countries a few months later. Again acting more on the basis of security policy and détente considerations, the Soviet Union encouraged the Federal Republic of Germany to host an international conference to discuss the question of reductions of emissions. At this 'Multilateral Conference on the Causes and Prevention of Damage to Forests and Waters by Air Pollution in Europe', held in Munich in June 1994 and co-sponsored by the ECE, the East Bloc countries signalled that they were prepared to reduce 'transboundary fluxes' of SO_2 (Boehmer-Christiansen and Skea, 1991, p.28). In so doing, they successfully isolated the US and UK within the international community while enhancing the image of the Soviet Union in Western Europe (Boehmer-Christiansen and Skea, 1991, p.28).

During the conference the '30 per cent club' was joined by eight other countries and an agreement on the reduction of annual national sulphur emissions was put on the agenda for the second session of the ECE regime Executive Body for the same year. Against the background of this cresting tide of support, the 30 per cent proposal was adopted as an official protocol to the LRTAP Convention at the third meeting of the Executive Board in Helsinki in July of 1985. The Protocol stipulated a reduction of emissions/transboundary fluxes of sulphur dioxide (SO_2) by at least 30 per cent as soon as possible, and by 1993 at the latest, with 1980 levels taken as the baseline. As it was immediately signed by the 21 members of the already existing '30 per cent Club', it obtained enough ratifications to bring it into force in October 1985. However, some major emitter states refused to join the agreement, the UK and Poland being the most important.

The NO_X Protocol

As soon as the Sulphur Protocol had been signed in Helsinki in July of 1985, the Executive Body established a Working Group on Nitrogen Oxides. There already was general agreement that NO_X was a major contributor to ecosystem damage although there continued to be disagreement over the extent of the damage it caused (Levy, 1993, p.94). However, negotiations on NO_X faced some additional problems compared with SO_2. First, the scientific consensus about the damaging effects and the information on control technologies had been much stronger in the case of SO_2, not least because SO_2 had been at the centre of the internally co-ordinated research under EMP. Second, a large part of the NO_X emissions came from mobile sources (especially automobiles) and was more difficult to monitor and to control. It was, however, clear that NO_X reductions of any significant magnitude would require strict automobile emission standards. A vigorous debate was in progress over how best to reduce auto emissions (the principal source of nitrogen oxide emissions) between proponents of catalytic converters and those pushing for lean burn motors. Depending on the position of the national auto industry, governments were favoured or opposed one or the other measure. Of the large Western countries, Germany supported a protocol while the UK, France and Italy opposed it. A third problem was that while SO_2 emissions in many countries had been declining since the mid-1970s, NO_X were still increasing.

The FRG had wanted to include the issue of NO_X emissions already in the Sulphur Protocol and stuck to its leadership position when it circulated elements of a draft proposal in 1986. With respect to stationary sources a 30 per cent reduction was proposed. With respect to mobile sources, the application of the best available technology was suggested, a measure that has always been favoured by the Germans. However, there were several other positions held by different countries, and no clear coalitions emerged as has been the case during the negotiations on CLTRAP and the SO_2 protocol. Since British decision-makers '...perceived a higher ecological vulnerability to nitrogen oxides as compared to sulphur oxides...' (Sprinz, 1992, p.61) and abatements costs for this pollutant were substantially lower, the UK was willing to support a 'freeze' protocol for nitrogen emissions. In this way it could avoid antagonising the other countries the way it had on the Sulphur Protocol.

It is interesting to note that in contrast to both the CLRTAP itself and the Sulphur Protocol negotiations no clear coalitions emerged in favour of a particular approach. But this seems not to have been a problem since, as Gehring has noted, the regime had created its own dynamic:

> The start of the negotiations in the absence of a well organised initiating group and the absence of sufficient technical information suggests that in the case of NO_X a larger part of the deliberation process proceeded within the institutional structure of the international regime. It also suggests that the existence of the institutional framework of an international regime considerably lowers the threshold for such negotiations (Gehring, 1994, p.163).

The 'Protocol to the 1979 Convention on LRTAP Concerning the Control of Emissions from Nitrogen Oxides or their Transboundary Fluxes' was finally adopted at the sixth session of the Executive Body in Sofia in 1988. The signatories pledged to freeze NO_X emissions at the 1987 level from 1994 onwards and to negotiate subsequent reductions at a later point in time. Twenty-five countries signed the protocol, including the UK and the United States. Moreover, 12 European signatories went a step further and signed an additional (and separate) joint declaration committing them to a 30 per cent reduction of emissions by 1998 taking as a baseline any chosen year between 1980 and 1985. EC countries not supporting this objective were the UK, Spain, Portugal, Greece, Luxembourg and Ireland.

Compared to the SO_2 Protocol, the Protocol on Nitrogen Oxides differs in important respects. It requires only a freeze on NO_X emissions or of their transboundary fluxes by the end of 1994, and instead of the proposed baseline year 1987, any other previous year could be chosen. However, further negotiations on reductions were institutionalised in the Protocol and were scheduled to begin six months after the Protocol entered into force. Furthermore, twelve countries pledged unilaterally to reduce NO_X emissions by 30 per cent by 1998. While in the case of sulphur the formation of a '30 per cent Club' preceded the Protocol, in the case of nitrogen oxides it was the other way around. Other innovations included a commitment to promote the transfer of emissions reduction technologies, a commitment to emission standards that are based on the 'best available technologies that are economically feasible', and a reference to critical loads as a possible concept for a later reduction protocol.

The Protocol on VOCs

Following agreement on the nitrogen oxide protocol negotiations began in 1989 on a protocol to regulated VOCs. That VOCs contributed significantly to European ozone formation, and should therefore be regulated, was not disputed. However, the issue of Volatile Organic Compounds is once again different from SO_2 and NO_X because these substances, as we have seen, do not cause acid rain but lead, together with NO_X to the formation of ground-level ozone in the presence of sunlight. In this sense, the 1991 Protocol Concerning the Control of Emissions of VOC or their Transboundary Fluxes, signed by 21 countries at the ninth session of the Executive Body in Geneva in 1991, can be regarded as an extension of the original regime (although attempts to include VOCs in a protocol date back to 1986, when the FRG proposed a joint regulation for VOCs and NO_X emissions).

Regulation of these compounds, however, presented a particularly challenging scientific problem in that VOCs are comprised of a large number of chemicals that behave in very different ways. They interact differently with meteorological and topographical conditions, in complex ways not yet fully understand, so that their ozone-creating potential varies widely. Furthermore, the number and variety of compounds used in different economic activities met there was very little information available. As of mid-1991, only five countries had conducted thorough inventories of their VOC emissions.

Nevertheless, in November 1991 21 parties signed a protocol which was, in the words of Levy, 'a rather awkward pastiche (of different national positions) that merely listed three entirely different forms of regulations (based on the three sets of national interests)' (Levy, 1993, p.100). The protocol calls for a reduction of 30 per cent in VOC emissions between 1988 and 1999, based on 1988 levels – either at the national level (for the country as a whole) or within specific 'tropospheric ozone management areas'. Countries are permitted to select their own base year. Some countries are allowed to opt for a freeze of 1988 emissions by 1999. Similar to the NO_X Protocol a mechanism for the negotiation of further reductions is set out in the Protocol. Moreover, for the first time, a mechanism for monitoring compliance was included, enabling countries to bring complaints before the Executive Body.

'Second Generation' Protocols: Initial Steps Toward Joint Problem Solving

Levy argues that it was a combination of collaborative science and adversarial diplomacy that drove the process that led to the initial protocols to the LRTAP convention – the agreement in Helsinki in 1985 to reduce SO_2 emissions and the Sofia protocol in 1988 to freeze emissions of NO_X (Levy, 1994, p.3). Under the auspices of LRTAP's Working Group on Effects and the monitoring activities of the EMEP, collaborative science was able to play an important role in advancing the state of consensual knowledge regarding the nature of transboundary flows and the extent of damage caused by acid rain. The data and analysis generated through these activities made it possible to develop estimates regarding who did what to whom. By shedding light on the nature and extent of damage suffered by the different countries, such information was crucial for determining how national actors perceived the problem the national governments arrived at the positions they took in the negotiations.

Building on the shared insights developed through such joint research activities looking at the severity of the problem and which countries were responsible for causing it, adversarial diplomacy was used to put pressure on states to reduce emissions. The focal points for this diplomacy were the SO_2 and NO_X protocols. It is interesting to note that this pressure did not move reluctant countries to undertake commitments beyond what they were planning to do in any case. In this sense, SO_2 and NO_X had in common that when initial negotiating positions were formulated prior domestic plans defined clear baselines beyond which governments did not budge (Levy, 1993, p.114).

Levy points out that in neither case did any country sign with the intention to use the protocol as a guide to revision in its domestic emission reduction policies. Indeed, he adds, '...there is no evidence that any state (with the possible exception of the Soviet Union) signed any protocol without first determining that already-planned policies would bring it into compliance more or less automatically' (Levy, 1994, p.4). Although the policy followed by the British government, of not committing itself to reduction targets unless it was sure it could meet them, would seem to support this argument, Lewanski argues convincingly that Italy made up its mind – at the last minute – to sign the SO_2 protocol without having clear ideas about how it would be able to fulfil its commitment.

But if countries did not intend to use the instrument as a guide to reducing national emissions, why did they sign? For different reasons. Some, for which acid rain already constituted a serious national problem had already undertaken far-reaching domestic measures on their own. These countries, with measures that were stricter than what the protocol called for, were pressing for further-going multilateral action. Another group of countries expected that the reductions required would be achieved as the result, for example, of changes in their energy policies. Finally, as Levy points out, there was a group, including many East block countries, that signed for opportunistic reasons, never intending to undertake serious reduction measures of their own (Levy, 1994, p.3-4). With regard to the reduction targets in the protocol, the first two groups were able to sign without second thoughts since their domestic measures exceeded the required 30 per cent. Likewise, the third group did not hesitate since these countries intended to cheat from the start (Levy, 1994, p.4). On the other hand, in contrast to the commitments on SO_2 reduction targets, there continues to be some doubt on the part of some of those countries that signed as to whether they will be able to achieve the targets agreed to.

In an important sense it was developments with regard to national emission policies of the countries involved that made it possible to reach agreement of the protocols in the first place. While, in general, this may be true, there is evidence that the negotiation process within the regime did influence the position individual countries took with regard to national reductions. Again Italy can serve as a good counter-example of a country that signed the SO_2 protocol as a consequence of this process. Indeed, a reading of the record of the decisions taken by different member countries supports the conclusion that in several cases the unilateral targets that these countries would have set on their own would probably have been lower than the reductions agreed to in the protocol. In some cases little or no action at all would have been taken without the process set in motion by the CLRTAP regime. Whether the adjustments made were the result of international 'pressure' or 'learning' is, of course, another matter. The bottom line, however, is that national acid rain policies would have been different without the CLRTAP process. Moreover, even though it may be true that the protocols did little more than formalise the existing state of affairs, 'under the right conditions' they may help advance that status quo. Such movement cannot be realised by forcing states to undertake measures they otherwise would not have. Rather it is the result of helping to shift countries' perceptions of their self-interest and the possibilities for realising it through joint actions (Levy, 1994, p5).

It is this kind of role that the international agreements reached within the acidification regime are beginning to play in the present phase of co-operation on this issue. It is for this reason observers claim that that LRTAP entered 'a new phase' after the 1988 signing of the protocol on nitrogen oxides (Levy, 1994, p.5). At present, the most recent protocols are less 'instruments of normative persuasion that influence parties during the negotiations but not after signature' and more '…genuine regulatory instruments that impose serious constraints on domestic policies after they are signed' (Levy, 1994, p.5).

This shift reflects changes in the way in which the problem is perceived politically. In this regard the most important change has been the conversion of those countries that previously opposed international controls for dealing with acid rain into supporters of the regime that has been set up. Here the most dramatic shift was the turn-about on the part of Great Britain in 1988. As a consequence of this reversal of the British position, there was less need for normative persuasion based on the previous brand of adversarial diplomacy. Under conditions of general consensus regarding the seriousness of the problem and the need for international regulatory action, it has become possible for the first time to engage in collective problem solving based on shared perceptions of the problem (Levy, 1994).

This change in style in the direction of collective problem solving can be seen in the agreements that have been reached since the NO_X protocol. For example, during the VOC negotiations, exchanges were less confrontational and acrimonious. Since no government had any reliable estimates of what abatement costs and damage estimates might be, there were no serious disputes on this point. Instead the debate focused on the kinds of instruments that might be most effectively employed in order to achieve the desired environmental results. Equally important, more attention was paid to how 'variations in national situations could be accommodated most productively' (Levy, 1994, p.6). In this way, an element of differentiation among the countries was introduced into the regulatory regime. Consequently, the VOC protocol commits signatory countries to 'emission reduction policies that in most cases go beyond what these states had earlier committed themselves to domestically'. Indeed, '...most states had no VOC regulations in place prior to the protocol' (Levy, 1994, p.5-6).

It is, however, the 1994 'Oslo Protocol' on Further Reductions of Sulphur Emissions, the first protocol based on the concept of critical loads, that earns the title of 'the first instrument of the second generation' (Gehring, 1994, p.181). Critical loads are defined as 'a quantitative estimate of an exposure to one or more pollutants below which significant harmful effects on specific sensitive elements of the environment do not occur according to present knowledge' (NO_X Protocol, Article 1). The relevance of the critical loads concept can hardly be overestimated, as it establishes targets as the ultimate goals of environmental diplomacy. More importantly, this approach provides an ecological rationale and formula for establishing *differentiated* targets. Integrated assessment models provide major assistance for negotiations in evaluating the environmental impacts of their joint reduction plans. The aim of this approach is that emissions reductions should be negotiated on the basis of the effects of air pollutants, rather than by choosing an equal percentage reduction target for all countries involved. With this tool, it is possible to overcome a number of limitations related to the use of flat rate reductions. For one thing, such flat rate reductions are inevitably arbitrary from a scientific point of view. Furthermore, they tend to ignore prior national efforts and overlook the fact that both sensitivity to pollution and the cost of reducing emissions vary markedly across and within the countries of Europe.

The use of this new management tool both reflects and, subsequently, has contributed to the dramatic transformation of the issue of acidification as a regulatory in Europe. While the first sulphur protocol may be seen as 'primarily a proselytising tool in which the terms were established early on by a small group of leaders who then campaigned for accession among a group of reluctant states', the revised sulphur protocol represents an effort to create a collective regulatory instrument, the terms of which were to be more than a recording of existing policies. On the contrary, now national policies were to be adjusted to internationally negotiated norms. It is this new approach to joint regulation that marks a radical departure for the convention. This change could not, of course, gave been realised without a far-reaching consensus with regard to the scientific and technical basis that can probably only be achieved through a long history of scientific co-operation such as characterises the LRTAP regime. Also crucial for the emergence of the critical loads approach was that the RAINS-model used for the implementation of the allocation of reductions among the countries, had been developed by a highly respected *international* research institute, the International Institute for Applied Systems Analysis (IIASA).

While the UNECE regime may have been weak on reaching agreement on mandatory measures, it has been effective in encouraging national measures, setting goals and collecting information. Indeed, as Boehmer-Christiansen and Skea argue, perhaps its main achievement has been the creation of a common information and science base (1991, p.29). The collaborative science, organised through LRTAP working groups, was important for advancing the state of consensual knowledge about the extent of acid rain damage and about the nature of transboundary flows. The politically most important science was probably that conducted under the auspices of the Working Group on Effects and the European Monitoring and Evaluation Programme (EMEP). The Working Group on Effects oversaw collaborative research on forests, materials, freshwater ecosystems, crops and integrated monitoring. This research helped solidify a consensus about the importance of the acid rain problem. In some cases this information led countries to discover domestic acid rain damage that they had not expected to find. (The UK probably represents the starkest example of this latter phenomenon, but at an earlier point in time the Netherlands, too, had become aware of domestic damage as a result of information acquired from regime-related activities.)

The growth of reliable knowledge regarding the severity of the problem and who was responsible for it, helped, according to Levy, put pressure on states to reduce emissions (Levy, 1994. p.3). Furthermore, as a result of the highly scientific nature of the workshops which underpin political activity within the UNECE regime, more sophisticated science-based concepts to emission control have been developed. In particular, the concept of critical loads for pollution burdens on ecosystems was developed within the framework of the UNECE regime.

The critical loads idea did shift 'the nature of the public debate, both internationally and in many domestic settings, away from who the bad guys were, and toward determining how vulnerable each party was to acid rain' (Levy, 1994, p.9). Within Britain, for example, the debate was thoroughly transformed when critical load maps revealed the extent to which the country was vulnerable to damage by acid rain. In the train of such revelations, 'the constituency supporting acid rain controls expanded from environmental activists and elites concerned with the standing of Britain abroad to include groups worried about the fate of the British countryside – a much larger group...' (Levy, 1994, p.9), which was committed to the protection of an object of high value in Britain, comparable to the forests in Germany.

Such changes in constellation of domestic political forces and the resulting shifts in the pattern of support and opposition with regard to international regulatory action to deal with the threats of acidification are particularly important factors shaping the evolution of an international regime. Evolving patterns of the domestic politics of acidification as the result of changes in the perceived interests of both problem-specific and more general internal actors filter the impacts of international developments on domestic political processes and the positions taken by national governments regarding international co-operation. At the same time, there were also other factors pushing the country in a 'greener' direction, thereby lending support to the use of and reinforcing the impact of critical loads and the LRTAP. As the research reported in the UK national case points out, a number of factors converged beginning in 1988 which contributed to the transformation of the British government from obstructionism on international environmental issues toward a more activist position.

The Role of the EC: Actor and Arena

In contrast to the LRTAP, the European Community is a body with powerful regulatory instruments at its disposal, which, however, can only be brought into action if the member states can agree to a course of action on a given issue. In comparison with the other international organisations involved in the formation of the acidification regime, the EC arrived relatively late on the scene. Although it was a signatory to the CLRTAP, the problem is not mentioned in official documents until the beginnings of the eighties. The appearance of this issue on the agenda of the EC is an interesting example of the way in which Community policy making was pressured by both the domestic politics of its member states and policy developments in the broader international community. When the concern for acid rain spread from Scandinavia to Central Europe with the discovery of forest dieback in Germany (and subsequently in other countries of the regime) in 1982, a political struggle over the issue erupted within the EC.

Initial concern on the part of the European Community with the regulation of substances responsible for acidification was expressed in the early seventies, when sulphur compounds, suspended particles, NO_x and CO were described in the First Environmental Action Program as pollutants deserving high priority; the first concrete measure dealing with air pollution Directive 75/716 regulating the sulphur content of light fuel oil. EC 'sulphur policy' addressed health and local air pollution problems and was justified in terms of the need to harmonise. Consequently, no mention was made of acidification as such to justify its policy measures. On the other hand, also at this time, the EC did undertake its first activity directly related to acidification, i.e. monitoring activities with regard to atmospheric pollution caused by sulphur compounds and suspended particles. This was done in support of the OECD's technical programme to measure long-range transport of air pollution that had been initiated in 1972 (Huber and Liberatore, 1995, p.8). In 1976 the Commission took a further air pollution measure, once again intended to tackle local pollution and harmonise the efforts of member states, in the form of ambient air quality standards. After significant delay – due to strong opposition from major SO_2 producers such as the UK, Germany and France – the directive was finally adopted in 1980. While this directive was being negotiated the EC member states were also participating – together with the EC as a whole – in the international negotiations leading to the formation of the CLRTAP regime. After having committed themselves to international collaboration on the same issue, it was difficult for the EC member states to keep postponing action on air quality control within the EC itself. In this way, as Huber and Liberatore argue, it can be said '...that sulphur policy and acid rain policy had come together through the interplay between the EC and the international levels' (Huber and Liberatore, 1995, p.9).

1983 was a 'banner year' for action on acidification on the part of the EC, and once again events at both the international and domestic levels converged to create pressure for action. The growing importance of the acidification issues was reflected in a number of major EC policy documents. In December of this year the Commission introduced its draft for a directive regulating emissions of acidifying substances by large combustion plants. An important factor behind the initiatives and activities in 1983 was the about face by Germany on the issue of long range air pollution. Once it had been discovered that acidification was damaging German forests, domestic political pressure led to strict national legislation to deal with industrial air pollution. Consequently, the German government was quite interested in achieving equally strong international measures dealing with acidification so that it could ensure that the economic costs of dealing with this problem would be shared among all countries. In this way German industry would not face unfair competition from countries which did not have equally strict regulations. Since Germany was one of the main emitters of SO_2 and one of the most powerful countries within the EC, its new strong stand on acidification proved influential in placing the issue on the Community's agenda and the development of the initiatives introduced by the Commission.

With Germany's conversion to the side of those seeking significant reductions in emissions of SO_2, the EC suddenly shifted from a body exhibiting a united front against ameliorative measures to one whose most powerful members were in open conflict over the issue – at the level of both the LRTAP and the EC itself. Germany sought EC-wide regulations that reflected its own domestic response to the damage caused by acidification to its forests, i.e. steep reductions in sulphur and nitrogen emissions from large combustion plants. On the other hand, Britain, and later Spain, supported initially by a number of smaller member sates, opposed these measures strongly until a compromise solution was agreed to in 1988.

The German about face on the issue of acid rain proved to be decisive for placing the matter on the EC agenda. Although there had been continuous pressure on the EC from the Scandinavian countries for action, these efforts had not been successful. Prior to 1982 there was little support for such measures within the Community. This all changed with the discovery of forest dieback in Germany. However, even the weight of Germany was not sufficient to force a quick closure on this issue within the EC. It was not until 1988 that the directive regulating SO_2 and NO_X emissions from large combustion plants was finally passed. At the same time the EC also concluded work on another source of acidifying pollutants, i.e. nitrogen oxides from auto emissions.

In its Fifth Environmental Action Program (1992) the EC states that its long-term goal concerning acidification is to ensure that the critical loads and levels are not exceeded. In 1995 the environmental ministers of the member states invited the European Commission to elaborate a Community strategy to combat acidification, identifying and proposing cost-effective measures for abating emissions of acidifying air pollutants. In December 1997, the European Commission presented its acidification strategy, in which it sets an interim environmental target of a 50 per cent gap closure, between the base level year (1990) and the critical loads levels, by the year 2010. In order to reach this goal, the Commission has since proposed a number of measures, including a directive establishing binding national emissions ceilings for SO_2, NO_X and particulates, a proposal for the revision of the 1988 directive on large combustion plants; a proposed further reduction in the sulphur content of certain liquid fuels; and a proposal for ambient air quality standards. The Commission has also recommended that the Oslo Protocol be ratified and that the member states play an active role in the negotiations on a new multi-polluntant protocol within the CLRTAP. The Commission has since presented an ozone strategy focusing on the regulation of VOCs.

Concluding Comments

As we have seen in chapter 1, a process consisting of a series of interrelated (for the most part, sequential) steps covering a wide range of activities and stretching over a long period of time. Each of these phases of international problem-solving activity can involve different sets of governmental/non-governmental actors acting under the auspices of or in the arenas created by different national or international authorities. In each phase of development there will emerge varying patterns of relationships between actors and arenas at the international and domestic levels.

In the case of LRTAP the cycle of international/domestic interactions begins with the 'discovery' by (a) national scientist(s) of the phenomenon 'acid rain' and the definition of the problem around the assumed causal links between local damage and pollutants emitted from other countries. At that time, in most West European countries air pollution was viewed in the first instance as a local, mostly urban problem, of concern because of its health consequences. It is in this form that the issue later also initially comes onto the agenda and is dealt with at the EC level. The long-range transport of air pollution, the problem of acid rain and, later acidification, had to be *perceived* as problems and *'sold'* to the international community as issues requiring co-operative action. The initial efforts at creating concern for the problem and a willingness to take action on it was carried by the Nordic countries, which throughout the acidification story played an important entrepreneurial role. In their efforts to gain the attention and commitment of other European governments, the Scandinavian countries undertook a march through the available international institutions: the UN Conference on the Human Environment, the OECD, the CSCE, and the UNECE. With the exception of the OECD and its programme of research on the long-range transport of air pollutants – and the role played by the World Meteorological Organisation and the UNEP in setting up the EMEP – these organisations played primarily important 'arena' roles. They brought together those countries most directly affected to discuss the problem. A crucial juncture in the development of international co-operation on the issue was the insertion of the problem into CSCE deliberations on quite unrelated matters. Linking it to the discussions on how to further East-West détente enabled the Nordic countries to push for action on the acid rain question itself.

During the pre-regime preparatory phase, therefore, a number of international institutions played important roles in bringing the issue of transboundary air pollution onto the international agenda and to the negotiation table. It is during this phase and the phase of the institutional bargaining itself that we can most clearly see the impact of such institutions in providing occasions and arenas for promoting international co-operation.

Once the international agreement had been signed, the third phase of international co-operation began, which involved both the implementation of the initial commitment and the starting up of activities that were to lead to further developments of the regime as a whole. The evolution of the regime led to changes in the nature of tasks that needed to be carried out – at both the international and the domestic levels – in order to realise the objectives of the regime.

The EC was an initial member of the regime as a signatory of the original convention. But it has been a member *sui generis* in that it could bind a sub-set of regime members to further-going international commitments. More importantly, however, it could also develop legislation of its own setting forth regulations with significant consequences for the achievements of the objectives of the UNECE regime. In an important sense these EC measures are a central component of the acidification regime. At the present time, the co-operation and co-ordination of actions within the context of the CLRTAP and the European Union is shaping the developing of the international regime as a whole.

Even when this co-operation stops far short of 'rules and regulations' (reduction objectives or other substantive and procedural commitments), international activities can have important impacts on the politics and policy of environmental regulation (Underdal, 1994). Of particular interest in this regard is the extent to which (and the mechanisms through which) national actors (both within the national context and collectively at the international level) 'learn' with regard to the nature of problem and the range of possible solutions. Above all, it is through an examination of the relationships between the phases or stages of intra-regime evolution that we can best trace the 'learning curve' of international co-operation within the institutional context of a given regime. With regard to the shift from what Levy calls 'tote-board diplomacy' to collaborative decision making we have seen how the redefinition of the political agenda of acidification was, in large part, the result of a better understanding of the nature of the problem and possibilities for dealing with it and how the input of new information and the changing context for international negotiation helped redefine the interest of participants and the range of acceptable options open to them for joint solutions. In turn, these developments modified existing patterns of conflict and co-operation at the domestic level, and thus the configuration of support and opposition with which national policy makers were confronted.

A focus on the domestic underpinning of international co-operation only tells part of the story. It is clear that events in the international context – or at the international level (changes in the state of the environment itself; decisions by other governments or international organisations; developments in technology or knowledge regarding the nature of a problem; and the range of possible ways of dealing with it) can be quite important inputs into the domestic policy process. Likewise, different international organisations – either separately or as constituent elements of an international regime – can play a number of different roles in facilitating co-operation among national governments. Most importantly, as a result of inputs from collaborative international science, the experiences gathered in carrying out earlier protocols, and the contribution made by this 'common history' to the normative institutionalisation of the regime on the basis of shared culture, familiarity and trust, the regime itself becomes a factor shaping domestic developments. An examination of the impact of the regime on the evolution of domestic policy and institutions needs to consider as well the role played by international actors and processes in legitimating concern and mobilising support for the problem within a country; the way in which international agreements and institutions can contribute to the empowering of groups of actors at national levels; and the way in which participation in and responsibility for the implementation of an international regime can feed into the organisational self-interest of bureaucratic actors in support of international co-operation as a 'piece of turf' to be promoted and defended.

References

Boehmer-Christiansen, S. and J. Skea (1991), *Acid Politics: Environmental and Energy Policies in Britain and Germany*, Belhaven, London.

Gehring, T. (1994), *Dynamic International Regimes. Institutions for International Environmental Governance*, Peter Lang, Frankfurt/Main.

Huber, M and Liberatore, A. (1995), 'A Regional Approach to the Management of Global Environmental Risk' (unpublished manuscript).

Levy, M.A. (1993), 'European Acid Rain: The Power of Tote-Board Diplomacy', in P. M. Haas, R. O. Keohane and M. A. Levy (eds), *Institutions for the Earth. Sources of Effective International Environmental Protection*, MIT Press, Cambridge, MA.

Levy, M.A. (1995), 'International Cooperation to Combat Acid Rain', *Green Globe YearBook 1995*, (Pre-publication manuscript, 1994).

Schwarzer, G. (1990), Weitraeumige grenzueberschreitende Luftverschmutzung. Konfliktanalyse eines internationalen Umweltproblems, Tuebingen: Tuebinger Arbeitspapiere zur internationalen *Politik und Friedensforschung*, nr. 15.

Sprinz, D. (1992), *Why Countries Support International Environmental Agreements: The Regulation of Acid Rain in Europe*, University of Michigan, Ann Arbor. Unpublished Ph.D. Dissertation.

Sprinz, D. and T. Vaahtoranta (1994), 'The Interest-Based Explanation of International Environmental Policy', *International Organization*, vol. 48, pp. 77-105.

Underdal, A. (1994), 'Progress in the Absence of Substantive Joint Decisions? Notes on the Dynamics of Regime Formation Processes', in T. Hanisch (ed), *Climate Change and the Agenda for Research*. Westview Press, Boulder CO.

Wettestad, J. (1999), *Designing Effective Environmental Regimes*, Edward Elgar, Cheltenham.

3 Conceptual Framework: Modelling Supply of and Demand for Environmental Regulation

ARILD UNDERDAL

Purpose and Scope

This chapter outlines the analytical framework guiding our empirical research. In the political science literature, answers to our basic research questions are sought by pursuing at least three different paths of research. Each path is defined by a set of assumptions about political processes and state–society relationships, from which distinct explanatory propositions and policy implications can be derived. We refer to one of these approaches as 'the unitary rational actor model' (or simply *model I*), another as the 'domestic politics model' (*model II*), and the third as that of 'social learning and policy diffusion' (*model III*). In the presentation below, we distinguish between *assumptions*, *propositions* and *hypotheses*. Assumptions are basic premises that are normally not considered objects of empirical testing in any specific study. However, these assumptions imply a number of more concrete propositions which can be examined in empirical research. Only a subset of these propositions will be addressed in this particular study. To distinguish this subset we refer to them as 'hypotheses'. Some of these hypotheses are hard to examine with the kind of data available to us; these are marked with an asterisk in the text. Below, also the basic assumptions and a number of important 'propositions' have been included in the overview simply to provide a more coherent statement of the argument. In some cases, they may serve as heuristic guides to empirical research.

In the presentation below as well as in the empirical analysis we will deal most extensively with model II. As pointed out in chapter 1, we will also pay more attention to the politics of implementation than to the formation of negotiating positions.

Dependent Variables

To repeat, the basic question addressed in this book may be formulated as follows: How can we best explain (variance in) the *positions* that national governments take in international negotiations about environmental (acid rain) regulations, and (variance in) their *implementation and compliance records*, i.e. the extent to which these regulations are in fact implemented and complied with? We thus have two dependent variables that call for elaboration and operational measurement: negotiating positions and implementation and compliance records.

Negotiating Positions

For phase 1 – the formation of national positions – our dependent variable is *the 'strength' of preferred regulations*. 'Strength' is measured simply as the relative reductions in emissions advocated as the appropriate target by the various governments in the international negotiations.

In order to rank national positions along this dimension we need to deal with several methodological problems and pitfalls. First, 'strength' of preferred regulations can be measured only in relation to a certain *baseline* and *deadline*. It is therefore necessary to specify the baseline (year) referred to and the deadline by which targets are to be achieved. For purposes of comparative analysis, we refer to the official base years used in the protocols. These reference years are: sulphur 1980, NO_x 1987, VOCs 1988. Corresponding deadlines are: end of 1993 for Sulphur I; 2000, 2005, and 2010 for Sulphur II; end of 1994 for NO_xI; and 1999 for VOCs. At this stage, implementation records can be evaluated only with regard to Sulphur I and (tentatively) NO_xI.

Second, actor positions may sometimes be hard to describe with precision, for at least two reasons. One is that, particularly in the early stages, some parties may formulate only vague or very tentative positions. The other is that a government may well change its position over time. Our rule-of-thumb has been to try to obtain information about position(s) at the opening and the concluding stages of the negotiations.

Third, as any student of negotiations will know, inferring *real preferences* from *official positions* is not always a straightforward exercise. Tactical considerations may lead a party to inflate or in other ways misrepresent its real preferences. An outside observer will usually have access only to official positions.

Fourth, 'strength' of preferred regulations should not be confused with willingness to pay. Since abatement costs differ substantially from one country to another, we have no basis for inferring a government's willingness to pay from the reduction target it advocates. In other words, the party advocating the most ambitious reduction target will not necessarily be the party expressing the greatest willingness to pay for environmental protection.

Compliance and Implementation Records

In the analysis of compliance and implementation records, our dependent variable is *the actual change (reduction) in emissions achieved, by the agreed upon deadline* (or, when the deadline has not yet expired, by the latest year for which we have confirmed data).

As indicated in chapter 1, a distinction is made between *outputs* (i.e. policy measures undertaken by governments in response to an IEA) and *outcomes* (i.e. relevant behaviour of the targets of government interventions, notably the polluters). We are interested in *both*, and particularly the extent to which government policies affect polluter behaviour. Accordingly, we are interested not only in compliance records but also in implementation *efforts* and the relationship between efforts and achievements (cf. Knoepfel, 1997, p.177f). In this study no attempt has been made to examine how changes in emissions affect the environment itself *(impact)*.

Explanatory Model I: The Unitary Rational Actor

This model focuses on the calculations that a single, rational decision-maker would perform in deciding whether or not to sign an agreement (phase 1) or to comply with the terms of an agreement that he has signed (phase 3). The model treats his decisions as a function of incentive structures. More specifically, it builds upon three basic assumptions:

A_{I1}: States are unitary, rational actors.

A_{I2}: Decision-makers evaluate options in terms of costs and benefits to their nation – and only in those term – and choose whichever option (is believed to) maximise(s) net national gain.

A_{I3}: States are in full control of 'their' societies.

No serious scholar will claim that these assumptions qualify as an accurate description of political systems or typical decision-maker behaviour. The main arguments usually submitted in favour of model I are (a) that it 'is peculiarly conducive to the development of theory' (Schelling, 1960, p.4), and (b) that its basic assumptions are at least as close to reality as any other set of equally simplistic assumptions. The parsimony, precision, and rigor permitted by the assumptions of unity and rationality make this model particularly attractive for purposes of deductive reasoning and extensive empirical studies.

Negotiating Position

The core assumption of model I is that states can, for analytical purposes, be conceived as unitary rational actors maximising net national welfare. Applied to environmental management, this assumption basically implies that a state will pursue environmental protection up to the point where its marginal costs of abatement equals its marginal damage costs, and no further. With regard to negotiating positions the model predicts that the most 'progressive' countries will be those suffering high damage costs relative to own abatement costs. Conversely, 'laggards' will be countries with high abatement costs relative to damage costs. In the case of *transfrontier* pollution the relationship between own damage and abatement costs will be a function not only of own emissions, ecological vulnerability to pollution, and the costs of reducing own emissions; it will be affected also by what might be called 'the balance of pollution exchange' vis-à-vis other countries. Translated into more specific hypotheses, the model predicts that

H_{11}: the *pushers* of international environmental regulations will be those countries that are (a) most (ecologically) vulnerable to pollution, (b) have the most 'negative' pollution exchange balances (i.e. are net importers), and/or (c) have the lowest abatement costs. Conversely, the *laggards* will be those that are (a) least vulnerable to pollution, (b) 'net exporters' of pollutants, and/or (c) have the highest abatement costs.

Compliance and Implementation

Model I focuses primarily on *compliance* rather than implementation. The main proposition that can be derived from the rational actor model is quite simple (cf. Young, 1979):

P_{11}: A party will comply with an international agreement if, and only as long as, its marginal costs of compliance are lower than (or at most equal to) the marginal benefits it expects from fulfilling its obligations.

In the most general version of this model, the nature of the costs and benefits are left unspecified. Once we want to apply it to specific cases, however, costs and benefits must be specified. In the context of environmental policy, the default option would be to conceive of the marginal costs of compliance with an international environmental accord in terms of the *abatement costs* incurred by honouring one's commitments (dAC), and marginal benefits in terms of the reduction achieved in environmental *damage costs* (dDC) by doing so. Thus, P_{11} can formally be restated as follows:

P_{11a}: An actor (i) will comply if $dAC_i < dDC_i$, and it *may* comply if $dAC_i = dDC_i$.

In applying this model to empirical cases, the assumption is usually made that abatement as well as damage costs can be derived from 'objective' characteristics of the country itself and the system of activities in question (see e.g. Sprinz and Vaahtoranta, 1994). The calculus, particularly with regard to damage costs, may certainly be complex and call for the most sophisticated methods of valuation. Even when we use the best information and most powerful techniques available, considerable margins of uncertainty will usually remain. One reason is that we typically have only an imperfect understanding of the causal relationship between human activities (such as emissions of polluting substances) and environmental damage. Moreover, it is important to recognise that abatement *efforts* sometimes have positive *side* effects – e.g. the development of new technology that may turn out to be commercially profitable. To the extent that such side effects can be predicted with reasonable confidence, they should be included in the equation.

In some cases compliance may bring benefits other than a reduction in environmental damage costs or the development of new technology. Additional benefits might come in the form of e.g. technical or economic assistance. Moreover, defectors may not only trigger the defection of others but also risk additional cost in the form of 'external' sanctions, loss of prestige or damage to their reputation. If we refer to such compliance-generated *side*-benefits as IB and the defection-generated costs as SC, a more comprehensive interpretation of P_{II} can be formulated as follows:

P_{IIb}: An actor (i) will comply as long as $dAC_i - dIB_i < dDC_i + dSC_i$,
and it *may* comply if $dAC_i - dIB_i = dDC_i + dSC_i$.

In model I it is usually assumed that any implementation bonus and defection costs will be provided by other members of the regime or by an international organisation. Except for regimes that are perfectly stable (see below), the more coercive leadership that is provided within the regime, the more compliance we should expect, everything else the same. Thus, what is known as the 'hegemonic stability' hypothesis suggests that international regimes will tend to decline in the absence of a 'hegemon' that is sufficiently powerful to provide positive incentives for compliance or impose punitive sanctions on defectors.[1] The basic logic of the model is, however, fully compatible with the notion of decision-makers as moral beings, incorporating the norm of *pacta sunt servanda* into their utility functions – as long as the norm is not considered a categorical imperative.

Now, some policy problems have a solution that is perfectly *stable* in the sense that once agreed upon, no actor – not even a pure egoist completely cynical about its own commitments – would have any incentive to defect.[2] The most trivial case would be one where compliance is a dominant strategy for all parties, meaning that no actor has any inventive to defect regardless of what others do. More relevant and interesting to the study of international agreements are cases where compliance is not the universally dominant strategy, but still the best option for each and every actor (at least) as long as (all) others comply. Here, no actor has any incentive to defect *unilaterally*. Agreements establishing joint information systems or pure 'co-ordination rules' (such as those governing air traffic or navigation at sea) often fit this description. When we are dealing with joint action to reduce pollution or conserve resources, however, some temptation to 'cheat' will often persist. An agreement is, by definition, 'an exchange of conditional promises' (Iklé, 1964, p.7), whereby one actor commits itself to a certain course of action in exchange for certain contributions from its partners. In most cases at least one party will have to settle for an 'exchange rate' that is less favourable than the one it would have chosen had it been able to make the decision all by itself; its most preferred solution would be one where the actor himself contributes less and/or others contribute more. Whenever the agreed solution is not considered to be the arrangement that maximises utility for each and every party involved, compliance can not be taken for granted.

The damage cost reduction stemming from compliance with an international environmental agreement results in part from changes in one's own behaviour, in part from changes in the behaviour of one's partners. In a system characterised by interdependence, one actor's net benefits from complying with the provisions of an agreement will depend upon what others do. The more important the benefits that result from the compliance of *others*, the more the incentives to comply will depend on the impact of an actor's own behaviour upon the behaviour of its partners. Thus, from P_{11} we can deduce several more specific propositions, including:

P_{12}: The weaker the causal link between each actor's own compliance and that actor's own benefits from the project, the weaker its incentives to comply.

P_{12a}: The less likely that (important) partners will detect defections, the weaker one's incentives are to comply.[3]

P_{12b}: The less sensitive other contracting partners are to one's own defection (even when detected), the weaker one's incentives are to comply.

The sensitivity of one partner to the behaviour of another is seen essentially as a function of the latter's contribution to the achievement of the project and its role in bringing about the agreement.[4] The less substantial a party's contribution to the project, the less sensitive its partners will be to its level of compliance. A small party stands a better chance of being tolerated as a free rider than a party who is supposed to make a substantial contribution. Similarly, if we assume that actors have preferences not only with regard to outcomes but also with respect to the paths leading to those outcomes, it seems reasonable to hypothesise that the greater a party's role in bringing about an agreement, the more sensitive its partners would be to its own efforts to meet its obligations. Other things being equal, the defection of a 'pusher' who has worked hard to persuade others to contribute will have a greater impact upon the motivations of its partners than the defection of a party who joined the agreement reluctantly, giving in to persuasion and pressure. Assuming that at least some partners are sensitive to one's own contribution and can monitor compliance, we can see that

P_{12c}: The more important the benefits that will accrue – directly or indirectly[5] – to oneself from the compliance of others, the stronger one's own incentives are to behave in a way that enhances or at least does not jeopardise their compliance.

The total amount of benefits that an actor (A) can expect from complying with the terms of an international agreement can, then, be summarised in the equation

$$U_a^a + (P_i^a S_i^a)(U_a^i + IB_a^i),^6$$

where U_a^a is A's net gain from its *own* implementation measures, P_i^a is the probability that A's partners will correctly distinguish between actual defection and compliance, S_i^a is a coefficient expressing the aggregate sensitivity of these partners to A's level of compliance, U_a^i is the aggregate utility to A of the compliance of its partners, and IB_a^i is any additional benefits or 'rewards' for successful implementation controlled by its partners.

We can now identify the 'perfect' agreement in two steps, as follows:

P_{13}: The agreement maximising incentives to comply will be one that is (a) strongly integrative (meaning that all partners reap net benefits from the project, and realise that they do), (b)

characterised by universal pivotality, i.e. an agreement in which
every partner is strictly pivotal and known by all to be so, and –
unless compliance is the dominant strategy for all parties – (c)
fully transparent, so that compliance can be mutually monitored
and everyone is assured that this is the case (see Hovi 1992,
chapter 15).

Being strictly pivotal means that full compliance is absolutely
necessary to achieve the compliance of others. Strict pivotality is achieved
only when the 'exchange rate' between one's own contribution and those
of one's partners is fixed and known to be fixed, so that no actor expects to
be able to modify it in his own favour through unilateral action. An
agreement that meets the requirements included in P_{13} needs no *external*
bribes or sanctions to survive.[7] It would, however, be extremely fragile,
since even a minor defection by one single party would lead to the
immediate and complete collapse of the agreement.[8] To avoid such a trap,
Chayes and Chayes (1993, p.176) argue that a regime 'should not be held
to a strict standard of compliance but to a level of overall compliance that
is "acceptable" in the light of the interests and concerns the treaty is
designed to safeguard'. In other words,

> P_{14}: To maximise overall long-term compliance, a regime must
> have a critical minimum of *robustness*. The optimal level of
> robustness is based on a delicate balance between
> deterrence/retaliation and forgiveness (see Axelrod, 1994).

The propositions above suggest several more specific implications for
the design of stable agreements. One suggestion is simply to make sure that
no (pivotal) party is brought down to its 'resistance point' (BATNA), or is
manipulated into signing an agreement that is not in its own real interests.
Thus, the desire to maximise success at the bargaining table needs to be
tempered with the reminder that, in order to survive, an agreement should
provide net benefits to all partners in actual operation; at the very least it
must leave no partner worse off. Second, the logic of the model suggests
that compliance will, in certain circumstances, be inversely related to the
amount of behavioural change required by regime rules. Everything else
being equal, the risk of deliberate defection and implementation failure
tends to increase the greater and more costly the behavioural change
prescribed by the regime (Downs et al., 1996). Third, the model suggests

that wherever the agreement does not give each and every party its most preferred solution, transparency and monitoring will enhance, and may even be necessary to ensure, compliance. 'Minimise the fear of deception' is a good rule to follow.[9] Fourth, model I offers some clues as to *who* are the most likely to defect. Some rules-of-thumb are: beware of those that have (a) the least to gain from general implementation of the agreement; (b) the most to gain from the compliance of *others* relative to the net benefits that they can reap from their own measures (notably net importers with high abatement costs); and (c) have the least important contributions to make in the eyes of important partners (presumably small, low-profile 'followers').

In the empirical analysis we will examine model I from two different perspectives. First, we will explore its heuristic value, i.e. whether it points us towards the main *mechanisms* shaping negotiating positions and compliance records. The important question here is whether and to what extent decision-makers *approached* the problem in terms of this kind of cost-benefit calculus, not how well they in fact succeeded in deriving precise or accurate conclusions from the exercise. Second, we ask whether the predictions derived from the model fit observed patterns of behaviour. The latter part of the analysis will be done in an integrated fashion in chapter 13.

Explanatory Model II: Domestic Politics

Implicit in model I is the assumption that if faced with the same problem – described in terms of abatement and damage costs – all actors would choose the same solution (with the possible exception of cases where actors are indifferent to the choice between options). Any *variance* in policy is attributed to differences in national cost/benefit balances.

The domestic politics model recognises that environmental policies will be substantially influenced by 'objective' damage and abatement costs. However, one basic assumption is that neither national positions nor implementation records can be understood *simply* as a derivative from some 'objective' calculation of national material interests. More specifically, model II introduces three complications: (1) Decision-makers as well as the general public perceive costs and benefits in subjective rather than objective terms, and values and beliefs may differ significantly within as well as across societies; (2) Actors are concerned not only (perhaps not even primarily) with maximising net national welfare, they typically have several more 'parochial' concerns that affect their behaviour; (3) Political systems distribute power and influence unequally, and domestic political processes tend to produce outputs that can deviate systematically – and in ways that can be predicted – from those that would maximise net national welfare as understood by the unitary rational actor model. In order to understand environmental policies we must therefore know not only the structure of the problem, but also the basic values and beliefs through which it is 'filtered', as well as the structure of the political system and the dynamics of the political processes through which policy is shaped and executed.

From what we have said above, it follows that models I and II differ also in terms of *complexity*. Instead of one single decision-maker guided by the 'national interest', model II depicts governments as complex organisations where sub-actors pursue multiple and to some extent conflicting objectives, and where policy decisions are a weighted aggregate of sub-actor preferences. Model II opens the black box of the unitary actor and relaxes – although does not abolish – the assumption of rationality. Furthermore, instead of an omnipotent state we have one which is influenced and constrained by the society it is supposed to govern. Accordingly, model II allows for the possibility that a government may be *unable* to fulfil its commitments because of failure to win domestic games of implementation (cf. Putnam, 1988).

To formulate the basic assumptions in general form:

A_{II1}: Governments are complex organisations, over which no single decision-maker has full control.

A_{II2}: Decision-makers evaluate options in terms of costs and benefits and choose the option (believed to) maximise(s) net benefits, but utility functions sometimes differ, and none will necessarily be fully consistent with that of the nation or state at large (the 'national interest').

A_{II21}: The perspectives and interests of decision-makers are to some extent shaped by role and position; 'where you stand depends on where you sit' (Allison, 1971, p.176).

A_{II3}: A state is itself influenced by and has only partial control over 'its' society; the state not only governs but is also influenced and constrained by society.

Let us move on to explore where these assumptions may lead.

Societal Demand and Support

Governance is always a relationship between 'government and the governed', to quote Crossman (1965). This suggests that domestic sources of variance in negotiating positions and implementation records can be found either in government itself, in society, or in the relationship between the two. Borrowing terms from systems analysis, we can see environmental policy as a function of two interrelated driving forces: societal demand for environmental quality and governmental supply of policies to protect the environment.[10] Both are presumably heavily influenced by 'objective' damage and abatement costs, but a basic assumption is that neither demand nor supply can be adequately predicted or understood simply in terms of some objective calculation of national interests.

Using a simple flow-chart, we can depict societal *demand* for environmental quality as being shaped by (a) 'objective' damage and abatement costs, (b) the values, interests, and beliefs of different segments of society, and (c) the configuration and capabilities of actors or agents (such as NGOs, political parties and media) available to articulate and promote different values and interests (see figure 3.1).

Figure 3.1   Determinants of societal demand and support

Damage/ Abatement costs	Values, interests and beliefs of various segments of society $_{(1...n)}$	Interests, power and influence of intermediate agents $_{(1...n)}$	'Net' societal demand and support
->	-> <-	->	

<-   **Model I**   ->   <-   **Model II**   ->

Note that *net* demand and support is the critical concept. This implies that, in our study, we have to balance 'negative' demand (from polluters and others worried about abatement costs) against 'positive' demand (from victims, environmental NGOs and others).

Values, interests and beliefs of the public   In the unitary rational actor model we assume that all societies are equal in all respects left unspecified by the model. Model II directs our attention to the possibility that societies may assign different values to the same environmental good:

> A_{II4}: Societies differ in terms of their valuation of environmental resources. This applies to the valuation of environmental quality *in general* as well as the valuation of *specific* resources (e.g. particular species of plants or animals, or a particular ecosystem).

Differences in valuation of environmental goods are likely to be found along at least two dimensions. First, in instrumental terms environmental goods that constitute the *resource base* for (important) economic activities are likely to be valued higher than those that are valued for some other reason, such as recreation or aesthetic qualities. Secondly, some environmental goods may have higher symbolic or cultural value than others. At this stage we have scant knowledge about such culture-specific values and beliefs as they relate to environmental resources. For purposes of modelling, we therefore will have to treat them as idiosyncratic factors to be included in the residual 'error term'.

There is a substantial amount of evidence suggesting that the valuation (as well as the 'consumption') of environmental goods tend to increase with affluence (see e.g. Jänicke and Weidner, 1997). Thus, we expect

> H_{II1}: The relative priority given to (the protection of) environmental quality tends to *in*crease with affluence and *de*crease when a society experiences negative growth rates (i.e. in circumstances of economic recession or some other major crisis).

Societies may differ not only in terms of basic values, but also with regard to public *concern* over environmental degradation in general or a specific environmental problem in particular. Variance in public concern may reflect different levels of knowledge about or awareness of environmental change:

> A$_{II5}$: Societies – and segments within societies – differ in terms of their knowledge about environmental deterioration and the causes of such degradation. More specifically, knowledge about environmental degradation and its causes tends to be higher (a) the higher the level of education, and (b) the more open and pluralist the political system and the mass media.

The general public can rarely be considered a political *actor*. For most issues only a small fraction of the public will in fact be sufficiently knowledgeable and interested enough to voice its concerns. The general public will in most cases be politically 'dormant' most of the time, and slow to respond to environmental threats. Moreover, some segments of the public are likely to be able to exercise greater influence in political processes than others.

> A$_{II6}$: All societies are to some extent socially stratified, giving some groups a better position from which to influence political agendas and decision. Societies differ, however, in terms of *degree* of social inequality, and also with regard to which groups belong to which strata.

The practical implication of this assumption for empirical research is that we need to determine not only damage and abatement costs at the aggregate national level, but – whenever costs are unequally distributed (as they normally are) – also their domestic *distribution*.

> H$_{II2}$: Damage or abatement costs that hit the social 'centre' of society will be better articulated and carry more weight than those that hit the social 'periphery' only, and the bias will be stronger the greater the distance in social and political resources between centre and periphery.

The distribution of costs and benefits should be described not only in terms of the social stratification of society, but also in terms of the degree to which costs and benefits are concentrated in particular economic sectors or social groups. The general hypothesis would be that:

H_{III3}: Under 'business-as-usual' circumstances, the amount of political energy infused into a particular concern tends to be higher when costs and/or benefits are concentrated to specific and important sectors of the economy than when they are widely dispersed throughout society or indeterminate. For issues generating substantial political mobilisation, the opposite 'bias' can be expected.

Finally, some *kinds* of consequences tend to generate more demand for action than others. The general pattern seems to be:

P_{III}: Consequences that are *certain, close in time, highly visible,* come as a *shock*, and affect *productive assets or important national treasures,* tend to generate more political energy than those that are less certain, belong to some distant future, are less visible, occur gradually or according to expectations, and affect non-salient recreational values or consumptive interests.

The propositions suggested above are summarised in table 3.1.

In a nutshell, we argue that the more the environmental damage resembles Type A, the stronger the public demand for environmental protection will be. We have also used the dimensions included in table 3.1 to categorise the consequences of measures taken to abate the damage. We argue that in political decision-making processes Type A consequences tend to carry more weight than Type B consequences. Moreover, we argue that in cases involving asymmetries between damage and abatement costs, type A consequences will carry even greater weight the greater the asymmetry.

Ideally, we would have liked to describe environmental damage in terms of the dimensions listed above. In other words, we would like to know *who* are affected (the most), *what values* are threatened, and whether the threat is perceived as *certain, immediate*, etc. Where applicable, we would like to have the same pieces of information about abatement costs. This is, however, a tall order. In this study we have therefore decided to focus on only two aspects: the domestic distribution of damage and abatement costs, and the kind of values threatened by pollution.

Table 3.1   Factors affecting public demand and support for environmental protection

Dimensions	Type A (strong)	Type B (Weak)
Distribution of damage/abatement costs in 'social space'	Affect 'centre' of society	Affect social 'periphery' only
	Concentrated in specific and important sectors of the economy	Widely dispersed or indeterminate
Kinds of values affected	Productive assets, human health, or important natural treasuries	Recreation or non-vital consumption
Damage seen as	Certain	Uncertain, hypothetical
	Close in time	Remote in time
	Highly visible	Non-visible
	Appearing suddenly ('shock')	Occurring gradually and as expected

*Intermediate actors (*agents*)*   Public opinion may very well influence decision-makers *directly* – often by holding out some implicit promises of rewards for positive responses, or threats of sanctions (e.g. at the next election) for those who fail to respond. However, public concern is most often picked up by intermediate agents, such as non-governmental organisations, political parties, or the media. These agents *articulate*, *amplify*, *aggregate* and to some extent also *shape* societal interests and concerns. Thus,

A_{II7}: The strength with which societal demand and support are articulated depends upon the existence, capabilities and incentives of intermediate agents, such as political parties, non-governmental organisations, ad hoc campaigns, and the media.

The availability and strength of such agents may clearly vary for reasons that are not at all related to the specific problem addressed by the LRTAP regime. Actors such as NGOs and political parties typically address broader issue-areas. Moreover, the configuration of NGOs and political parties at any particular point in time tends to reflect conflicts that were salient and coalitions that were formed at critical junctures in the earlier history of that society. *Today's* ecology of organisations tends to a large extent to reflect *yesterday's* concerns. History is with us not only in the form of cognitive 'lessons' but also in the form of institutional relics.

A_{II8}: The political *capabilities* of intermediate agents is a function of their:

- financial resources
- (professional) staff
- membership size
- capacity to mobilise members and its broader constituency, and
- access to and role in relevant decision-making processes (including representation on committees and links to government agencies, political parties or other actors).

Actual *influence* is a function of capabilities and behaviour (skills and energy invested).

A_{II9}: Demand for and support of strict environmental policies will typically show the following pattern:

- Victims of environmental damage demand stricter emission reductions than any other group;
- Environmental NGOs and 'green' political parties will take positions close to those of the 'victims', while industrial organisations and parties with close links to labour or industry tend to give more weight to the (legitimate) interests of the polluters;
- Producers/providers of environment-friendly technology tend to demand less abatement than victims, but more than polluters. The technologically most advanced producers will accept stricter regulations than those that are less advanced; and
- Polluters tend to accept (but not actively promote)

environmental regulations to the extent that they are compensated for abatement costs.

H_{II4}: The greater the relative political capabilities of environmental NGOs, 'green' political parties, authorities of any geographical regions particularly affected by 'acid rain', and producers/providers of abatement- or environment-friendly production technologies, the more 'progressive' national positions will be, and the more effective the domestic implementation of international environmental agreements.

Governmental Supply

Even in democratic political systems, public policy is rarely if ever driven merely, and often not even primarily, by societal demand. Although embedded in society – and probably to a large extent sharing its values and concerns – governments do not merely react to social pressure; they have policy preferences of their own and political capabilities to pursue what they consider to be sound policies.

In the domestic politics model, governments are seen as complex organisations, over which no single decision-maker has complete control. Different political parties and branches of government may have different perceptions of the problem and different policy preferences, and none of these will necessarily be fully consistent with the 'national interest'. According to this perspective, the environmental policy pursued by a government can be seen as a function of four major determinants:[11]

(1)   the ideological profile of the cabinet in power;
(2)   the distribution of power and influence among different branches of government;
(3)   the extent to which the government (particularly the cabinet) controls state policy;
(4)   the extent to which the state in effect governs and controls 'its' society.

While factors 1 and 2 seem relevant at all stages of the policy process, the latter two are relevant primarily to the implementation phase.

Ideological profile of the government It is well known that different political parties and leaders subscribe to different world-views or ideologies. What is less clear is how the conventional ideological distinctions relate to environmental policies. Previous research indicates that there are only small systematic differences in this particular issue-area between moderate conservative, liberal, and social-democratic governments (see e.g. Jänicke, 1977). We nevertheless suggest:

> H_{II5}: Governments subscribing to a strong free-market ideology and cabinets of the extreme left (including communist parties) tend to give lower priority to environmental protection than governments formed by moderate conservative, liberal or social-democratic parties.

The distribution of power and influence among various branches of government Since the perspectives and preferences of decision-makers are to some extent shaped by role and position (cf. A_{II2}), the distribution of power and influence among ministries and agencies may affect negotiating positions as well as implementation records.

> A_{II10}: Within the governmental system, environmental ministries or agencies tend to be the champions of environmental regulations, while ministries/agencies in charge of the polluting sectors of the economy tend to be the most reluctant. In between we typically find major actors such as the Ministry of Finance and the Foreign Ministry (the latter probably preferring positions that can at least to some extent accommodate the prevailing view in the conference).

The extent to which a particular branch of government will be able to influence public policy depends upon several factors. One is its *institutional capacity*, measured in terms of, e.g., budget (per cent of total, and relative to ministries/agencies in charge of polluting activities), and staff (measured as for budget). Another important factor is the *political clout* of its leadership, measured in terms of the 'political standing' of the cabinet member/official currently in charge.[12] A third factor is *involvement* in relevant decision-making processes. Involvement may be measured in terms of formal authority as well as actual participation. Since environmental damage is typically a *side*-effect of other activities – such as industrial production, transportation of people and goods – environmental policy must, in order to achieve its objectives, succeed in 'penetrating' these other policy spheres. Accordingly, 'relevant' decision-making processes are not only those that deal with environmental protection in a narrow sense.

> H_{II6}: The greater the relative institutional capacity and involvement of the environmental branch of government, and the greater the 'political clout' of its leadership, the greater its influence over policy formation and implementation, and the more 'progressive' will be national negotiating positions and the better the record of implementation.

The extent to which government (the cabinet) controls state policy As pointed out above, this dimension is important primarily for implementation.

The extent of governmental control over state policy seems to depend on several factors, including:

The degree of *internal unity* of the cabinet itself (single party vs. coalition government; internal policy distance within ruling party or coalition), hence:

> H_{II7}: The more unified a cabinet, the greater its capacity to deliver on its promises, and the better its implementation record will be.

Cabinet stability and control over the legislature, hence:

> H_{II8}: The more stable the cabinet and the firmer its control over the legislature, the greater its capacity and the better its performance in implementing international agreements.

Policy distance to major opposition party or bloc. Thus,

> H_{II9a}: The shorter the policy distance between the cabinet and the opposition, the more likely that international commitments will be met even under a new cabinet. (*)

> H_{II9b}: Implementation of international environmental agreements at the national level will be more effective to the extent that domestic consensus was reached among important political actors – and with target groups – *before* and *during* the negotiation phase (cf. Victor, Raustiala and Skolnikoff, 1998).

Personal political authority of the *head* of government:

> H_{III0}: The stronger the personal political authority of the head of cabinet, the greater the government's capacity to fulfil its commitments and the better its implementation record. (*)

The extent to which government controls society (policy outcomes) This dimension is relevant primarily, perhaps only, to the analysis of implementation. Important factors include:

The extent to which (formal) authority over environmental policy is *centralised* at the national (federal) level, hence:

> H_{III1}: Centralisation of authority will tend to enhance implementation capacity but not necessarily improve records of compliance. Decentralisation of authority is likely to increase the range of *variance* in implementation performance, the overall outcome depending on the extent to which political domains are 'congruent' with the ecosystems or systems of activities concerned – in particular the extent to which victims and polluters are located within the same administrative unit (region).

The *extractive capacity* of the state. This is a complex and elusive concept. One rough indicator is fiscal capacity, measured as central government income in per cent of the country's GDP, but also legal provisions and judicial systems may be important.

> H_{III2}: The greater the extractive capacity of the state, the more capable it is of implementing its own decisions, and the better its implementation record will be. (*)

The extent to which the target groups (i.e. the polluters whose behaviour is to be modified) belong(s) to the public sector, or are *financially dependent* upon funding from central government sources.

> H_{III3}: Implementation is easier and will be more effective in relation to public sector targets and targets who are financially dependent upon the national government. (*)

Relationship of supply to demand   As pointed out above, a basic assumption of the domestic politics model is that supply and demand are interrelated driving forces. Thus,

> A_{III1}: Governments tend to respond to societal demand by taking some step(s) to provide what is demanded, but in politics there is no straightforward relationship between demand and supply.

> P_{II2}: Governments in democratic political systems tend to be more responsive to societal demands (in general, and demands for environmental policies in particular) than governments in non-democratic systems, and politically 'vulnerable' governments tend to be more responsive than those that are less 'vulnerable'.

In practical terms, we will explore to what extent national policies were driven mainly by societal demand or by governmental leadership, and how the two interacted to produce national positions and determine implementation outcomes. We should realise, however, that such assessments will have to be based on informed judgement rather than systematic analysis of 'hard' data.

The Domestic Politics of Implementation: A Closer Look

Another fundamental premise of the domestic politics model may be summarised as follows:

> A_{III2}: The policy process consists of a series of (partly overlapping) games. Implementation games tend to differ systematically from those of earlier phases in terms of arenas, patterns of participation, and the distribution of influence (see Allison, 1971; Allison and Halperin, 1972).

This assumption can be further specified as follows:

> A_{III2a}: As the policy process moves from the initial stage of problem identification and diagnosis to the stage of implementation, the policy options considered tend to become more specific.

A_{III2b}: The more specific the policy measure that comes up for decision, the more determinate and differentiated its impact tends to be on society.

A_{III2c}: As subgroup concentration of policy impact increases, the involvement of sector agencies and institutions (the 'organisational–corporate channel') tends to increase, while the role of the 'numerical–democratic channel' – including the role of Parliament, political parties, and 'generalists' such as the Ministry of Foreign Affairs – tends to decline.

A_{III2d}: The less symmetrical the impact of a problem or a policy on society, the more conflict it tends to generate; the more conflict it generates, the more likely there will be some mobilisation of non-governmental actors, such as environmental groups, industrial organisations, ad hoc campaigns, media, etc.

A_{III2e}: The 'organisational–corporate' channel tends to amplify the interests and concerns of specific sectors of the economy or segments of society, while 'numerical–democratic' processes tend to shift the balance in favour of concerns that are widespread among the attentive public.

The major propositions that can be deduced from these assumptions point towards outcomes that deviate systematically (but not necessarily dramatically) from those that are predicted by the unitary rational actor model. Thus,

H_{III4a} (cf. H_{II2}): In 'business-as-usual' circumstances, the policy measures that are most easily implemented will be those which offer tangible benefits to some specific sector of the economy or segment of society while costs are widely dispersed throughout society (cell 4 in figure 3.1). Conversely, the commitments that are hardest to implement are those where costs are concentrated to specific sectors or segments while benefits are indeterminate or widely dispersed (cell 1).

H_{III4b}: For issues generating substantial political mobilisation, the opposite 'bias' can be expected. Political campaigns tend to be most successful when they can target one or a few distinct

'culprits' (rather than mass behaviour) and a small number of
discrete events (rather than a continuous stream of activities).

Taken together, these propositions suggest that implementation is
likely to be most successful for policies that (a) offer net benefits to
specific sectors of the economy or segments of society, while costs are
indeterminate or widely distributed, and (b) conform to core values
subscribed to by the attentive public. In other words, there is nothing as
attractive as a measure that *combines* private benefits (profit) for organised
interests with public virtue.

Figure 3.2   **Policies distinguished according to the distribution of
costs and benefits (based on Wilson 1973)**

Costs

Benefits

	Concentrated	Distributed
Distributed	1	2
Concentrated	3	4

The misfortune of environmental policy is that it is relatively poor in
such measures. Environmental damage typically occurs as *side effects* of
other perfectly legitimate activities, such as industrial production or
transportation of people and goods. Accordingly, most environmental
policies cannot simply supplement other policy commitments; in order to
succeed they somehow have to *penetrate* the activities that cause damage
to the environment in the first place. As policy ideas are further developed
and specified, it will become increasingly clear that many of them will
have substantial consequences for particular sectors of the economy or
segments of society. Regulatory policy tends to impose costs upon those
whose behaviour is to be changed, and the targets of environmental
regulations are normally societal actors (producers, consumers). They are
also the ones who in effect 'control' the activities that are to be modified.

Accordingly, governments are dependent upon their co-operation – voluntary or coerced – to ensure compliance.

In fortunate (though perhaps rare) circumstances, a company or even an entire branch of industry will reap substantial profits from environmental regulations. For example, banning the emission of a certain substance most often puts a premium on benign substitutes. A company that is ahead of its competitors in developing such substitutes may find that a ban can be a powerful device to strengthen its competitive edge. Regulations to phase out CFCs may be one case in point. One reason why the major US producer changed its mind in support of global regulations may have been that it was a pioneer in developing environmentally benign substitutes. International regulations prohibiting the use of CFCs could therefore serve as an instrument for enhancing its own share of the global market. In other cases, producers of new abatement or production technology can see environmental regulations as a vehicle for attracting customers. Even when it is opposed to strict regulations, a particular branch of industry will often want to see its foreign competitors subject to at least equally strict measures, and hence join 'its' government in pushing for *international* standards. Finally, it should be added that the toolbox of environmental policy does include instruments other than regulatory restrictions. Financial support for the development of renewable energy or more energy-efficient technologies are examples of instruments moving policy towards cell 4. More generally, governments have at least some leeway in designing policies and measures so that they generate politically benign configurations of interests.[13]

Environmental policies also tend to suffer from the existence of certain basic asymmetries between costs and benefits. In particular, the costs of measures to protect the environment tend to be fairly certain and immediate, while benefits tend to be uncertain, perhaps even hypothetical, and accrue only in some more or less distant future (see table 3.1).

All this leads us to suggest:

> P_{II3}: Environmental policy faces a serious risk of *vertical disintegration*, i.e. a state of affairs where the aggregate thrust of 'micro-decisions' deviates more or less significantly from what policy doctrines or principles would lead us to expect.

Note that the logic of politics is such that *neither* macro-principles nor micro-decisions will necessarily accurately reflect the 'true' preferences of government or the 'national interest'. General doctrines and principles – often developed in the early stages of the policy process – are likely to be biased in favour of *collective* values and concerns (including environmental values), while specific micro-decisions – many of which will have to be taken in the implementation phase – tend to be biased in favour of *subgroup* interests, particularly the interests of producers.

So far we have focused on basic political mechanisms that are assumed to be present in all or at least many political systems. However, model II leads us to expect that the strength of these mechanisms will depend on, inter alia, characteristics of the political system itself and the relationship between state and society. Two of the most general propositions may be stated as follows:

> P_{II3a}: The mechanisms leading to inadvertent vertical disintegration of policy tend to be stronger in open, pluralist political systems than in those that are more closed and centralised (cf. Lundqvist, 1990, p.34).[14]

> P_{II3b}: The weaker a state's control over 'its' society, the greater the risk of implementation failure.

Finally, the domestic politics model suggests that *signing* (and ratifying) an international agreement can make a significant difference. However, while model I attributes any impact on actor behaviour to some implementation bonus or sanction costs, model II focuses, first, upon the impact the agreement may have upon the domestic distribution of power and influence, and, second, upon incentives generated by the *act* of signing:

> P_{II4}: Signing (and ratifying) an international agreement tends to *empower* the governmental agencies (and also the IGO) in charge vis-à-vis other branches of government (or IGOs), and also vis-à-vis the societal actors to whom the regulations apply. For the sub-actors in charge, it also tends to generate enhanced incentives to 'deliver'.

'Empowerment' typically occurs through mechanisms such as the formal approval of a more comprehensive mandate, informal attribution of legitimacy and status, and the transfer of resources. A similar effect may occur for non-governmental organisations associated with the agreement or the general policy field as well.

Explanatory Model III: Social Learning and Policy Diffusion

Our third model differs from the other two primarily in its assumptions about the nature of the policy-making and policy-implementation processes. While most applications of the unitary rational actor model treat the problem and available 'cures' as exogenously determined, model III assumes costs and benefits to be concepts that leave considerable scope for interpretation, and solutions as something that will, to a significant extent, have to be discovered or invented through the policy-making process itself. In principle, this applies to all issue-areas, but environmental protection is – in comparative terms – a new policy field characterised by substantial uncertainties about the nature and magnitude of problems and few firmly established policy strategies. Moreover, while models I and II both see *decision-making* as the essence of the policy process, our third model focuses primarily on processes of *searching, learning* and transnational *diffusion* of knowledge and ideas. In brief, it directs our attention to what might be called the 'ideational' basis of public policy, as opposed to interests and power (cf. Hasenclever, Mayer and Rittberger, 1997). As outlined below, it combines elements of what is often referred to as 'cognitive' (Jönsson, 1993) and 'contructivist' (Adler, 1997; Ruggie, 1997) approaches.

Its basic assumptions can be summarised as follows:

A_{III1}: States and governments are complex organisations, embedded in and influenced by (even more) complex societies.

A_{III2}: Decision-makers typically enter policy processes with imperfect information and tentative preferences. Accordingly, they engage not only in persuasion but also in active search for information and ideas (cf. Haas, 1992)

A_{III3}: Policies develop to a large extent through *learning* – through the development and adoption of new knowledge, norms and ideas – and are maintained through *internalisation* and *routinisation*, i.e. by becoming incorporated into a party's own belief system and 'standard operating procedures'.

The implications of the latter assumption are more profound than they might seem at first. In the jargon of constructivist theory, model III leads us to see international regimes not only as regulatory devices but also as 'constitutive' entities – meaning that they also *shape* actor preferences and beliefs. Moreover, it leads us to view implementation as being subject to what March and Olsen (1989) call 'the logic of appropriateness' – i.e. guided by a sense of duties and obligations – rather than the 'logic of consequence'.

Learning and the 'social construction' of ideas may be reciprocal processes, but they will rarely if ever be fully symmetrical ones. For example, several studies of the development of EC environmental policies (e.g. Liefferink, 1996; Heretiér, Knill and Mingers, 1996) indicate that the typical pattern is one where one or a few member countries, in collaboration with the Commission, tries to persuade the EC to adopt Community policies in line with the those that the most 'advanced' member country has already adopted for itself. More generally, to the extent that processes of learning and policy diffusion are at work, we would expect to find a particular pattern of policy development:

> P_{III1}: The problem diagnosis and policy ideas submitted by those actors who enter the process with superior knowledge (i.e. candidates for what Young (1991) refers to as 'intellectual leadership') or higher social status, will to a large extent be adopted by actors entering with inferior knowledge, less well-prepared positions, or lower status.

> P_{III2}: Transnational learning and policy diffusion are most likely to occur between actors who are ideologically and culturally 'close' and with regard to politically 'benign' problems.

> P_{III3}: International agreements as well as informal norms and ideas will be adopted and implemented to the extent that they offer compelling models that powerful actors (a) accept on their own substantive merits, and (b) can use to convince or persuade others (cf. Strang and Chang, 1993).

P_{III4}: Compliance tends to reinforce compliance. In other words, environmental policies develop in part by imitation, and the build-up of political 'momentum' is an important mechanism in the establishment as well as the maintenance of international regimes.

One important implication of P_{III1}-P_{III3} is that substantial progress towards solving international problems can be made even in the *absence* of any formal, substantive agreement (see Underdal 1994). Studying the effectiveness of formal international environmental accords should, therefore, only be part of our agenda. If we want to understand how governments respond to common problems, we also need insight into the processes and mechanisms whereby ideas become adopted and *unilateral* adjustments occur in response to new knowledge or ideas.

Some of the contributions – particularly those belonging to what Hasenclever, Mayer and Rittberger (1997) refer to as the 'strong cognitivist' camp – describe the formation and diffusion of ideas mainly in terms of 'natural' or 'organic' processes. Others – including Haas (1990, 1992) – have more to say about the agents and arenas involved:

A_{III4}: Transnational networks of experts ('epistemic communities') are the principal *agents* of learning and policy diffusion, and international negotiations and organisations – non-governmental as well as intergovernmental – are major *arenas* where these processes occur.

Finally, model III also has something to say about factors enhancing compliance and facilitating implementation. The essence may be summarised as follows:

P_{III5}: Compliance with an international agreement will increase (a) the more comprehensive and deeper the base of 'consensual knowledge' and shared policy norms upon which the agreement is built; (b) the more tightly integrated and stronger the transnational epistemic community that serves the regime; and (c) to the extent that the international regulations become incorporated into the standard operating procedures of domestic bureaucracies.

At this point we will not pursue model III further. In the empirical analysis we will focus mainly on models I and II. We do so partly because this third model is as yet less developed and specified than the other two. Scholars working with this latter approach have demonstrated that learning and diffusion of knowledge and policy norms can, at least under certain circumstances, be important processes shaping national policies as well as international regimes (see e.g. Andersen and Liefferink, 1997; Jänicke and Weidner, 1997). They have also demonstrated that epistemic communities can be major agents developing and disseminating consensual knowledge and thus help foster substantive international co-operation (cf. Haas, 1990). However, in its present form the model does not yield conclusive predictions in terms of the substantive *content* of the agreements that are most likely to be adopted or successfully implemented. It identifies some of the mechanisms that serve to enhance compliance, but has yet less to say about the circumstances under which we can expect these mechanisms to be sufficient to ensure that commitments are honoured. This should not necessarily be taken as criticism; processes such as learning, diffusion and the 'social construction' of beliefs seem inherently more difficult to model than processes of choice. Some scholars explicitly see it primarily as a *supplement* rather than a full-fledged alternative standing on its own feet only (Jönsson, 1993, p.203). Finally, model III is arguably more relevant to the initial stages of regime formation than to the 'final bargains' and the stage of implementation (see Young and Osherenko, 1993, p.236). That being said, however, we hasten to add that it does also have important implications for the study of compliance and implementation, particularly in issue-areas characterised by high uncertainty about the nature and magnitude of the problems and the effectiveness of alternative 'cures'. Environmental protection fits this description quite well.

Modelling Open Systems and Dynamic Processes

The description of model III implicitly points to a problem with the unitary rational actor and domestic politics models as outlined above. So far, they have both been described largely with reference to a closed system and in static terms. Let us now very briefly consider how these two models might be extended to deal with *open* systems and *dynamic* processes.

Open Systems

Extending these models to open systems requires essentially two kinds of adjustments. First, we have to allow for inputs from the outside into 'domestic' political processes. Second, we have to allow for the possibility that an issue can be considered not only on its own substantive merits but also in the context of some other issues or policy concerns. We must, in other words, be sensitive to issue linkages and *policy context.*

The unitary rational actor model responds to the former challenge by conceptualising negotiating behaviour and compliance in *strategic* terms, i.e. as based on assumptions or information about what others do. For example, what constitutes the optimal level of compliance for a particular actor is seen as contingent upon what one or more of the other parties do.[15] Adapting model II to the analysis of open systems involves at least three major steps. One is to consider the possibility of social learning and transnational diffusion of knowledge and ideas, as we have done in model III. The other is to allow for the possibility of transnational or transgovernmental coalitions. One example may be environmental ministries and NGOs working together to overcome the resistance of ministries and associations of industry. The third step is to examine the impact of international organisations as arenas and actors.[16]

Moving to the functional dimension, the point of departure is the observation that any issue will be handled in the context of a particular 'policy space' that exists in that particular political system at that particular time. Moreover, political processes often produce issue-linkages based on strategic and tactical considerations rather than inherent functional inter-connectedness (for an interesting example, see Boehmer-Christiansen, 1992). The context into which a particular problem is framed can make a significant difference with regard to, inter alia, the kinds of perspectives and concerns that are considered relevant or appropriate, and what actors will have access to the policy-making process. The general implication of these observations is that 'acid rain' policies may be significantly affected by other concerns, and can – as the motives behind Mr. Brezhnev's initiative remind us – serve purposes that are in no way substantively related to the problem itself. Similarly, exogenous events – such as the need to form a new governing coalition – may involve some kind of package deal in which certain environmental policies are 'traded' for something else. No attempt will be made in this project to develop a model for predicting such linkages or determining their impact. All we have been able to do here is to try to determine, on an ad hoc basis, the extent to which issue-linkages and policy context have influenced national positions and implementation behaviour. Where the policy context seems to be important, we will also try to determine *which* exogenous (non-environmental) policy concerns seem to have been particularly important in that particular case.

Dynamic Processes

We consider the LRTAP Convention and subsequent protocols to be distinct but related events. One important question becomes, then, *to what extent and how did events occurring at the beginning of this chain affect what happened later in the process?* As indicated in chapter I, the LRTAP regime has evolved through different iterations of policy-making along two separate, but interrelated trajectories – one dealing with the same pollutant (e.g. SO_2), the other expanding the scope of the regime to include new pollutants. Relevant literature offers at least two different perspectives for analysing such (causal) paths. One stems from the neo-functionalist notion that – under certain circumstances – processes of international co-operation gather momentum, through functional spill-over, elite socialisation and other mechanisms (see e.g. Nye, 1970). In essence, the proposition is that successful co-operation tends to generate various kinds of positive feedback facilitating further steps. An alternative perspective stems from the economists' 'law of diminishing returns'. Applying this 'law' to the study of environmental policy, we would suggest that the first steps taken are likely to be the 'easy' ones – those from which marginal utility clearly exceeds marginal costs. As attempts are made to extend the regime to new areas or to reach more ambitious environmental targets, however, marginal abatement costs tend to increase and/or returns in the form of environmental quality tend to decrease. To the extent that the cost-benefit balance changes as described above, extension tends to become increasingly difficult. The competing propositions can be formulated as follows:

> P_{IV1a}: Governments that have signed (an implemented) IEAs in a particular domain tend to increase their policy ambitions over time and will be more eager than others to negotiate more ambitious follow-up agreements.

P_{IVIb}: Governments tend to become increasingly reluctant to follow up initial commitments with more ambitious measures – at least in so far as marginal abatement costs increase or marginal benefits decrease.

A basic assumption, common to all our three models, is that

A_{IVI}: The existence of an international regime tends to enhance rule-consistent behaviour of member countries. The degree of rule-consistent behaviour tends to increase as regimes *mature*, but may decline again if a regime becomes 'obsolete'.

The *mechanisms* assumed to enhance rule-consistent behaviour are different, though. Model I attributes the impact to some implementation bonus or sanction costs. Model II focuses upon the impact that the regime may have upon the domestic distribution of formal authority and political influence (empowerment of the environmental branch of government and other agencies in charge of implementation) and incentives generated by the act of signing. Model III points to internalisation of knowledge, norms and ideas.

Finally, we may point out that all models lead us to expect links between the different stages of the process. The most obvious proposition is:

P_{IV2}: The closer the terms of an IEA to a state's preferred solution, the more likely that implementation will be successful.

Exogenous Factors

There is, of course, a very real possibility that any change in actual emissions may be caused by factors exogenous to all three models – notably by changes in the level or structure of economic activities or in available technologies for pollution abatement and/or production (see e.g. Weiss and Jacobson, 1998; Victor, Raustiala and Skolnikoff, 1998). Both would, of course, in turn show up in estimates of abatement costs, but they are potentially so important to the understanding of the mechanisms shaping the final outcomes that we will try to explore their impact in our case studies.

Concluding Remarks

Our point of departure was that all three models have important insights and policy 'implications' to offer. The question before us is not which model is 'true' but rather *how much* of the variance observed can each of them account for, and which provide the most *useful* approach given a particular purpose of analysis. Let us conclude by offering a few rules-of-thumb that may help us answer the latter question.

First, the unitary rational actor model provides a picture of what might be called the 'essential (material) structure' of the problem. As indicated in figure 3.1), this essential structure can be seen as a basis from which other and more complex models proceed. Although strong reasons can often be given for moving beyond the confines of model I, the burden of proof seems to rest with anyone who would argue that it can be discarded completely.

Second, it seems to me that the more *specific* our dependent variable, the stronger the case for moving beyond the unitary rational actor model. The finer the nuances in behaviour that we want to understand, the more likely it seems that domestic politics and/or cognitive processes will leave a significant imprint. Conversely, the cruder our distinctions in terms of positions or levels of compliance/implementation, the more likely it seems that model I would account for most of the variance observed.

Third, the more *extensive* the analysis (i.e. the larger the number of cases to be covered) relative to the resources available for research, the stronger the case for the most parsimonious model (i.e. model I). The amount and range of additional data required as we move from model I to model II can be quite substantial. Thus, one important challenge facing those of us who work with the more complex models is to demonstrate not only that politics 'matter' (which, in unspecified form, is a rather trivial observation), but also that the marginal returns of the time and energy invested in developing and applying complex models exceed the marginal costs involved.

Fourth, while (stochastic) generalisations may serve the needs of the scholar himself the practitioner would – in the absence of universal laws – almost invariably be more concerned about the specifics of his particular case(s). This suggests that to the policy-maker model I provides a necessary, but rarely if ever a sufficient basis for action; he will most often demand also the kind of insights that can be provided by model II and – probably to a lesser extent – those that flow from model III.

Finally, as we have pointed out above, model III seems to be most relevant for issue-areas characterised by high uncertainties and few firmly established policies serving as default options. We have also suggested that learning and policy diffusion are more likely to occur with regard to *politically 'benign'* problems and in relationships among 'friends'. One might also argue that model III provides a useful conceptual framework for studying the evolution of environmental policies *over time*, in response to changing perceptions of the problem and the development of policy ideas. Overall, model III may yield the most useful insights if it is *combined* with one (or both) of the other two – and the domestic politics model may be the more compatible 'partner'.

Acknowledgements

This chapter builds on my article 'Explaining Compliance and Defection: Three Models', which appeared in *European Journal of International Relations*, vol. 4 (1998), pp. 5-30. The material is used here with the permission of the editor and Sage Publications. I gratefully acknowledge useful comments and suggestions to earlier drafts from other members of the project team as well as from Elaine Carlin, Jon Hovi, Peter M. Haas, and two anonymous reviewers for EJIR. Thanks are due also to Lynn P. Nygaard for editorial assistance.

Notes

1. There is also a more 'benign' version of the hegemonic stability hypothesis, focusing on the hegemon as provider of collective goods through *unilateral* action (see e.g. Kindleberger, 1981; Snidal, 1985).

2. This does not mean that defection can be completely ruled out, but if it occurs it would be due to lack of capacity or human *error* rather than to incentive structures. Model I focuses on the latter.

3. $P_{12a} - P_{12b}$ apply only under the assumption that others will comply at least as long as we do. An actor considering whether or not to defect *in response to* defection by others may, in fact, find that the reverse would apply in his case.

4. The overall relationship between the parties may also be important. Other things being equal, a party is likely to be more sensitive to the defection of a rival or enemy than to the 'implementation failure' of a friend. Note that the greater the costs of coercive action to

enforce the compliance of a partner, the less likely it is that others will respond to its defection by imposing external sanctions.

5. The word 'indirectly' is inserted to indicate that the defection of a partner in response to an actor's own failure to deliver may trigger the defection of *others* as well. A rational actor would certainly take into account such 'chains of falling dominoes'.

6. Given that A has signed the agreement and that at least some partners would react to defection by imposing some kind of sanctions *beyond* defecting themselves, the sanction costs *avoided* could be seen as a benefit of compliance, implying that the *difference* in utility to A between complying and defecting can be calculated according to the equation $U_a^a + (P_i^a S_i^a)(U_a^i + IB_a^i + SC_a^i)$.

7. It may, however, depend critically upon the credibility of threats of 'internal' retaliation (i.e. upon the willingness of others to respond to any defection by defecting themselves).

8. I am grateful to Duncan Snidal for helping me see this point clearly.

9. Note that the importance of transparency and monitoring tends to increase the more costly the behavioural change prescribed by the regime. Several practical suggestions for minimising the fear of deception can be found in Bilder (1981).

10. The distinction between *societal* demand and *governmental* supply is a crude organising device, adopted mainly for heuristic purposes. Clearly, environmental protection can in many cases be undertaken by societal actors themselves, and governmental actors often enter the equation not only on the supply side but also as generators or articulators of demands. In empirical accounts of the policy process, demand and supply factors are likely to be intertwined; governmental supply will at least to some extent respond to demand, while societal demand and support are likely to be influenced by governmental policies.

11. For a different and broader specification of overall 'capacity' for environmental protection, see Jänicke (1997).

12. Admittedly, this is an elusive variable. The only feasible way to measure political clout would probably be to use what is often referred to as the 'reputational' method, i.e. determining whether the minister in charge is a junior or senior member of cabinet – according to the conventional wisdom of informed observers, such as political journalists.

13. For an interesting illustration at the international level, see Mitchell (1994).

14. This is certainly *not* to suggest that closed political systems with high concentration of power will have a better record of environmental protection or compliance with international environmental agreements than open, pluralist systems. In fact, the empirical evidence that we have seems to point strongly towards the opposite conclusion. The point to be made here is that, other things being equal, concentration of power tends to enhance implementation *capacity* – but, of course, also the capacity to cheat or defect. What P_{II3a} suggests is that the *mechanisms* leading to non-compliance, if any, are likely to be different: authoritarian and totalitarian systems defecting by deliberate choice, while open, pluralist systems may be *unable* to deliver on their promises.

15. The model does, of course, allow for *dominant* strategies, i.e. a course of action that is superior to all other options in at least one situation and under no circumstances inferior to any other option.

16. The latter has been done in a follow-up study, see Hanf, Sprinz et al., (1997).

References

Allison, G. T. (1971), *Essence of Decision: Explaining the Cuban Missile Crisis*, Boston: Little, Brown & Co.

Allison, G. T. and Halperin, M.H. (1972), 'Bureaucratic Politics: A Paradigm and Some Policy Implications', in R. Tanter and R.H. Ullman (eds.) *Theory and Policy in International Relations*, pp. 40-79, Princeton, N.J.: Princeton University Press.

Andersen, M. S. and Liefferink, D. (1997), 'Introduction: the Impact of the Pioneers on EU Environmental Policy', in M. S. Andersen and D. Liefferink (eds.), *European Environmental Policy: The Pioneers*, Manchester: Manchester University Press.

Axelrod, R. (1984), *The Evolution of Cooperation*, New York: Basic Books.

Bilder, R. B. (1981), *Managing the Risks of International Agreement*, Madison: University of Wisconsin Press.

Boehmer-Christiansen, S. (1992), 'Anglo-American Contrasts in Environmental Policy-Making and their Impacts in the Case of Acid Rain Abatement', *International Environmental Affairs*, 4, pp.295-322.

Chayes, A. and Chayes, A.H. (1993), 'On Compliance', *International Organization*, 47, pp.175-205.

Crossman, R.H.S. (1965), *Government and the Governed*, London: Chatto & Windus.

Haas, P. M. (1990), *Saving the Mediterranean: The Politics of International Environmental Protection*, New York: Columbia University Press.

Haas, P.M. (1992), 'Introduction: Epistemic Communities and International Policy Co-ordination', *International Organization*, 46, pp.1-35.

Hanf, K., Sprinz, D. et al. (1997), *International Governmental Organizations and National Participation in Environmental Regimes: The Organizational Components of the Acidification Regime*, Unpublished report. Rotterdam: Erasmus University, Department of Public Administration.

Hasenclever, A.; Mayer, P. and Rittberger, V. (1997), *Theories of International Regimes*, Cambridge: Cambridge University Press.

Heretiér, A., Knill, C. and Mingers, S. (1996), *Ringing the Changes in Europe. Regulatory Competition and the Transformation of the State. Britain, France, Germany*, Berlin: Walter de Gruyter.

Hovi, J. (1992), *Spillmodeller og internasjonalt samarbeid: oppgaver, mekanismer og institusjoner*. Unpublished Dr.Philos.-dissertation, Department of Political Science, University of Oslo.

Iklé, F.C. (1964), *How Nations Negotiate*. New York: Harper & Row.

Jacobson, H.K. and Weiss, E.B. (1995), 'Strengthening Compliance with International Environmental Accords: Some Preliminary Observations from a Collaborative Project', *Global Governance*, 1, pp.119-148.

Jänicke, M. (1997), 'The Political System's Capacity for Environmental Policy', in M. Jänicke and H. Weidner (eds.), *National Environmental Policies: A Comparative Study of Capacity-Building*, Berlin: Springer.

Jänicke, M. and Weidner, H. (1997), 'Summary: Global Environmental Policy Learning', in M. Jänicke and H. Weidner (eds.), *National Environmental Policies: A Comparative Study of Capacity-Building*, Berlin: Springer.

Jönsson, C. (1993), 'Cognitive Factors in Explaining Regime Dynamics', in Volker Rittberger (ed.) *Regime Theory and International Relations*, pp. 202-222. Oxford: Clarendon Press.

Kindleberger, C. P. (1981), 'Dominance and Leadership in the International Economy', *International Studies Quarterly*, 25, pp.242-254.

Knoepfel, P. (1997), 'Switzerland', in M. Jänicke and H. Weidner (eds.), *National Environmental Policies. A Comparative Study of Capacity-Building*, Berlin: Springer.

Liefferink, D. (1996), *Environment and the Nation State*, Manchester: Manchester University Press.

Lundqvist, L. J. (1980), *The Hare and the Tortoise. Clean Air Policies in the United States and Sweden*, Ann Arbor: University of Michigan Press.

March, J.G. and Olsen, J.P. (1989), *Rediscovering Institutions*, New York: Free Press.

Mitchell, R.B. (1994), 'Regime Design Matters: Intentional Oil Pollution and Treaty Compliance', *International Organization*, 48, pp.425-458.

Nye, J.S. (1970), 'Comparing Common Markets: A Revised Neo-Functional Model', *International Organization*, 24, pp.796-835.

Putnam, R.D. (1988), 'Diplomacy and Domestic Politics: the Logic of Two-Level Games', *International Organization*, 42, pp.427-260.

Ruggie, J.G. (1997), *Constructing the World Polity*, London: Routledge.

Schelling, T. C. (1960), *The Strategy of Conflict*, Cambridge, MA: Harvard University Press.

Snidal, D. C. (1985), 'The Limits of Hegemonic Stability Theory', *International Organization*, 39, pp.579-614.

Sprinz, D. and Vaahtoranta, T. (1994), 'The Interest-based Explanation of International Environmental Policy', *International Organization*, 48, pp.77-105.

Strang, D. and Chang, P.M.Y (1993), 'The ILO and the Welfare State', *International Organization*, 47, pp.235-262.

Underdal, A. (1994),'Progress in the Absence of Substantive Joint Decisions?', in T. Hanisch (ed.), *Climate Change and the Agenda for Research*, Boulder, CO: Westview Press.

Victor, D.G., Raustiala, K; and Skolnikoff, E.B. (eds.) (1998), *The Implementation and Effectiveness of International Environmental Agreements*, Cambridge, MA: The MIT Press.

Weiss, E.B. and Jacobson, H.K. (eds.) (1998), *Engaging Countries: Strengthening Compliance with International Environmental Agreements*, Cambridge, MA: MIT Press.

Wendt, A. (1992), 'Anarchy is What States Make of It: The Social Construction of Power Politics', *International Organization*, 46, pp.391-425.

Wilson, J.Q. (1973), *Political Organizations*, New York: Basic Books.

Young, O.R. (1979), *Compliance and Public Authority*, Baltimore: The Johns Hopkins University Press for Resources for the Future.

Young, O.R. and Osherenko, G. (1993), 'Testing Theories of Regime Formation', in V. Rittberger (ed.) *Regime Theory and International Relations*, Oxford: Clarendon Press.

4 Sweden - A Pioneer of Acidification Abatement

KARIN BÄCKSTRAND AND HENRIK SELIN

Introduction

In Sweden, the combination of the country's natural ecological vulnerability to acidification and its geographical location have been domestic focal points for the treatment of the 'acid rain' issue. Environmental damage varies across regions, depending on geological and meteorological factors. The greatest impact is found in the South-western part of Sweden, where the acidity of the soil is high (i.e. low pH), precipitation high, and winds carrying acidifying substances from Europe blow in.

Political interest in the acidification issue in Sweden emerged in the late 1960s, when the problems of long-range transport of sulphuric compounds and their harmful environmental effects were first highlighted by scientists. In the 1970s and 1980s, public and political interest in the acidifying effects of sulphur emissions reached its peak. As the sulphur problem was gradually mitigated, the focus partly shifted in the late 1980s towards the more multifaceted and complex character of NO_X emissions. The early discussions on NO_X were related mainly to eutrophication, but as the acidifying effects of these pollutants became clearer, increased attention was given to NO_X and its contribution to the problem of acidification. Moreover, as sulphur emissions have been reduced, NO_X emissions have come to account for a larger proportion of Sweden's overall national acidifying depositions. The VOC issue per se has received only marginal attention, but is conceptualised as being closely intertwined with the NO_X problem. Actions aimed at reducing NO_X emissions are also viewed in terms of their effects on VOC emissions and vice versa.

In the Swedish debate on SO_2, NO_X, and VOC emissions, actual and potential domestic damage costs have been influential factors. These

87

damage costs have been high in contrast to national abatement costs that have generally been lower, albeit varying with time and pollutant. In addition to the public and private costs of mitigating already existing environmental damages, future costs scenarios have been presented. While the principal focus has been on the forest and paper industry, negative effects for lakes and fish stocks, affecting both the professional fishing sector and the tourist industry, have also been highlighted.

Sweden is one of the largest net importers of acidifying substances in the UNECE region. Prevailing air currents are a major transport medium for acidifying substances, originating from sources located in central Europe and the United Kingdom. In an attempt to reduce the amount of pollution that is carried into the country, Sweden has consistently been a strong advocate of strict and legally binding international emission control schemes.

Domestic Actions

Following the discovery that the burning of coal and oil result in SO_2 and NO_X emissions, related issues were initially treated as local air pollution problems, perceived to affect mainly people living in crowded cities.[1] In 1966, the first major government report on air pollution concluded that the high concentrations of SO_2 emissions in the larger cities were hazardous for the human health, causing respiratory diseases (Swedish Official Investigations, 1966). In addition, local effects such as plant damage and corrosion were also discussed.

A turning point in the framing of the SO_2 issue came in 1967, when the Swedish scientist, Svante Odén, reported in the prestigious daily newspaper *Dagens Nyheter* that the long range transport of acid air pollutants had serious adverse effects on ecosystems. In the wake of Odén's article, a wide-spread national debate followed and the issue was now transformed from being a local human health issue into a national environmental problem. Odén's findings were not fully accepted among all of his scientific colleges (Naturvårdsverket, 1991). Virtually the whole research community agreed on the basic connection between SO_2 emissions and acid precipitation, but scientists differed in their views on the effects of acidification (Lundgren, 1991, p.26). Although Odén's theory was contested, the Swedish government responded quickly and forcefully to his findings. In the decade following Odén's article, a series of national legislative measures was introduced with the aim of reducing national sulphur emissions.

In 1967, the newly established Swedish Environmental Protection Agency (SEPA) submitted a plan to the Social Democratic government for lowering the sulphur content in fuel oil to 2.5 per cent weight. In addition, SEPA proposed a long-term program for reducing the sulphur content in fuel oil to 1 per cent weight for the whole country by 1979, emphasising the importance of starting the program immediately because of an oversupply of low-sulphur oil. In 1968, the government presented the plan in a bill to Parliament, where it, in part, was supported by the Communists and the Liberals. However, the Liberals also pointed to several questionable aspects of the proposal. Possible future scarcity of low-sulphur oil could lead to price increases with detrimental effects for the economy. Further, it was argued that stack gas regulation was a better alternative. Interest organisations, such as the Swedish Steel Producer's Association and the Swedish Petroleum Institute were against the bill. They argued that there was not sufficient evidence justifying the introduction of costly national measures. National action, it was felt, should be postponed until other countries also adopted policy measures to reduce acid emissions (Lundgren, 1991, p.26). As a result of this opposition, the cabinet backtracked and proposed that only the first part of the plan should be adopted, i.e. the part that prescribed a maximal permissible sulphur content in fuel oil of 2.5 per cent weight. The long-term plan for further reductions was postponed until the technical possibilities and economic costs of further restrictions were investigated. After ten days of debate in the Parliament, the bill was adopted (Lundqvist, 1980, pp.46-47, 162-169).

The government and SEPA also disagreed with regards to the pace at which low-sulphur fuel should be introduced. SEPA argued that a market for low-sulphur fuel oil had to be created by introducing stricter legislation.[2] The Ministry of Agriculture, pressured by the Ministry of Finance and Industry, claimed that far-reaching legislation could not be introduced until the market supply of low-sulphur oil was guaranteed. Not until 1976, did the government become convinced that a shift to low-sulphur oil should be made while this oil was in abundant supply. This change of position stemmed from a government commission that concluded that the effects of sulphur pollution were in the long-term damaging to Sweden's economy, and particularly to the vital forest and paper industry (Departementserien, 1976). As a response, an action plan was outlined to continue the stepwise reduction of the sulphur content of fuel oil.

In 1976, the first national goal for sulphur emissions was also formulated; by 1985 the sulphur emissions were not to exceed the levels of the 1950s (Cabinet Bill, 1976). Although all parties supported the five-year lake liming program that was introduced simultaneously with the bill, the Swedish Federation of Industries and the Royal Academy of Engineering were sceptical about the need to introduce stricter regulations with respect to sulphur emissions. A recurrent argument was that far-reaching unilateral actions should not be introduced before international commitments were enacted and taken.

Continuing with its work on acidification, SEPA presented the first comprehensive national action program to combat acidification in August 1984 (SEPA, 1984). The goal was to reduce SO_2 emissions by 65 per cent (including a 50 per cent reduction of emissions from industrial processes), and NO_x emissions by 30 per cent by 1995, with 1980 as the base year in both cases. The plan was presented by the government to Parliament in February 1985, and was enacted into law in May of that year (Cabinet Bill, 1984, p.6). In 1987, a second reduction goal for SO_2 was set, aiming at reducing sulphur emissions with 80 per cent from the 1980 levels by 2000. In 1990, an overall long-term objective for both SO_2 and NO_x emissions was formulated: atmospheric deposition of sulphur and nitrogen should be reduced to the point where the depositions fell below the critical load for each substance.

VOCs was addressed formally by SEPA for the first time in 1987. It was estimated then that VOC emissions would decrease by 30 per cent by 1995 without any legislative measures. However, by 1990 this optimistic prognosis had been revised and an action program for VOCs was proposed. The environmental bill of the cabinet of 1991 included the goal of reducing VOC emissions by 50 per cent by the year 2000, with 1988 taken as the base year (Bäckstrand, 1995).

International Performance and Negotiating Positions

Parallel to the gradual introduction of domestic regulations, Sweden has played a very active role in the international arena. From the discovery of long range transport in the late 1960s to the integration of the critical loads approach in the 1994 Second Sulphur Protocol, Sweden has played the role of a pusher for a LRTAP regime.

The strong early Swedish position was driven by an urge to put long range transports of sulphur emissions on the international political agenda. The efforts to mobilise the international community behind this goal were carried out primarily by an active group of scientists and bureaucrats within, or closely associated with, SEPA. They had strong national political support, mainly from the Ministry of Agriculture and for several years they lobbied for international actions to reduce emissions. At the 1972 United Nations Conference on the Human Environment in Stockholm, a Swedish case study on transboundary pollution was presented. This was followed by activities through the Air Management Group of OECD and the OECD's Scientific Committee. The activities of these bodies were important for the establishment of the Co-operative Technical Programme to Measure Long-Range Transport of Air Pollutants, a precursor to EMEP. Sweden was also one of the prime movers behind the creation of LRTAP (Larsson, 1996).

In 1983, when the LRTAP Convention entered into force, Norway, Finland and Sweden submitted a proposal that called for a sulphur agreement stipulating a 30 per cent reduction of SO_2 emissions between 1980 and 1993. Gradually, the 30 per cent Club increased its membership and in 1984 the Executive Body of LRTAP decided to initiate negotiations aiming at a legally binding agreement in the form of a negotiated protocol. Sweden has strongly endorsed the creation of all LRTAP protocols, as well as signed the Declaration of Intent on NO_x. In addition, the Swedish research and policy community, in conjunction with the other Nordic countries, has promoted the critical loads concept as a method for additional reduction of sulphur compounds. The 1994 Second Sulphur Protocol represents the fruit of these efforts.

Overall, Sweden has implemented domestic reduction schemes independently from the negotiations in LRTAP. National reduction goals have been set prior to the adoption of the various LRTAP protocols. These national reduction goals have always been the same or higher than the emission reductions stipulated in the LRTAP protocols. Sweden's role in LRTAP has been defined primarily by the effort to convince other parties of the urgency and seriousness of the acidification problem. The second task has been to lobby for the adoption of new stringent protocols inducing other countries to adopt more ambitious national reduction targets.

Implementation and Compliance Records

Based on the requirements stipulated in the First and Second Sulphur Protocol, and the Nitrogen Protocol, Sweden's implementation and compliance records have been successful. Sweden has fulfilled the binding obligations enshrined in all these three protocols. For VOCs, the situation is more uncertain.

When comparing Swedish national reduction goals for SO_2 and the reductions stipulated in the 1985 First Sulphur Protocol, Sweden exhibits a pattern of over-compliance. When Sweden signed the First Sulphur Protocol in 1985, the provisions of the protocol had already been fulfilled. By then Sweden had reduced SO_2 emissions by 48 per cent compared with the figures for 1980.[3] At the time of the signing of the 1994 Second Sulphur Protocol, Sweden had once again already achieved the stipulated target.[4]

With regard to the NO_X, Sweden has fulfilled the obligations set out in the 1988 Nitrogen Protocol which demanded a freeze in NO_X emissions. By 1995, Sweden had reduced its national NO_X emissions by 20 per cent.[5] However, the nationally adopted goal of a 30 per cent reduction by 1998, derived from the Declaration of Intent that Sweden signed, has not yet been achieved. The main reason for not being able to cut NO_X emissions further is the growth in traffic volume. Despite measures such as the introduction of catalytic converters and exhaust emission controls, there has been an increase of NO_X emissions emanating from the traffic sector. In 1980, 69 per cent of the total NO_X emissions originated from mobile sources in the traffic sector; in 1995 this figure had reached 81 per cent.[6]

It is not certain that Sweden will be able to fulfil the emission reduction commitments demanded by the 1991 VOC protocol. By 1994, with 1988 as a base year, Sweden had managed to reduce VOC emissions by 16 per cent (Cabinet Bill, 1990; SEPA, 1992a). This is still well below the 30 per cent reduction required by 1999. The fulfilment of the unilaterally set goal of a 50 per cent reduction of VOC emissions by year 2000 seems even more unlikely.

Sweden as a Unitary Rational Actor

Import / Export

Sweden is one of the nine out of 38 European countries that are net importers of SO_2 emissions. Only a very small amount of the SO_2 emissions affecting Sweden originates from national sources. Moreover, the relative import/export ratio has increased over time. In the 1970s and the early 1980s more than 20 per cent of the national depositions originated from domestic sources. Between 1985 and 1994, they had decreased to 8 per cent, meaning that 92 per cent of the national sulphur deposition originated from foreign sources. At this time Sweden exported 77 per cent of its sulphur emissions, but import was three to four times larger than the export (EMEP, 1996, pp.66 and 69). The largest proportion of imported emission originates from Germany, followed by Poland and Great Britain.

The budgeting of nitrogen, according to EMEP's principles, is divided into reduced nitrogen[7] and oxidised nitrogen.[8] Whereas Sweden is a net importer of reduced nitrogen, the country is a net exporter of oxidised nitrogen. In the period 1985-1994, 69 per cent of the total national depositions of reduced nitrogen were imported, while 48 per cent of national emissions were exported. With regard to oxidised nitrogen, during the same period Sweden imported about 87 per cent of the total national depositions but was also a net exporter of emissions (EMEP, 1996, pp.70-71). On VOC, no national import and export figures are available.

National Damage Costs and Abatement Costs

National sulphur damage costs are high, especially those associated with the forest and paper industry. This makes Sweden not only a net importer of sulphur emissions but also a country where national sulphur damage costs clearly exceed national abatement costs. There has been no comprehensive and integrated calculation of the total national damage costs, but disparate accounts suggest that the accumulated costs of corrosion, the decline in harvest and fish populations, acidified forest and drinking water add up to several billions SEK (Bäckstrand, 1995, p.37).

The introduction of sulphur regulation in Sweden has not directly affected any central national industries, as for instance, was the case with the coal and mining industry in the United Kingdom. This has helped to keep down national abatement costs. Moreover, abatement costs in the 1980s benefited from the major investments in energy saving technology in the industry sector that took place in the economically expansive period before the oil crisis in the 1970s. The highest abatement cost has been the costs of the large-scale lake-liming programs that were introduced in the 1970s and subsequently extended to forest soil liming. Together they added up to a cost of almost one billion SEK (Staaf and Bertills, 1992, p.415).

National NO_X damage costs were originally thought to stem from the effects of eutrophication. As the link between NO_X and acidification became clearer, national NO_X damage costs became intertwined with SO_2 costs. However, NO_X exhibits higher implementation costs than SO_2. Abatement costs for NO_X are related primarily to the traffic sector, thereby affecting a nationally important export-oriented motor industry which also represents a strong vested interest. Because of its ties to deep-seated consumption patterns, NO_X emission cannot be reduced primarily through an industrial clean-up as was the case with SO_2 in the 1970s. In this respect the reduction of NO_X emission emanating from the traffic sector represents a major challenge. The socio-economic context in which the NO_X controls were introduced was very different from that of SO_2. After having benefited from the economic boom of the 1980s, Sweden in the early 1990s entered into a deep economic recession with growing unemployment. As a result, political priorities shifted from environmental issues towards more bread-and-butter economic issues. The VOC problem in Sweden has to a large extent been connected with the NO_X issue, and the two issues have often been treated concurrently. No calculations of the national damage costs and abatement costs for VOC as a separate class of pollutants have been made.

The URA model is most useful in explaining the sulphur case, in which factors such as the natural sensitivity to acidification, the high net import of sulphur deposition and the high national damage cost/low national abatement cost convincingly explain the Swedish position. Still, model I reveals only a limited part of the whole story, and even less in the cases of NO_x and VOC. A more detailed account can be gained by looking at the domestic politics of acidification, which is where we turn next.

The Domestic Politics of Acidification

In the following section, the domestic politics of acidification are examined. The main national actors involved in the SO_2, NO_x and VOC cases, as well as their different roles in shaping national policies and international negotiating positions are elucidated. As noted above, national damage costs generally outweigh the national abatement costs in both the SO_2, NO_x and VOC cases, but interesting variations exist.

Societal Demand and Support

Values, beliefs and attitudes of the public   Since the late 1970s, the acidification problem has been one of the most prominent environmental issues on the Swedish public agenda. The early strong public support for SO_2 reductions was stimulated by a large media exposure of the forest damage in the early 1980s that quickly became a pivotal issue that caught the attention of the public. This coincided with a steadily increasing public interest in environmental issues in general during the 1980s. In 1988, national environmental interest in Sweden rose to a record high and in the general election that same year the environment was ranked as the most important political issue. As the environment 'went global' in the late 1980s, public interest in the acidification issue diminished as other environmental issues such as the depletion of stratospheric ozone layer and climate change became more salient. In the 1991 general election, the environment declined to second place (Bäckstrand, 1995). The general decline in interest in environmental issues continued as the economic recession deepened in the 1990s.

NO_x and VOC have not gained the same public symbolic significance as SO_2. In the more complicated NO_x issue, the public has been both more differentiated in terms of its positions and less aware of the issue as such. The severe economic crisis in the early 1990s resulted in public resistance to NO_x regulations that would impose higher costs on driving. The VOC problem has been a marginal issue which has not been able to generate significant public support.

Presence and strength of intermediate actors   Concern for acidification was expressed by a number of intermediate actors. One of the first NGOs to react to the detrimental effects of acidification was the Swedish National Sport Fishing Federation.[9] In the early 1980s this organisation actively lobbied both central and local governments, advocating the liming of lakes as well as preventive measures aimed at reducing emissions. The actions taken by the Swedish National Sport Fishing Federation at that time reflected both a growing public concern for environmental issues and a direct stake in minimising the effects of acidification on lakes and fish stocks.

Swedish national affiliates of international environmental NGOs, such as Greenpeace and the WWF, have not been very active on the acidification issue. The same is true for the largest national environmental NGO, the Swedish Society for Nature Conservation, which has remained focused primarily on conservation issues. Instead, the most active environmental NGO has been the Secretariat on Acid Rain. The Secretariat was set up in 1982 after an international conference on air pollution and acidification in Gothenburg the year before. It functions as an umbrella organisation for five (originally four) major national NGOs, including the Swedish Society for Nature Conservation, WWF and Friends of the Earth. The Secretariat often acts on behalf of its affiliated organisations which at least in part explain the relatively low degree of direct involvement by these organisations in the acidification debate. The Secretariat was funded mainly by SEPA until 1992, when the Ministry for the Environment replaced SEPA as the largest financier.

On the international level the Secretariat has initiated information campaigns as well as conducted lobbying. The campaigns have been targeting foreign governments both directly and indirectly through various foreign national NGOs. The Secretariat's status as an NGO has not prevented it from working closely with SEPA. In many instances the interests of these two organisations have converged. The Secretariat, in its role as an NGO, has been able to act in places where the Swedish government does not have a mandate. This has created a situation in which the Secretariat, in many international contexts, has acted in parallel with the Swedish government.

Three aspects of domestic NGO involvement are striking. Firstly, all NGOs did not become involved until after the scientific and bureaucratic community had formulated the issue and brought it on the political agenda.

Secondly, the role of NGOs was largely confined to the function of information dissemination. This was in line with the Swedish political and legal system, in which NGO participation was limited to national or international information campaigns. Participation in the implementation and litigation process was constrained by the fact that environmental NGOs are not considered 'affected parties'. However, the new Environmental Code that entered into force 1 January, 1999 increases the scope for participation by NGOs in the policy process by granting them the status of ombudsmen for the environment (Bäckstrand, Kronsell and Söderholm, 1996, p.223). Finally, many NGOs are largely dependent on the government. The Swedish NGO Secretariat on Acid Rain, for example, is financed by the government and the SEPA. Moreover, there are several inter-organisational ties between Secretariat, the SEPA and the Ministry of Environment. Partly as a result of that key staff members have moved between these different organisational contexts.

No single political party has acted as a significant intermediate actor. Since the national political recognition of acidification as a major environmental threat in the late 1960s, there has been a high degree of political consensus on policy measures. Similarly, the ratification of LRTAP and subsequent protocols have not been controversial issues. The entrance of the Green Party in the Parliament in 1988 has not led to any significant change in acidification policies.

Kinds of values affected Since Odén's discovery, the SO_2 issue has been perceived as a serious environmental problem contributing to wide-spread acidification of forests and lakes. In the SO_2 case, there was initially an inherent conflict between two competing goals: on one hand, there was the desire to prevent large-scale damages to forests and lakes while, on the other hand, it was deemed important to keep energy prices low. This conflict particularly affected the paper and forest industry. Many Swedish paper companies own large forests and have an immediate economic interest in preventing forest damage. At the same time, the forest and paper industry is a highly energy intensive sector that utilises large amounts of fossil fuel in its manufacturing processes, thereby making it responsible for the bulk of the country's SO_2 emissions.

The problem of forest death vis-à-vis low energy prices was, in the end, largely solved by political decisions, prompted by the oil crisis and aimed at transforming the entire energy sector. A central element in this reform was to reduce dependence on fossil fuel which was achieved by large-scale investments in nuclear power. Thirty two per cent of the SO_2 emission reductions result from general energy savings. Only 24 per cent of the total reduction in SO_2 emission in Sweden can be explained by more stringent environmental regulations directly targeting the sulphur emissions (Bäckstrand, 1995). In a short-term perspective, the transition to nuclear power resolved the two apparently conflicting goals. Calculations show that 44 per cent of the achieved SO_2 emission reductions can be attributed to an increased reliance on nuclear and hydroelectric power (Bäckstrand, 1995). Between 1970 and 1990 the dependency on oil decreased from 70 to 20 per cent due to investment in nuclear energy and continued use of hydroelectric power. These two sources also account for the bulk current electricity production in Sweden. However, with the recent political decision to start phasing out nuclear power, the conflict is arising anew.

While SO_2 emissions largely come from stationary sources as a result of the burning of fossil fuels, NO_x emissions to a much higher extent originate from mobile sources, with the traffic sector accounting for 80 per cent of national emissions. The conflict with the energy intensive industry over SO_2 regulations has thus in the case of NO_x been replaced by resistance from the motor industry and transport sector against imposing stricter exhaust emission controls than in the rest of Europe. As a result, domestic abatement costs are higher for NO_x in political as well as economic terms. The motor industry is not only a corner stone in the important export-oriented sector; it also represents a strong vested interest. In addition, the NO_x issue lies at the heart of deep-seated consumption patterns, where the role of the car as a status symbol and an expression of individual freedom should not be underestimated. The fact that major parts of Sweden are sparsely populated regions makes people highly dependent on their cars. Moreover, the socio-economic context in which regulations were introduced differs significantly. While the SO_2 regulations were introduced in a period of economic growth, the issue of NO_x regulations appeared on the agenda at a time when Sweden was facing the worst economic recession since the 1930s.

The single largest source of VOC emissions in Sweden are mobile sources, which in 1994 contributed 37 per cent of the total amount of emissions (SEPA, 1992b, p.9). The VOC problem has largely benefited from actions taken in connection with NO_x. The VOC issue as such has not caused much national controversy since many of the actions undertaken in the traffic and energy sector to reduce NO_X emissions, already have helped, and will also in the future help, to reduce VOC emissions.

Governmental Supply

The initial Swedish demand for national and international action to combat SO_2, NO_X, and VOC emissions originated in a scientific-bureaucratic network located within SEPA. The Swedish government, through SEPA with its well-developed links to the scientific community, took on a leading role in putting the acidification issue on the agenda and initiating policy proposals.

Relative strength of the environmental branch of the government The Swedish Civil Service is characterised by comparatively small ministries and larger, considerably independent administrative boards or agencies. In the environmental administrative system the division of policy functions between the Ministry of Environment and SEPA varies with the environmental issue at hand. Acidification stands out as the issue where the Ministry of Environment has carried out the most far-reaching delegation of policy-making power to SEPA.[10] SEPA is the first central environmental administrative authority established in the world (Nordic Council of Ministers, 1991). However, Sweden was comparatively late in establishing a separate Ministry of the Environment. Not until 1987, after the Brundtland Report had appeared, was the Ministry of Environment and Energy established. Previously, the Ministry of Agriculture, and to some extent the Ministry of Industry, were responsible for environmental issues. The new Ministry was expected to play 'an offensive and co-ordinating role within the Cabinet'. In order to strengthen its role as co-ordinator a Special Office of Co-ordination was set up for environmental issues (Lundqvist, 1996, p.270). However, the scope of the activities of the Ministry of Environment was initially determined largely by decisions by the Ministry of Finance. This dependence was heightened with the onset of the economic recession in 1991.

The amount of financial resources allocated to environmental policy has varied with the general socio-economic situation. In the 1970s, the heydays of what has been called the 'major industrial cleanup', the expenditures of the Ministry of Agriculture for environmental measures was 5 per cent of the total expenditures of the national government. In the early 1980s the proportion of the expenditure on environmental policy and research fell to 1.4 per cent. With the increased focus on the global dimension of environmental issues in the late 1980s more than 2 per cent of the national budget expenditures were allocated to environmental research and policy. In the early 1990s, with the pressing problems of the economic recession, priorities again shifted from environmental to economic issues, resulting in cuts in public expenditures for environmental protection. Recent promises of increased funding for environmental policy and research indicate, however, that the pendulum is swinging back again.

SEPA has been granted a decision making, advisory, preparatory and expert role under the Environmental Protection Act from 1969. About half of the people in the central environmental bureaucracy are employed by SEPA, that currently has a staff of 400 employees. SEPA's responsibilities include environmental monitoring, research, information gathering and dissemination, and formulation of national plans. SEPA also plays an important role in co-ordinating a large number of administrative agencies. Finally, SEPA has a supervisory function with regard to the Environmental Protection Act that mandates the agency act to ensure that that activities that could be damaging to the environment are carried out in accordance with national laws and regulations. The activity of the National Licensing Board for Environmental Protection also plays an important role in the central environmental regulatory system. It functions like a court of law and its main role is to examine applications for licenses for activities that are hazardous to the environment. The acidification issue is the prime example of an issue in which many policy functions have been transferred to SEPA. In the acidification case as well as in many other environmental issues, the political initiative was taken by SEPA. Since 1984, the agency has initiated action plans to combat acidification every third year. Subsequently, these action plans have served as the most important basis for cabinet bills.

By and large there have been few conflicts between the Ministry of Environment and SEPA with regard to acidification policies. This has been especially true in the case of SO_2, where there has been a broad consensus on Sweden's position in the LRTAP negotiations. However, in the NO_x case, SEPA and the Ministry of Environment have differed in their judgements of the possibilities for achieving reductions of NO_x emissions. While the Ministry of the Environment pushed forcefully for a 30 per cent reduction goal during the negotiations of the NO_x protocol, officials at

SEPA argued that such an ambitious target would undermine the fragile consensus for a NO_x protocol.[11] In addition, SEPA pointed out that such a goal was inappropriate since Sweden would not be able to meet is commitment to reduce its own NO_x emissions by 30 per cent. In hindsight, this seems to have been a correct estimation (Bäckstrand, 1995).

Governmental control over state policy The environmental policy and administrative system in Sweden has traditionally been marked by both vertical and horizontal centralisation. This was particularly true for the environmental policy system from the 1960s to the mid-1980s. The vertical centralisation could be seen in the dominance of the central level as compared to the powers granted to the regional level (county administrations) and the local level (municipalities). This era was also marked by horizontal centralisation, since most of the central environmental policy functions were concentrated within SEPA (Lundqvist, 1996). As a result, SEPA played in this period a central role – within a traditional framework of command and control regulations – in introducing large-scale programmes for industrial clean-up of sulphuric emissions, effectively dealing with the problem.

In general, the trend in the 1990s has been towards a greater dispersion of policy functions across more sectors and levels of the administrative system. This development was prompted by financial challenges as well as by new ways of conceptualising environmental governance. The net result has been less control by SEPA and more responsibilities to county administrations and municipalities. In the Local Community Act of 1991, the municipalities were granted the authority to decide on their own organisational framework. Another buzzword in the wake of the Brundtland Commission report was 'sectorisation', or 'sector responsibility', implying that all sectors have 'a responsibility to prevent environmental damage and to solve existing environmental problems' (Cabinet Bill, 1987, p.35f). The adoption of holistic approaches is stressed also in the new Environmental Code that integrates previous pieces of environmental legislation (Swedish Official Investigations, 1996). The development within the administrative system is reflected also in the environmental research system. In the past, SEPA's Research Council was the centre around which environmental research revolved. Today, funding of environmental research is more widely dispersed.

In a quick assessment it may be tempting to attribute the less successful implementation records for NO_x and VOC compared to that of SO_2 at least in part to the increasing fragmentation and decentralisation of the environment policy system. There are, however, at least three reasons for being cautious about this interpretation. First, the NO_x issue in particular poses fundamentally different regulatory challenges, with multiple and interacting effects (acidification, eutrophication, and low-level ozone). Second, the sources of NO_x and VOC emissions are to a higher degree related to vital production as well as consumption interests. Third, the severe economic recession in the early 1990s created conditions less favourable for advocating and enforcing a progressive environmental policy.

Social Learning and Policy Diffusion

The third model focuses on ideational factors as a basis for the demand for international environmental agreements. It is an actor-oriented account that highlights the importance of networks of policy makers and scientists in illuminating and diffusing new ideas and policy innovations beyond national boundaries.

As previously pointed out, the scientific community in Sweden was an important actor, both in terms of bringing national and international attention to the acidification problem, and in providing scientific input into the LRTAP work. Various accounts of the Swedish political system point to the importance of close channels of communication between the policy making apparatus and the scientific community (Lundqvist, 1971; Lundgren, 1991). Equally important, the policy community is easily identified in such a small country and cultivates close relationships with the scientific elite. The prime example of this is the speed with which Svante Odén's 'discovery' of the acidification problem was accepted by the government. Many actors, such as industry, political parties and groups within the scientific community were initially sceptical (Öberg and Bäckstrand, 1996). Nonetheless, the government launched a policy proposal that enforced a reduction of the sulphur content in oil only a couple of weeks after Odén's discovery was made public (Lundgren, 1991).

The platform for the scientific-bureaucratic dialogue on acidification and subsequently critical loads was the Research and Development (R&D)

Unit within SEPA. The main responsibilities of the unit have been the administration of environmental research funds, preparations of project proposals, initiation of new research programs as well as the maintaining of contacts with the scientific community (SEPA, 1990). The various heads of the R&D Unit have lobbied actively for the domestic recognition of the acidification problem and pushed for further international commitments in the form of more stringent protocols.[12] In order to convince recalcitrant states of the seriousness of the acidification problem, active use of international arenas was emphasised from the onset. Staff from the R&D Unit and the International Unit at SEPA have been active in the various working groups and task forces, first within OECD and later in the organisational context of LRTAP. In addition, they have been an integral part of transnational alliances, involving different institutions such as IIASA, LRTAP and the UN-ECE Secretariat. The network also had close contacts with NGOs, including the Swedish NGO Secretariat on Acid Rain. The concept of critical loads, that spurred the development of a new generation of international environmental agreements, was originally developed by Swedish scientists affiliated with SEPA, and further developed during a series of Nordic and international workshops. In all this, the Nordic Council has been an important strategic platform where cross-national alliances of bureaucrats from national EPA's and scientists developed a consensus and co-ordinated their policies on critical loads.

The scientific-bureaucratic community has been behind the initiation of several national research activities. In 1977/1978 a permanent research program on the effects of acidification on soil, surface, water, flora and fauna was established by SEPA. A decade later another large permanent research project on measures to counter surface water acidification was initiated under the auspices of SEPA (Naturvårdsverket, 1991, p.93). More recently, a new research program was launched in 1993 under the heading 'Acidifying substances and tropospheric ozone – Effects and Critical Loads' (Bertills and Hannenberg, 1995). Since 1977 SEK 250 million has been spent on investigating the effects of acidification. During the same time period 950 million SEK has been spent on monitoring and liming, especially the liming of lakes (Staaf and Bertills, 1992, p.415).

The close relationship between the scientific and policy elites is institutionalised in the research policy system. While the Parliament and the Government formulate the overall research policy, the research councils are granted an extensive independence in financing and allocating research funds (Nordic Council of Ministers, 1991, pp.55-56). For a long time the most important actor in allocating funds for environmental research was the Research Council within SEPA. Already in 1968, when SEPA was established, it received a grant for environmental research and the same year the Research Council was established. A quarter of the public funding for environmental research (SEK 100 million, 1989) has been channelled through the Research Council while SEPA's own laboratories received a grant half of the size of the Research Council. However, as previously pointed out, the environmental research system is today undergoing a fundamental change. SEPA's laboratories have been dismantled and located within the universities. Since 1997, SEPA has had its funds for environmental research cut severely. As a consequence, the heavily reduced R&D Unit within SEPA has a more marginal role today. The Foundation for Strategic Environmental Research (MISTRA) has taken over the previous role of SEPA and is now the main financier of environmental research in Sweden.

In summary, the social learning and policy diffusion model highlights how the acidification issue in Sweden was very early defined by a scientific-bureaucratic network that succeeded in conveying the urgency of the issue to the government and the Parliament. This network, which was located in the institutional context of SEPA, spread its scientific ideas to the international arena, such as the OECD and, later, LRTAP. The network had the institutional support from the government and various large-scale research projects on acidification. More recently, policy-related research on critical loads has catalysed second-generation international agreements in transboundary air pollution. The process of developing critical loads as a viable policy instrument is very much a story explained by the demand for policy innovation by a research policy community in the Swedish and Nordic context.

Conclusions

Sweden has throughout the relevant time period played the part of a 'pusher' and actively lobbied for stringent binding international agreements. With regard to all three pollutants, Sweden has unilaterally adopted more ambitious national emission reduction goals than those demanded by the LRTAP agreements. Currently, Sweden has successfully implemented the 1985 First Sulphur Protocol, the 1988 Nitrogen Protocol and the 1994 Second Sulphur Protocol. There has been an over compliance in both the SO_2 case (80 per cent reduction already by 1994) and the NO_X case (20 per cent reduction by 1995). However, Sweden has failed to implement both the non-binding Declaration of Intent demanding 30 per cent reduction in NO_X emissions and the corresponding national reduction goal. The prospect of reaching the 30 per cent reduction by 1999 as stipulated in the VOC protocol seems highly uncertain. The long-term goal of 50 per cent reduction by year 2000 seems even more unlikely.

During the course of this study we have found that the *Unitary Rational Actor Model* best explains the sulphur case. The national sulphur *damage* costs were high, especially for the domestically important forest and paper industry, and the fishing sector with its associated commercial and recreational values. In addition, the national sulphur *abatement* costs were comparatively low in the 1980s, since many of the policy measures that were introduced in the in the late 1960s and early 1970s reduced emission levels (for example decreasing sulphur content in fuel oil, stack gas regulation, and stricter emission limits). Furthermore, the transformation of the energy sector in the 1970s gradually replaced fossil fuel by nuclear energy and significantly reduced sulphur emissions.

The mixed success in implementing NO_X and VOC reductions is better explained by the *Domestic Politics Model*. Despite the difficulties involved in estimating and comparing damage and abatement costs for the NO_X and SO_2 cases, it can be argued that they are higher in the NO_X case. Effective regulation of NO_X emissions requires a major restructuring of the transport system, which basically means putting into question the lifestyles, habits and private choices of millions of people. In addition, the industries that will be affected by such regulation are central to the Swedish economy. While the damage and abatement cost ratio may have been the same in the NO_X and in the SO_2 case, the factor responsible for the relative lack of success in reducing NO_X emissions have been the perceived *political* costs associated with the paradigm change that far-reaching NO_X regulation will require. The deep economic recession in the early 1990s also hampered the development of progressive environmental policies and generally shifted the interest from environmental issues to economic/welfare issues. The cross-sectoral nature of the NO_X and VOC issue with their many linkages to important sectors, such as the traffic, the industrial and the agricultural sector, rendered the implementation process more difficult. Both industrial interests (such as the car lobby) and consumers (car owners) protested against policy measures that would impose high costs on individuals and/or companies.

The national actors most in favour of far-reaching abatement measures were SEPA, the Ministry of Environment and the Swedish NGO Secretariat on Acid Rain. In general, the supply side actors, i.e. the governmental actors in conjunction with the scientific community, were most active in putting the issue on the agenda in the late 1960s and demanding national action plans in the 1970s. The demand side actors, i.e. the public and the NGOs, were involved in the acidification issue at a much later stage. We argue that SEPA, with its network of advisory scientists, has been the most important actor in bringing the acidification issue to the forefront. But at the same time the governmental control over state policy has decreased since the mid-1980s as the environmental policy and research systems have become more fragmented. Decentralisation, sectorisation, fragmentation, and reduction in funding to environmental protection and research are factors that have precluded that SEPA has been able to monopolise the NO_X issue as it did in the sulphur case. In other words, the institutional background for regulating NO_X is very different in comparison with the regulatory context for sulphur. Yet, the institutional factor should not be overestimated. The cross-sectoral nature of the NO_X problem and following from that, the high political costs of enforcing stringent regulation are key factors in understanding the difference between the SO_2 and the NO_X case. Another key is found in the socio-economic context of regulation. The protracted economic recession in Sweden and the challenges to the Swedish welfare system brought about by the globalised market economy have reduced the scope for enacting stringent NO_X regulation.

This brings us to the third model, *Social Learning and Policy Diffusion*. It is very clear that the knowledge, beliefs and ideas embraced by policy community network remain important factors for explaining the development of domestic policy and international regimes. As indicated by the Domestic Politics Model, the officials at SEPA, with their associated network of advisory scientists, were key actors in the national and international context. In the Swedish case, a bureaucratic-scientific network within SEPA invested efforts to convince other national actors and recalcitrant states about the urgency of the acidification problem. It should be noted that key persons in the network, such as the heads of the R&D

Unit within SEPA, were active for almost 25 years in the acidification issue, both at the national and international level. In other words, a core group has been following the issue for long and lobbied for the adoption of the framework convention and for the subsequent protocols. The network was stable until the 1990s, when some of the senior people retired or left for higher paid jobs outside SEPA. At that time the organisation was also beginning to face severe budget cuts. The notion of critical loads, on which the Second Sulphur Protocol is based, was also a result of the efforts of a network of closely intertwined scientists and policy makers. Sweden's role as a pioneer for acidification abatement is very much defined by the effort of this long-lasting network of bureaucrats and scientists.

Notes

1. Lake liming occurred already in the 1950s in Sweden. However, the deterioration of fish life then was not connected to the sulphur problem. Jan Nilsson, interview Stockholm 1994 and 1997.
2. Göran Persson, 1994, interview.
3. Statistiska Meddelanden, Serie NA18, 1996. From 508 000 tons to 266 000 tons.
4. Statistiska Meddelanden, Serie NA18, 1996. From 508 000 tons to 97 000 tons.
5. Statistiska Meddelanden, Serie NA18, 1996. From 454 000 tons to 362 000 tons.
6. Statistiska Meddelanden, Serie NA18, 1996. In actual figures, from 313 000 tons to 294 000 tons.
7. Mainly in the form of ammonium nitrate, NH3.
8. NO_X, mainly in the form of NO2 nitrate oxides.
9. Christer Ågren, 1997, telephone interview.
10. Peringe Grennfelt, 1997, interview.
11. Lars Björkbom, 1997, interview.
12. Göran Persson and Jan Nilsson, who currently occupy central positions within MISTRA, were previously heads of the R&D Unit within SEPA.

References

Bäckstrand, K. (1995), 'The Politics of Acid Rain Policy in Sweden', Report to the project *The Domestic Basis of International Environmental Agreements: Modelling National-International Linkages.*

Bäckstrand, K., Kronsell, A. and Söderholm, P. (1996), 'Organisational Challenges to Sustainable Development', *Environmental Politics*, vol. 5, pp. 209-30.

Bertills, U. and Hannenberg, P. (eds) (1995), *Acidification in Sweden. What Do We Know Today?*, Report 4422, SEPA, Stockholm.

Cabinet Bill (1976), 1976/77:3 *Om åtgärder för att motverka negativa effekter av svavelutsläpp*, Jordbruksdepartementet.

Cabinet Bill (1984), 1984/85:127 *Om program mot luftföroreningar och försurning*, Jordbruksdepartementet.

Cabinet Bill (1987), 1987/88:85 *Om miljöpolitiken inför 90-talet.*

Cabinet Bill (1990), Prop. 1990/91:90 *En god livsmiljö.*

Departementserien (1976), *Mindre svavel – bättre miljö.*, *Betänkande från utredningen om åtgärder för att motverka de negativa effekterna av svavelutsläpp*, 1976:2, Jordbruksdepartmentet.

EMEP (1996), *Transboundary Air Pollution in Europe*, MSW Status Report 1996, Part 1.

Larsson, P. (1996), *Regimförhandlingar på miljöområdet. En studie av förhandlingar av LRTAP-konventionen*, Dissertation, Studentlitteratur, Lund.

Lundgren, L. (1991), *Försurningen på dagordningen. En bild av ett händelseförlopp 1966-68*, Forskningsrådsnämnden och Naturvårdsverket, Solna.

Lundqvist, L. J. (1971), *Miljövårdsförvaltning och politisk struktur*, Prisma, Stockholm.

Lundqvist, L. J. (1980), *The Hare and the Tortoise. Clean Air Policies in the United States and Sweden*, University of Michigan, Ann Arbor.

Lundqvist, L. J. (1996), *Governing the Environment. Politics, Policy and Organization in the Nordic Countries*, NORD 1996:5, Nordic Council of Ministers, Copenhagen.

Naturvårdsverk (1991), *Miljövård i själva verket. Festskrift till Valfrid Paulsson*, SNV, Solna.

Nordic Council of Ministers (1991), *A Nordic Environmental Research Programme for 1993-1997*, NORD 1991:41.

Öberg, G. and Bäckstrand, K. (1996), 'Conceptualization of the Acidification Theory in Swedish Environmental Research', *Environmental Reviews*, vol. 4, pp.123-132

Schneider, T. (ed.) (1992), *Acidification Research. Evaluation and Policy Research*, Elsevier Science Publishers.

SEPA (1984), *Action Programme Against Air Pollution and Acidification*, The National Protection Board Informs, Naturvårdsverket, Solna.

SEPA (1990), *Forskningsnämnden - dess kommittéer och projektgrupper*, Naturvårdsverket informerar, Naturvårdsverket, Solna.

SEPA (1992a), *Sverige mot Minskad Klimatpåverkan*, NV Rapport 4459, Naturvårdsverket, Solna.

SEPA (1992b), *Utsläpp till luft av flyktiga organiska ämnen*, Naturvårdsverket Rapport 4312, Naturvårdsverket, Solna.

Staaf, H. and Bertills, U. (1992), 'Acidification Research in Sweden', in T. Schneider (ed.), *Acidification Research. Evaluation and Policy Research*, Elsevier Science Publishers.

Statistiska Meddelanden (1996), Serie NA18, 1996.

Swedish Official Investigations (1966), *Luftföroreningar, buller och andra imissioner*, SOU 1966:65, Justitiedepartmentet.

Swedish Official Investigations (1996) *Miljöbalken (Environmental Code). En skärpt och samordnad miljölagstiftning för en hållbar utveckling*, SOU 1996:10, Miljödepartementet.

5 The Reluctant Pusher: Norway and the Acid Rain Convention

TORUNN LAUGEN

Norway – A Downstream Country with a Vulnerable Natural Environment

What has determined Norway's acid rain policy is the fact that solutions must be sought *outside* Norwegian borders. Due to its geographical position downstream of major emitters such as Great Britain and Germany, Norway finds itself a large net importer of air pollution. As much as 95 per cent of total SO_2 fallout and 86 per cent of total NO_X fallout in Norway comes from abroad. Moreover, the geology of Norway is characterised by a variety of rocks vulnerable to the effects of acids; in addition, soil-layers are generally thin. As a result, the ability to neutralise sulphurous precipitation is poor. Moreover, a relatively large proportion of the precipitation falls in the form of snow. Acids accumulate over the winter, and then, as the snow melts, large quantities are released into the ground and water, causing 'acid shocks' and dramatic large fishkills (Rosenkrantz and Wetstone, 1993, p.60).

Norwegian scientists have been concerned by the decimation of fish stocks caused by acidification of fresh waters since the 1920s. Gradual acidification of precipitation in the 1950s was connected to the fish deaths, but the international link was not made clear until 1968, when the Swedish scientist Svante Oden published his pioneering report.[1] This report triggered a wider debate about the damaging effects of long-range transboundary air pollution. In the Norwegian government's next long-term programme (1970–73), acid precipitation was for the first time blamed on emissions from industry in Western Europe (*St.meld.* no.55, 1968–69, p.92).

Damage Costs

The most visible effect of acid rain in Norway has been the more than doubling of so-called 'dead lakes' where fish have been totally extinct over the last 20 years. Damage to fish-stocks has been registered in an area of 86,000 square kilometres in Southern Norway, equivalent to 26 per cent of total land-area (SFT, 1994, p.16). Exact figures as to the costs of this damage are difficult to produce, as fishing in these waters is not commercial but mainly for recreational purposes. However, estimates based on both the value of the existence of sport fishing and on the willingness of people to pay for its maintenance yield a figure of some NOK 400 million in damage-costs each year.[2] In addition the value of lost catches has been set at NOK 10 million, and loss of recreational opportunities at NOK 85 million per year (Brendemoen, Glomsrød and Aaserud, 1992, p.17). Furthermore, in 1994 Norway spent NOK 93 million on liming of lakes in order to neutralise damage caused by acid rain.

Acid rain is also held to be one of several long-term stress factors that might lead to serious forest damage; however, since nitrogen also has a fertilising effect, the extent of this damage is still somewhat controversial. On the whole, the long-term effects are assumed to be damaging, since increasing amounts of nitrogen in the soil will eventually lead to the leaching of other plant nutrients like potassium, calcium and magnesium. Acidification is now assumed to have reached the threshold level where damage starts.[3] In 1988 it was estimated that air pollution led to a timber-growth loss of 1–2 million m^3 in Norwegian forests, representing an estimated value-lost of NOK 300–600 million. As the buffering capacity of the soil diminishes, the costs are assumed to increase up to NOK 700–1000 million a year (Brendemoen et al., 1992, p.16). In addition, of course, the recreational value of the forest is also diminished.

Another problem, although less immediately apparent, is local health damage. Heavy concentrations of SO_2 and NO_x increase the risk of respiratory infections. Estimates made by Statistics Norway (formerly known as the Central Bureau of Statistics of Norway) indicate that the largest costs involve damage to health. Finally, acid precipitation also corrodes buildings and other materials exposed to it. It has been estimated that these damages cost at least NOK 220 million per year in directly increased maintenance expenses (Brendemoen et al., 1992, p.18).

Abatement Costs

Costs connected to reducing polluting emissions may be roughly divided into two types:

- first, there are the *immediate* costs of installing cleaning equipment, modernising industry, etc.
- second, there are *indirect* costs such measures might involve, such as loss of competitiveness compared to other countries (Claes, 1992, p.8). Production costs would be comparatively higher in a country with strict environmental regulation or high environmental taxes.

Both types are difficult to measure, and no attempts will be made to estimate at what point the damage costs to Norway would exceed abatement costs. However, it is a simple matter to calculate abatement costs connected to compliance with the SO_2 Protocol in the Norwegian case. The 30 per cent reduction required by the Protocol had already been achieved at the time of signing in 1985, and no further measures were needed. Any damage costs would therefore outweigh the abatement costs, which would be zero. For NO_x, on the other hand, the situation was quite different. Emissions were increasing, and no measures had been introduced before the signing of the Protocol. It was not clear exactly how reductions could be achieved, but it was evident that the costs would be extensive. The same can be said of VOC emissions, which were estimated to be on the increase at the time the Protocol was signed and are still rising.

Domestic Policy Prior to the LRTAP Regime

The First Joint Plan to Control Pollution – 1975

In 1975 the Ministry of the Environment (MoE), at the time headed by Gro Harlem Brundtland, introduced the first joint plan to regulate all types of pollution (*St.meld.*no. 44, 1975–1976). In its assessment of the acidification problem, the government noted that there were few local problems caused by acid rain, but that acidification of freshwater and lakes caused fish death, and emissions of SO_2 caused corrosion on materials. The problem was caused largely by foreign sources, *but Norway needed to* *reduce its own emissions in order to induce other countries to do the same.* The plan contained few fixed targets and no timetables for reductions. On the contrary, it was explicitly stated that reductions should be implemented to achieve the objectives within 'reasonable' limits. The following measures were introduced:

- *General regulations*
(1) Imposed use of low-sulphur (1 per cent) heavy crude oil in the southern parts of the country. For the rest of Norway, emission permits would be granted individually.
(2) Cleaning of industrial emissions, if technologically and economically feasible, by the use of individual concessions.
(3) Reduction of SO_2 emissions from industrial processes, if technologically and economically feasible. It was not specified how this was to be achieved.

- *Fiscal instruments*
Generous state programmes were instrumental in averting potential resistance from industry. Financial support from the state to industry contradicted the polluter pays principle, but since very little of the older industry was able to undertake the necessary changes, the government provided interim financial support. In 1975 it was estimated that some form of state assistance financed 72 per cent of the investments made by established industry to comply with environmental regulations. 'Environmental investment' was broadly defined, which helped to dissuade resistance from the industrial sector. At a later stage the government also provided financial support for general energy-saving measures in trade, industry, state buildings and the domestic sector. Part of the explanation behind the generous state programmes was a fiscal policy based on the belief that economic recessions could be countered by increased governmental spending. Keynesian anti-cyclical fiscal policy held a strong position in the Norwegian Ministry of Finance in the late 1970s and early 1980s, and environmental programmes for countering industrial pollution were seen as a good way to disburse governmental funds.

A general tax on mineral oil was introduced as early as 1971 and adjusted several times later. The tax was differentiated according to the sulphur content of the oil. But the tax had very little effect since low-sulphur oil was still more expensive than high-sulphur oil, even with the tax. Taxing the product, rather than the emissions, meant a lack of incentive to install cleaning equipment. For this reason, a tax rebate to reward reduced sulphur emissions through end-of-pipe cleaning measures was introduced in 1976.

Regulation of polluting emissions had thus started before the Long-Range Transboundary Air Pollution regime (LRTAP) had been established, and well ahead of the signing of the first protocol. However, regulation was not particularly strict, and care was taken to safeguard other concerns, like the competitiveness of Norwegian industry.

Norway's Role in the LRTAP Negotiations

The SO_2 Protocol – Pushing the Others as Far as Possible

Ever since the discovery of the transboundary nature of the acid rain problem in the late 1960s, the Norwegian view had been that the sources, and therefore also the solutions, to Norway's problem were located outside Norwegian jurisdiction. The primary motivation for Norway to join the international agreement was therefore not to reduce its own emissions, but to induce others to reduce theirs. Even though no specific cost-benefit calculations were made, it was clear that the 30 per cent reduction target was already within reach for Norway. What Norway therefore sought was to pressure other states into committing themselves to a fixed percentage reduction. And it is in the same manner that the national goal of 50 per cent reductions must be understood: the purpose was to set an international example by making substantial domestic reductions.

The NO_X Protocol – from Pusher, to Dragger, to Pusher Again

Behind the formulation of the NO_X goal lay a rather different story. Reduction of NO_X emissions was first mentioned in a Norwegian draft to the SO_2 Protocol as early as September 1984. It was recommended that '...additional actions for the purpose of achieving *quantified, substantial reductions* of emissions or of transboundary fluxes of nitrogen oxides' should be included in the SO_2 Protocol.[4] But this position was to change substantially during the following years.

When the formal NO_X negotiations started in 1988, Norway's position had changed to one of *stabilisation and reduction of NO_X as soon as possible.*[5] Further reductions were to be based on scientific calculations of critical loads.[6] In other words, Norway did not want any fixed reductions, and not even a fixed time limit for stabilisation. All the same, Norway ended up signing both the Protocol and the Declaration of intent requiring a 30 per cent reduction. The decision to sign was made at the highest political level in the government, and at the last moment of the negotiations.[7]

The VOC Protocol – Demanding Exemptions

When negotiations on the VOC Protocol started, Norway was no longer in the forefront in urging the highest possible percentage cuts. Norwegian emissions were high and expected to rise, since they were caused largely by the expanding offshore petroleum industry. Still, Norway again decided to commit itself to the ambitious 30 per cent reduction goal. However, Norway managed to obtain acceptance for exemption of emissions north of the 62nd parallel (the so-called TOMA areas, in practice in Norway and Canada). These emissions were allowed to increase as long as the overall country target was reached. However, this insistence on a more differentiated protocol meant that Norway was no longer seen as a pusher by the other participants in the negotiations.

The Second SO₂ Protocol – Siding With the Enemy

In the previous protocols all signatories had committed themselves to the same percentage reductions. With the VOC Protocol there had come a shift away from this principle with the acceptance of the TOMA areas for Norway and Canada. This trend was further developed in the position taken by Norway on the new sulphur protocol. Rather than setting a uniform reduction target for all the signatories, the new protocol should – in Norway's view – be based on the concept of 'critical loads' – the maximum level of sulphur deposition that soils, waters and other ecosystems can tolerate in the long term without suffering damage.[8] These new principles had several consequences for the Norwegian position in the negotiations. When the negotiations started in January 1992, the Norwegian delegation was clearly opposed to the German proposal, which required use of the Best Available Technology (the BAT principle). Norway held that countries should be free to choose whatever methods would involve the lowest cost to achieve their reductions. This position was criticised both by other countries and by Norwegian NGOs. Nor did it help matters that Norway here had the support of its old 'enemy' Britain. In August 1993 the Norwegian delegation finally bowed to pressure, and accepted the BAT obligation – making it clear, however, that Norway was still against the BAT principle, but had wished to signal its will to compromise as well as wanting to be able to 'close the deal' within the set time-limit. Progress in negotiations was deemed more important than taking an unyielding stand on principles.

Implementation and Compliance Records

The SO₂ Protocol

The first SO_2 Protocol and the national 50 per cent reduction goal have been implemented without hitches. In 1980, Norway's sulphur emissions were 141,000 tons. By 1985, they had already been cut to 98,000 tons – a reduction of 43,000 tons and more than 30 per cent. Even though the national goal of 50 per cent reduction was reached in 1987–1988, emissions continued to decrease every year; by 1995 they were down to 34,600 tons. Here, in other words, Norway has 'over-complied' with its international commitments.

The NOₓ Protocol and Declaration

Norway's emissions of nitrogen oxides increased annually, peaking in 1987 at 236,000 tons. Emissions then dropped to 223,000 tons in 1992 (SFT, 1995, p.4), so the stabilisation goal has been achieved. The 30 per cent reduction target is, however, not within reach. Emissions have been reduced only 5 per cent. Further measures are needed, but their introduction has been delayed several times. In January 1994, Thorbjørn Berntsen, the Minister of the Environment, first announced that the 30 per cent goal might have to be abandoned, and the Ministry of the Environment (MoE) has since then admitted that the 30 per cent reduction will not be reached.

The VOC Protocol

Emissions of volatile organic compounds in Norway have shown no decline since 1989. Instead they increased by 14 per cent between 1989 and 1995, and are expected to increase even further before the year 2000.

Summing up, Norway has continued to commit itself to ambitious international goals through the LRTAP negotiations. However, it is clear that Norway is no longer the country most eager to advocate the highest percentage reductions, having been outstripped by Sweden, the Netherlands and Germany in this regard. As to compliance with its commitments, Norway has gone from over-compliance with the SO_2 Protocol, to officially abandoning the NO_X Declaration, and being unlikely to reach the VOC target.

Model I: The State as a Unitary Rational Actor

Despite differing cost-benefit calculations, Norway's formal goals ended up quite similar for the three pollutants. Norway committed itself to a 30 per cent reduction in the first three rounds of negotiations, even though cost-benefit calculations might have indicated that more divergent goals could be expected.

When it comes to meeting these targets, however, Norwegian policy seems to accord better with realist predictions. For a long time, SO_2 emissions were reduced more than required by the international agreement. On the other hand, Norway has not been willing to assume the costs that would results from following up the NO_X Declaration and the VOC Protocol.

So far, the policy adopted by Norway does not seem to contradict the predictions of the URA-model. A vulnerable natural environment and net import of pollution are incentives for Norway to comply with the Agreement. In the SO_2 case, this did not cause any problems since abatement costs were virtually non-existent. With NO_X and VOC, however, abatement costs were substantial and outweighed damage-costs. But this explanation leaves two questions unanswered: 1) Why did Norway reduce its SO_2 emissions 40 per cent below the international commitment and 20 per cent below the national goal? 2) Why did Norway sign the NO_X Declaration and the VOC Protocol in the first place? This brings up another question: How is the decision about signing, and the decision about complying connected?

Connection Between Negotiation Position and Compliance, and the Costs of Defection

Why did Norway sign the NO_X Declaration and the VOC Protocol, knowing that the abatement costs would be substantial? One explanation might be that domestic factors other than material cost-benefit calculations determined the process. A second possibility is that Norwegian decision-makers underestimated the abatement costs at the time the commitments were made. Both these options will be examined in relation with the Domestic Politics Model further below. Yet another possibility should be mentioned: that Norway never had any intentions of fulfilling its obligation in the first place and was indifferent to the policy formulation. This could explain Norway's non-compliance with the NO_X and VOC targets in accordance with the URA model.

International Commitments as Deliberate Deception

Realists assume that states will always act in accordance with their national interest. The extreme consequence of this view is that, when the time comes to fulfil commitments, the respective national authorities will be obliged to consider the situation again, and decide whether it is in the state's interest to live up to pledges made earlier (Hovi, 1991, p.22).

Realists have several reasons for assuming that there might be some incongruence between the signing of a protocol and compliance with it. First, signing an international environmental agreement usually wins votes for politicians, while compliance could involve introducing costly and unpopular measures. (Most people are supportive of environmental issues as long as they are free of charge.)

Secondly, the motivation behind signing might be something totally different from the motivation behind compliance. Especially for small net-importers, the temptation of free riding can be strong (Underdal, 1993, p.2). A small net-importing country would need the international agreement and would therefore be in the forefront in the establishment phase. However, reductions of own emissions would make only an insignificant contribution to solving that country's own domestic problem, so when the question of compliance arises, the rational choice might well be to defect. It may also be that different mechanisms are at work when attempting to establish an agreement, and with implementation of it. In the establishment phase, the critical factor is whether states are willing to accept the relatively high costs of being in the forefront of the development. Once the co-operation has become established and the number of contributing states is growing, the costs will be more evenly distributed among the participants. But at the same time, the contribution of one individual state now becomes less vital for the realisation of the good, and the temptation of free-riding will surface (Claes, 1992, p.11).

There are several reasons for assuming that intentional non-compliance would be Norway's dominant strategy. It was important for Norway to establish the agreements since the solution to Norway's problems could be found only through international co-operation. At the same time, however, for Norway to cut its own emissions would hardly contribute to less acidification of Norwegian soil and water. Therefore Norway could be expected to be a 'pusher' in the establishment of the agreement, but a 'laggard' in compliance if the abatement costs are high. This analysis, however, does not consider the costs that may be incurred by defection.

Incentives for Compliance within a Realist Perspective

Two mechanisms suggest that Norway might have had incentives for complying with its commitments despite substantial abatement costs (Hovi, 1991). Firstly, a state may choose to comply out of fear that defection would lead to a breakdown of the whole agreement. But considering that the refusal of Great Britain (one of Europe's major emitters) to sign the SO_2 Protocol in 1985 did not halt the process, non-compliance by Norway would seem unlikely to have such severe consequences. Secondly, states may decide to comply if the costs incurred by defection might be increased through 'external mechanisms'. Two such external mechanisms that could be relevant here are *issue linkages* to other policy areas, and *maintaining one's reputation* as a reliable partner.[9]

Issue linkages　Compliance with one protocol has been linked to compliance with the other protocols through the LRTAP Convention. If Norway were to defect, that would make it difficult for Norway to demand that other countries comply with all the other protocols. Free riding would be tempting if the likelihood of being detected was fairly low. But because of the existing monitoring and reporting system, this would be difficult.

Maintaining one's reputation Norway is a small country and is thus dependent on international co-operation in many fields. It is therefore natural to assume that being regarded as a reliable partner will be more important for Norway than for larger and more independent states. Moreover, within this particular issue area, Norway has another reputation to maintain. The fact that the World Commission on Environment and Development was headed by the then Norwegian Prime Minister, Gro Harlem Brundtland, in addition to Norway's active international policy in environmental issues, has contributed to Norway's reputation as a leading country in environmental politics. At least some members of the Norwegian government would like to preserve the image of Norway as an environmental 'pusher'.

Summing up, then, Norway in some respects would benefit from defecting and becoming a free rider concerning the NO_x and VOC protocols. But this would be detected, and Norway's ability to put forward valid demands towards other countries would be weakened. Also Norway's reputation as a reliable partner and its image as a 'green' country could be damaged. All this makes the rationality of defection more questionable.

On the basis of the URA model, we may conclude as follows: Norway committed itself to ambitious goals since it needed a binding international agreement to solve a domestic problem. In practice, however, Norway complied with only one of its commitments, the others proving too costly. However, this explanation leaves some questions unanswered. Signing the NO_x Declaration and VOC Protocol would be rational only if Norway's intention was to defect (since abatement costs would be immediate and substantial). But our analysis has shown that defecting could also undermine Norwegian interests. The introduction of linkages between all the protocols under LRTAP makes it hard to determine exactly what the URA model would predict.

Model II: Domestic Politics

To find a better explanation of why Norway committed itself to the NO_X and VOC goals, and the different compliance records, we will need to take a closer look at the domestic policy formulation process. Which interests were reflected in the national positions?

Formulation of the National Position

The first SO2 protocol – Common focus on foreign sources In light of the later implementation process, it seems somewhat ironic that Norway's positions in the NO_X and VOC negotiations were in many ways more carefully prepared than was its position in the first SO_2 negotiations. As mentioned above, it was not really necessary to make any precise estimates of SO_2 reductions since the target had already been achieved at the time the protocol was signed. Consequently, the 30 per cent reduction goal for SO_2 did not give rise to any internal conflict since the reductions were already within reach, and no one's interests were threatened. At that time, the environmental NGOs were on the government's team, so to speak, even receiving financial support for activities directed toward Great Britain and toward Norwegian opinion.[10] The industrial sector, which represented the opposite side in this matter, may not have been an enthusiastic supporter of the Convention, but saw no reason to object to it, as the economic consequences would be minor.[11] In fact, external actors participated only to a very limited extent – not because they were excluded, but because they were indifferent.[12] Consequently, Norwegian preparations prior to the initial SO_2 negotiations were dominated by a small and close network of natural scientists and senior civil servants from the International Department in the MoE.

The NOX Protocol and Declaration – 'The national championship in percentages and chemical formulas' The NO_X negotiations were much more thoroughly prepared. The Norwegian government wanted a stabilisation of emissions, and this was supported by the sector ministries. The Ministry of the Environment issued a White Paper report (St.meld. no. 47, 1987–88) on how reductions could be achieved; here it was recommended that emissions should be stabilised. Outright reduction would be too costly because of the consequences for domestic and coastal transport. The damaging effects of NO_X were also somewhat controversial. This report went through regular circulation procedures in all the ministries, and no serious objections were raised.[13] However, when the final decision was to be made, political pressure made this careful preparatory process largely irrelevant.

In 1989, the NO_X goal was the subject of heated political debate, in which the government (the social democratic Labour Party) was criticised both by its own representatives and by the other parties for the 'low ambitions' of its environmental policy.[14] Sissel Rönbeck, Minister of Environment at the time, characterised it as a 'national championship in percentages and chemical formulas'.[15] This debate came about as a result of increasing public attention to environmental issues. Opinion polls carried out in connection with the general elections indicate the growing public attention to environmental issues in the late 1980s:

Table 5.1   Environment as the most/second most important election issue 1981–1989 (figures in percentages)

	1981	1985	1989
Most important	1.3	2.1	18.8
Second most imp.	1.0	1.0	11.6

Source: Kiberg and Strømnes (1992), Norwegian Social Science Data Service Report no. 94.

By the time of the 1989 elections, the environment had emerged as the single most important issue. In light of this development, the government felt that the political costs of abandoning Norway's previous position as an international 'pusher' would be too high.[16] The NO_X goal was in this respect not so much a result of anyone's interest, but as a result of a competitive political game for popular support.

The VOC Protocol – Including new actors and new principles When it came to the formulation of the Norwegian position on VOC emissions, a distinct change had taken place in the internal preparation process. The reason was most likely the realisation of the cost of fulfilling the NO_X

commitments, and a desire to prevent the same from happening again. Thus, the whole process was better planned, and several new participants were included. A preliminary list of abatement measures was used as basis for discussions in an inter-ministerial group established well ahead of the completion of the international negotiations. The group included representatives from the Ministries of Petroleum and Energy, Transport, Industry, Foreign Affairs and Finance.

External organisations were also officially drawn into the process. The MoE arranged seminars to discuss the Norwegian negotiation strategy and possible abatement measures in 1990. In a letter from the Federation of Norwegian Chemical Industries to the MoE, the organisation corrected several estimates made by the Norwegian Pollution Control Authority (the SFT) on the size of the emissions and the costs of various abatement measures. This time the affected actors were not waiting for the implementation phase, but sought actively to influence the definition of the problem and its possible solutions. They chose to attack not the goals, which probably would have made them very unpopular, but rather the basis for calculation and the effects of the suggested abatement measures.

The result of having an expanded policy formulation team can be seen in the final protocol, where Norway's demand for exemption for emissions north of the 62nd parallel was accepted. Behind Norway's demand for special treatment lay both an economic and an environmental reason. The economic reason was advocated by the petroleum sector, which argued that a situation could arise where Norway would have to choose between opening a new oilfield (on the so far unexploited coastline of northern Norway) and fulfilling its commitments to the VOC Protocol. On the environmental side, calculations from the Meteorological Institute showed that environmental damage by VOC emissions in the northern part of Norway was minor.[17] Even though all participants in the Norwegian group agreed on this principle, there was some disagreement on how important it was to get the exemption into the protocol. The MoE assumed that there would be no need for the exemption, and that Norway would not be forced to choose between fulfilling its commitments to the protocol and opening up a new oilfield in the far north. They favoured a basic common reduction and were concerned about the impact such an exemption would have on other countries. This must also be seen in the context of Norway's desire to be seen as a leading country in these matters. In the final round, it was the economic interests that proved decisive. Nevertheless, the calculations of future emissions from the Norwegian offshore petroleum sector were to prove too low, as technological advances in production technology and new estimates of the extent of the oil reserves led to an unexpected increase in production.

The most striking new aspect of the VOC process was the formalised participation of the sector ministries in preparations for the negotiations. The strong influence of the oil-producing sector was demonstrated by Norway's insistence on exemption for emissions off the northern coast. Another new aspect was that external actors questioned the scientific basis for the formulation of the Norwegian position. Still, the process remained relatively closed, as the skirmishes between environmental and sector interests were fought within the administration.

The second SO2 Protocol – Increasing inter-ministerial tensions In preparations for the second sulphur protocol, the development towards less control by the MoE and more interference from the other ministries continued and reached new levels. The most significant change was that representatives of other ministries now also regularly took part in the actual negotiations at the international level.

Preparations for a strategy on a new sulphur protocol started as early as the summer of 1991. An informal meeting was held where Norway decided to oppose the German demand for a commitment to the use of Best Available Technology (BAT) and advocate more cost-efficient solutions. An inter-ministerial group on strategy was also established. In a letter from the Ministry of Transport to the MoE, the former objected to the organisation of the preparations to the negotiations. Instead of convening prior to international meetings, the Transport Ministry wanted regular meetings throughout the period, since informal consultations could take place between the formal negotiations. The Ministry also objected to the sole focus on sulphur, arguing that the whole problem of pollution in the LRTAP co-operation should be discussed. This probably reflected the divergence of views on the BAT principle. A commitment to BAT would not have any consequences for Norway in the case of sulphur emissions. But Norway feared that any acceptance of the principle in the Sulphur Protocol would make it difficult to exclude it in the protocols for other pollutants like NO_X, which would have economic consequences for Norway. In a letter from the Ministry of Industry, concern was also expressed regarding the organisation of the Norwegian preparations. The

Ministry further expressed a desire to participate in the actual negotiations since the results would affect Norwegian industry. And lastly, they wanted economic analyses of the consequences for several branches of industry in Norway. However, this time the environmental side managed to win the inter-ministerial battle, and Norway did commit itself to the BAT principle.

Summing up the policy-formulation process, it seems clear that the first two goals were more a result of political considerations than of analytic cost-benefit calculations. The SO_2 goal was intended to put pressure on other states, while the NO_X goals were mainly the result of political over-bidding in the quest for popular support. The actors were largely the same in both processes. The most important difference between them was the new public concern with environmental issues that emerged in the late 1980s. In the NO_X case, this resulted in disregard for the expert advice. In the preparation of the VOC Protocol, sector interests entered the policy formulation process, having realised that they might be affected by the outcome. In the final SO_2 negotiations, the level of conflict within the administration had reached a stage where the sector ministries demanded that they be present at the actual negotiations. By then, Norwegian policy formulation process had developed from one involving a small group of higher civil servants and scientists who advocated binding international commitments, to a complex and time-consuming internal negotiation process, mainly concerned with the domestic consequences of any commitments that Norway might make. This in turn indicates totally different points of departure for the implementation process.

The Implementation Process

SO_2 reductions – Measures that implied no costs or were unrelated to the international protocol The Norwegian SO_2 emissions were reduced by 75 per cent between 1980 and 1995 (SFT, 1996:41). The SFT (the Norwegian Pollution Control Authority) explains the reductions by the following factors:[18]

- regulation of the industry
- reduced emissions from industry in addition to regulation
- closing of the copper mines in Sulitjelma
- forced transition from use of normal to low sulphur content in heavy crude oil as a result of regulation and a differentiated tax

- voluntary transition from use of normal to low sulphur content in heavy crude oil
- reduced consumption of fuel oil
- reduced sulphur content in fuel oil
- reduced emissions from coastal shipping.

A closer look at these explanations reveals that only a few of them can be said to have involved any costs for Norway. Indeed, measures that implied costs were not introduced as a result of the international agreement, nor motivated by acidification problems. I will start by looking at reductions that came 'free of charge'.

The following measures did not inflict any additional costs upon Norwegian society:

- The *closing of the mines in Sulitjelma* in 1987 alone led a 10–15 per cent cut in Norwegian emissions alone (*Econ Report* 45/93, p.19). However, the mines had for many years relied on government subsidies, and were highly unprofitable. Proposals for shutdown had been brought up in Parliament earlier, but failed to gain sufficient support as the mines provided precious jobs which would be difficult to replace in that northerly area.
- Some of the *reduced emissions from industry* came as the result of generally reduced activity, due to recession and factory closure. Energy economising measures had also been initiated, and this meant less energy consumption in the industrial sector. These measures were intended to save costs, not increase them. Emissions were also reduced by modernisation of industrial processes, but again these changes did not imply increased costs.
- The *reduced consumption of fuel oil* was mainly a result of good access to cheap excess hydroelectric power, in combination with rising oil-prices.
- In the same way, the *reduced sulphur content in oil* was partly due to easy access to crude oil from the North Sea with a naturally low sulphur content.
- Finally, the reduction in *emissions from coastal shipping* was due to reduced activity and the lower sulphur content in oil.[19]

Norway also implemented several measures that did involve some costs, though few of them can be traced back to the international agreement.[20] *Three* types of measures have been implemented. The first relates to regulation of industrial emissions by individual concessions. These measures were initiated in 1975 by the industrial clean-up programme 'Measures against Pollution' (*St.meld.* no. 44, 1975–76). Although concerns about acidification may have played a role here, improvement of local air quality was probably the most important motivation.[21]

The two other measures involving costs were connected to the changeover from using normal sulphur oil (NS oil) to low sulphur oil (LS oil). This transition was encouraged by a differentiated tax on mineral oil and by regulation of the sulphur content in oil.

A differentiated tax on mineral oil was introduced in 1971, but 18 years were to pass before it was regulated in a way that made LS oil cheaper than NS oil (*NOU* 1992:3). The inefficiency of the tax was pointed out several times, but no adjustments were made. This led to speculation as to whether the underlying motive really was environmental improvement or simply to increase state revenues. In any case, the tax made no contribution whatsoever towards reducing emissions until the last few years.

Regulations on maximum sulphur content in oil were introduced in 1976, and have probably been more effective. The regulations, amended in 1985, prohibit use of combustion oil containing more than 1.0 per cent sulphur in 13 counties in southern Norway. However, the switch from NS to LS oil started before the regulation was introduced. The price-differential between NS and LS oil came as a result of new developments in refining technology. Secondly, new international standards led to a decreased demand for high-sulphur oil and associated products (*Econ Report* 45/93, p.29).

Thus, fully 60 per cent of the reductions were achieved by measures that in fact involved no additional costs to Norway. This is the main explanation for the 'over-achievement' in Norwegian SO_2 reductions. The 30 per cent and 50 per cent reduction can be said to have been 'free of charge'.[22] The measures that did imply some costs were directed primarily at industry, which was subject to direct regulation and was also the chief consumer of mineral oil. However, the primary motivation behind these measures was local pollution problems, and not concern for international agreements.

Measures to reduce NO$_X$ emissions – Or the lack of them   The implementation process of the NO$_X$ Declaration has not proved quite as smooth. First of all, the emitters of NO$_X$ are not the same as the emitters of SO$_2$. More than 70 per cent of NO$_X$ emissions stem from mobile sources – 36 per cent from road traffic and 35 per cent from ships (the fishing fleet, domestic coastal transport, mobile drilling rigs). The second largest emitter is the petroleum sector, representing 13 per cent of total emissions (SFT, 1994:18). This means an entirely different situation compared to SO$_2$. Industry, with its large and concentrated stationary sources, is not the major emitter: instead, it is the communications and transport sector, with its many small mobile sources, that poses the biggest problem. I will start by looking at what has been done to regulate these sources.

The process of NO$_X$ regulation started in 1987 when the Norwegian government prepared a report focusing exclusively on measures to reduce NO$_X$ emissions (*St.meld.* no. 47, 1987–88). Most of the measures proposed were rather vague. What should be done was not clearly defined, no time limits were set, and consequently little concrete action came out of it. A second major attempt was concluded in October 1991, when the SFT completed a new report on how to achieve a 30 per cent reduction of NO$_X$ emissions. The report was long kept secret, probably due to the resistance it was met with in the other ministries involved. The Ministry of the Environment concentrated on 'selling' the plan to the ministries and the directorate which would be responsible for its implementation (the Ministries of Transport and Energy, and the Directorate of Navigation and Shipping), and the government was unwilling to expose this internal struggle to media attention.[23] In August 1992 the environmental NGOs obtained access to the report.[24] The plan included measures to regulate emissions from all the major sources. The most effective measures were within energy production in the North Sea and requirements concerning exhaust gases from ships. However, so far very few of these measures have actually been introduced.

To date, the following measures have been introduced:

- *Requirements concerning exhaust gas from all types of vehicles* were introduced from the late 1980s and onwards (private petrol-fuelled cars in 1989, private diesel-fuelled cars in 1990, lorries in 1992 and heavy lorries in 1993), and these measures are expected to reduce emissions from road traffic by almost 50 per cent (SFT, 1994:19). According to Audun Rosland, leader of the Transport and Energy Section at the SFT, the major motivation behind these measures came from local pollution problems, but acidification problems have also been of some importance.[25]

- *Industrial measures* have mainly been directed towards the production of fertilisers in the highly industrialised Grenland area in southern Norway. Here the impetus has come from local pollution problems. Some reductions in the ferro-alloy industry have been registered, but they are due to a temporary recession.

Important sources like oil and gas production and ship engines have been exposed to no regulation whatsoever.[26] According to Rosland, no measures have been introduced as a direct consequence of the international agreement. Sissel Rönbeck, former Minister of the Environment, holds the same view.[27]

An interdepartmental committee – The Joint Group on Climate and Acid Rain Issues – is now responsible for developing new measures to reduce NO_x emissions.[28] Its members come from the following ministries: the Environment, Finance, Foreign Affairs, Agriculture, Transport and Communication, Industry and Energy. Internal conflict within the group has delayed plans for abatement measures repeatedly. On one hand the Ministry of the Environment has argued that Norway must fulfil its international obligations in order to legitimately expect the same from others. On the other hand, the Ministry of Finance has opposed more government spending, the Ministry of Transport has opposed measures to reduce traffic, the Ministry of Energy has opposed measures directed towards oil production and the Directorate for Shipping has opposed regulation of coastal traffic. So far the groups opposed to the MoE's agreement have managed to block effectively the introduction of any new measures.

VOC reductions The SFT finally published its plan for reduction of VOC emission in 1997. According to the SFT, Norway can reach the 30 per cent reduction goal, at the cost of NOK160 million (approximately UK £ 16 million). It is still too early to say whether this will be done. However, some of Norway's poor implementation record is most likely caused by growth in the offshore petroleum sector, a development that was not foreseen at the time of the VOC negotiations.

Comparing the Processes

Undoubtedly the main explanation of the difference in reductions lies in the different cost-levels connected to the two protocols. With the SO_2 reductions, 60 per cent were 'free', whereas NO_X and VOC reductions would imply substantial costs. On the other hand, this is probably not the only explanation, as will be further elaborated below.

Groups exposed to regulation – 'Paying off' the industry   With all abatement measures, implementation touches on conflicting interests. In the Norwegian case, as we have seen the cost level and therefore also the conflict level, is much higher in connection with NO_X and VOC measures. In the SO_2 case, the abatement group – industry – was well organised and resourceful, as well as being directly affected economically. Still, the government managed to implement several measures to reduce emissions from this group (St.meld.no. 44, 1975–76). The main explanation is most likely that much of the cost inflicted upon industry was transferred to the taxpayers through beneficial state support programmes. Since very little of the older industry was able to make the necessary changes, interim financial support was provided by the government. An evaluation conducted in 1982 concluded that on average the support programmes covered a little less than 70 per cent of the investment costs for the industry (NOU 1982:23). The definition of 'environmental investment' was practised in a broad manner, and prevented strong resistance from the industrial sector. Later on the government also provided financial support for general energy-saving measures in trade, industry, state buildings and the domestic sector.

As regards NO_X emissions, no measures have been introduced, due to internal struggles in the central administration. Thus no societal groups have so far been exposed to regulation. It is, however, obvious that conflicts of interests between different societal sectors are what underlie the inter-ministerial disagreement. Similarly, the petroleum sector, which is Norway's largest emitter of VOC, is unlikely to welcome expensive abatement measures in the offshore industry. The ministries seem to be loyal to their respective sectors, and firmly oppose the introduction of measures that would affect their clients.

An effective administrative apparatus? How was the implementation process organised? Are there any substantial differences between the organisation of the two implementation processes? The Norwegian Ministry of the Environment is the major administrative agency responsible for regulation of pollution. This Ministry was established in 1972, and two years later a sub-agency, the Norwegian Pollution Control Authority (SFT), was established as an expert organisation. At first, legal regulation was dispersed in several laws, but in 1981 the Pollution Control Act assembled all regulation of stationary sources of pollution into one comprehensive law.[29] The administrative system is relatively centralised, though a decentralisation process commenced some few years ago involving some delegation of power to the municipal level. However, air pollution issues are still subject to central regulation.

The implementation of the SO_2 Protocol was mainly the responsibility of these two central administration agencies. The Ministry was in charge of developing the measures, since most of the sources were stationary and therefore subject to regulation by the 1981 Pollution Control Act. Implementation of the regulations was largely delegated to the SFT. The administration of the NO_X regulations was much more widely dispersed. The fact that mobile sources were not covered by the Pollution Control Act made regulation by law more difficult. Implementation of measures in the transport sector was the responsibility of the Ministry of Transport. The Directorate of Navigation and Shipping and the Ministry of Fisheries were first to elucidate and later to implement measures in the maritime sector. The Ministry of Energy was responsible for the further development of energy-saving initiatives and a more efficient energy market. The environmental authorities were mainly responsible for co-ordinating the various measures (*St.meld.* no. 47, 1987–1988).

In other words, there was no effective central administrative apparatus ready to implement the international commitments that Norway assumed in 1989, nor is there any such apparatus today. The same can be said to apply in the VOC case. The Ministry of the Environment has been dependent on co-operation from the various dispersed implementation agencies, and so far they have not managed to work together. The conflicts described in the previous sections were therefore reinforced by the way in which the implementation process was organised. Conflicts of interest between societal groups became institutionalised in the administrative apparatus itself.

Public opinion and the connection between the phases The analysis of the policy-formulation process has shown that growing public concern about environmental issues was an important factor in the determination of national positions. And the fact that the impetus behind Norway's signing of the NO_X Declaration was internal political pressure is likely to have had consequences for its implementation. If the primary motivation behind signing were to win popular support, an equal motivation in the implementation phase would not necessarily result in the introduction of unpopular measures. On the other hand, non-compliance could also prove unpopular. The effect of public opinion could therefore point in both directions with regard to implementation. Another matter is the fact that environmental issues no longer head the political agenda today.

Model III: Social Learning and Policy Diffusion

The co-operation between the Norwegian government and the natural science community has been particularly close throughout the period under study. This close collaboration started in 1972 when the Royal Norwegian Council for Scientific and Industrial Research and the Agricultural Research Council of Norway established a joint project, 'Effects of acid rain on forest and fish'. The original title was 'Air pollution – effects on soil, vegetation and water', and the change of name reflected a change of focus due to the increasing political involvement in the project. In 1973 the MoE began to contribute funds to the project and soon became a larger contributor than the two research councils combined. As the project expanded, the overarching objective became the production of scientific background material for the international negotiations, which the MoE was preparing for at that time. Focus shifted from forest damage to damage to fish stocks because fish were presumed to be affected at a faster rate than the forest, and the MoE needed to document damage in its international work. In 1975 the MoE in a document to the Parliament (*St.meld.* no. 172, 1974-1975) explicitly stated this change. By this time the project was totally under the control of the MoE.

Criticism was raised against the replacement of scientific guidelines by political ones and against the heavy influence of the MoE. By 1977 tensions had escalated into an open conflict with broad media coverage. It became clear the project had found no evidence of forest damage as a result of acid rain. The politicians in charge of the international negotiations were not pleased, as this would confirm that the fear of damage had been exaggerated. The scientists in the project felt pressured to 'stretch' their results and conclude that forest damage could not be totally dismissed. However, by the time the project ended in 1980, acid rain was a well-acknowledged phenomenon in Norwegian opinion. The next step would be to gain the same degree of acceptance abroad. Thus, over the next five year Norway exported its store of scientific evidence of damage caused by long- range air pollution.

The close contact between scientists and the bureaucracy has continued to influence the Norwegian policy.[30] However, the research continues to be focused on damage caused by foreign sources, while much less has been done to develop abatement measures. In this regard, Norway has been surpassed by other countries, perhaps most notably by Germany, and has now clearly become an importer of new technology and policy ideas. The overall Norwegian focus on the foreign sources of pollution is therefore also reflected in the domestic research agenda.

Conclusions

Has Norwegian acid rain policy changed in recent years? Not surprisingly, the answer is both yes and no. *With regard to the policymaking process* the answer is clearly yes. The actors involved at all levels of the process are more numerous today than 20 years ago. When the process commenced in the 1970s, only the Ministry of the Environment and the natural science community seemed to show any interest in the acid rain issue. The other ministries and the external actors were basically indifferent to the international negotiations. The change came when these groups realised that Norway's commitment to a 30 per cent reduction of NO_x emissions would have serious consequences for large sectors of Norwegian society. Several new actors demanded to be involved in all stages of the policymaking process, from the formulation of positions and goals to their final implementation.

In regard to the policy content, however, the answer is no. The aim of Norwegian acid rain policy has remained the same for two decades: to reduce damage to the Norwegian environment caused by acidification. This can be achieved only by reductions in foreign sources. But there are varying views in Norway as to the best way of going about this. The Ministry of the Environment has argued that in order to persuade others to adopt costly measures, it is necessary for Norway to set a good example by introducing own reductions. Other sectors oppose this view, and point out that domestic emissions contribute only marginally to the domestic problems. As new actors have entered the process, these different views have had notable consequences for Norway's policy. The strategy of reducing own emissions in order to induce others to reduce theirs is still controversial within the Norwegian central administration. The main goal remains the same, but the means to achieve it are contested.

Theoretical Implications

At first glance, the URA model seemed to explain the Norwegian policy quite well. The cost-benefit-analysis showed that there were significantly higher costs connected to compliance with the NO_X and VOC goals than with the SO_2 goal. This difference in cost-levels was undoubtedly an important explanation for Norway's non-compliance with the NO_X and VOC goals. However, there were also some questions that could not be answered by a cost-benefit analysis. One such question was whether Norway's commitment to the NO_X and VOC goals was an act of deliberate deception, a miscalculation of future abatement costs, or perhaps the result of other domestic concerns. The answer to this question could be found only through closer analysis of the process of domestic policy formulation and implementation.

The analysis of the domestic politics contested both assumptions of the URA model: that is, that states act as unitary actors, and that policy choices are based on rational economic cost-benefit calculations. However, in the case of the first SO_2 Protocol, Norway based its policy on rational calculations undertaken by a relatively determined group of higher civil servants and scientists. In the NO_X case the most important driving force behind the Norwegian policy was a competitive game between political parties at a time when public interest and concern for environmental issues had peaked. Despite clear expert warnings of the abatement costs, the political costs which Norway would incur by abandoning its international role as a pusher became too high for the governing Labour Party. The Norwegian position in the two last protocols was determined by a tug-of-war between the ministries affected by the Norwegian commitments (the Environment, Finance, Energy and Transport). Public interest had dropped, but the sector interests had learned their lesson from the NO_x process and demanded extensive participation (through their respective ministries) from the earliest possible stage in the policy formulation process and straight on through to the international negotiations.

When it comes to implementation of the goals, sector interests are also significant. The SO_2 reductions were largely free of cost, as well as being unrelated to the international commitments. The commitment to 30 per cent NO_x reduction was made against better knowledge; the lack of implementation here has been a result of strong resistance from sector ministries and the Ministry of Finance. In the VOC case, the increase in Norwegian emissions was underestimated when the commitments were made, and this is probably some of the explanation behind Norway's failure to comply. However, resistance towards the introduction of costly abatement measures remains strong.

The URA model's focus on material interests undoubtedly pointed to an important explanation of the Norwegian policy. But in order to understand some of the decisions that were made, it has proven necessary to widen the concept of national interest beyond maximising net national welfare in strictly economic terms. For instance, a state's concern about maintaining its reputation as a trustworthy partner or preserving its 'green' image should be included in the notion of 'national interests'. Expanding the concept of national interests does not mean rejecting the URA model as such. However, when expanded into an all-inclusive preference list, the model loses much of its parsimony and conclusiveness, which are seen as its greatest advantages compared to other approaches. Even more important perhaps, is the question of whether application of the URA model can be more delusive than clarifying. Is, then, the model riddled with deficiencies so fundamental that it must be discarded altogether?

Here, we need to distinguish between the URA model as an analytical tool, and its basic assumptions about what determines state behaviour. Applied as an analytical tool for predicting policy choices through quantification of national interests, the model has proven of only limited value. In fact, if an imperfect match between the predictions of the model and the actual behaviour of a state (as in the Norwegian case) should lead us to conclude that the country acted irrationally or that its policy was not interest-based, the model has indeed been more delusive than clarifying. However, these deficiencies result largely from the use of quantitative methods in social sciences. What then of the URA model's basic assumptions about state behaviour? The observation that states are not unitary actors is trivial. More important is whether domestic politics take primacy over international commitments. This seems to be confirmed by the Norwegian case. The question of rationality is more complex, but also less interesting here, since the problem of the all-inclusive preference list will make it difficult to determine what is rational behaviour. Our examination of the domestic policy process showed that the policy adopted by Norway could be seen as 'rational' according to a broad range of criteria and that it was always based mainly on narrow self-interest. But it would not have been possible to reach this conclusion by applying the URA model.

Notes

1. See the chapter by Bäckstrand and Selin in this volume (chapter 4).
2. The estimates are made in 1987 NOK.
3. Interview with Sverre Thoresen, Director of the Norwegian Forest Owners' Federation, June 1994.
4. Undated Norwegian draft to the SO_2-negotiations.
5. Interview with Sissel Rönbeck, May 1994. In a draft for a speech held by the Minister of the Environment at the Sofia meeting in November 1988 it is evident that this long remained the dominant view. 'Norway has had considerable problems during the negotiations. At one stage it seems impossible even to accept a freeze of the total NO_X emissions, far worse to accept any requirements of significant short-term reductions.' The actual statement made at the meeting was slightly different: '....the protocol obviously does not prevent individual countries from going further than our joint commitment here. .... Norway is faced with major challenges in our effort to reduce NO_X emissions, challenges which we are prepared to meet'.
6. 'Critical loads' are defined as the highest load of pollution that will not cause chemical changes leading to harmful effects on the most sensitive ecological systems.
7. Interview with Sissel Rönbeck, May 1994.
8. Minister of the Environment, Liv Hille Valla, to *Aftenposten,* 30 June , 1990.
9. Hovi (1991) also suggests several other mechanisms connected to refraining from action: denying a state the opportunity to act, holding hostages, holding land-areas, guarantees by third parties and irreversible costs – none of which are very relevant to the present case.

10. Interview with Guro Fjellanger, The Norwegian Society for the Conservation of Nature, January 1994. She is currently (February 1998) the Minister of the Environment.
11. Interview with Erik Lykke, former Head of the International Department in the MoE, January 1994.
12. Interview with Erik Lykke, January 1994.
13. Interview with Sissel Rönbeck, former Minister of the Environment, May 1994.
14. During the debate, the pro-agrarian Centre Party suggested that plans were made for a 75 per cent reduction of NO_X emissions by the year 2000. The Left Socialist Party wanted a 30 per cent reduction by 1995 and the Conservative Party also expressed 'disappointment' over the low ambitions of the Labour government.
15. Interview with Sissel Rönbeck, May 1994.
16. Interview with Sissel Rönbeck, May 1994.
17. VOC emissions are harmful when they are combined with NO_X and exposed to sunlight. Since there is very little of both NO_X and sunlight north of the 62nd parallel the damaging effect was 30 times lower than from emissions in Central Europe.
18. Undated and unofficial letter from the SFT to the MoE, from 1992.
19. To go further into a description of these reductions would be beside the point here, since they are not connected to the LRTAP Convention at all.
20. These measures are described and evaluated in several official documents: *Stortingsmelding* no.44 (1975-76), 'Tiltak mot forurensninger', *Stortingsmelding* no.51 (1984-85), 'Tiltak mot forurensning', *Stortingsmelding* no.47 (1987-88), 'Om reduksjon av nitrogenutslippene i Norge', *NOU* 1976:47, 'Miljøvernpolitikken. Funksjonsfordeling og administrasjonsordninger', *NOU* 1982:23, 'Utslippsavgifter', *NOU* 1992:3, 'Mot en mer kostnadseffektiv miljøpolitikk i 1990-årene', *NOU* 1977:11, 'Tiltak mot forurensninger', *Econ report* 45/93 for the Ministry of the Environment, 1993, 'Virkemiddelbruk overfor utslipp av svoveldioksyd'.
21. 'The demands arose as a result of local environmental problems in the vicinity of the individual plants due to the concentrated emissions' (*Econ Report* 45/93, p.19).
22. This conclusion is reached by the SFT in the same undated letter referred to in note 18.
23. *Natur og Miljø Bulletin*, 28 August 1992.
24. *Natur og Miljø Bulletin*, 14 August 1992.
25. Interview with Audun Rosland, June 1994.
26. However, new areas of research have been developed. The Norwegian strategy when it comes to emissions from ships has again been to find international solutions and persuade other maritime countries to agree to international regulation. This has been done in order to avoid the introduction of national coastal state requirements that might distort competition. On this background a research programme, 'Environmentally Friendly Ships', has been undertaken by the Norwegian Centre for Industrial Research (SINTEF Annual Report 1992, p.13). The purpose is to develop the necessary technology to lower NO_X emissions from ships (which today represent 36 per cent of total NO_X emissions). The programme covers fuel technology, new engine technology concepts and post-combustion treatment of gases. Another programme ('Environmentally Friendly Diesel Engines') also involves major industrial companies. The project co-ordinates activities with a total budget of NOK 60 million.
27. Interviews with Audun Rosland (June 1994) and Sissel Rönbeck (May 1994).

28. The group also has one representative from the Office of the Prime Minister and is headed by the Ministry of the Environment.
29. *Lov om vern mot forurensninger og om avfall*, 13 March 1981, no.6.
30. The Norwegian emphasis on critical loads was the final stage in the development of an increasingly central role of the scientific community in the policy formulation. Even though the scientists were more visible in the 1970s, partly because they were the only actors besides the MoE, they have become more important in setting the premises for the political process in the 1990s. The Ministry of Finance was originally opposed to the critical load approach, but was finally convinced by the scientists that 'critical loads' was consistent with their favourite concept – 'cost efficiency'.

References

Brendemoen, A., Glomsrød, S. and Aaserud, M. (1992), *Miljøkostnader i makroperspektiv*, Central Bureau of Statisitcs of Norway report 92/17.

Claes, D.H. (1991), *Internasjonale klimaforhandlinger og teorier om kollektiv handling*, Energy, Environment and Development Publication no.6, 1991, The Fridtjof Nansen Institute, Oslo.

Engaas, B.R., Jervan, H. and Ribeiro, T. (1988), *Miljøreguleringer i industrien*, Resource Policy Group report no. GRS-720, Oslo.

Hestmark, G. and Roll-Hansen, N. (1990), *Miljøforskning mellom vitenskap og politikk*, NAVFs utredningsintitutts Melding 1990:2, Oslo.

Hovi, J. (1991), 'Pacta Sunt Servanda - bare en frase?', *Norsk Statvitenskapelig Tidsskrift* 1991 (7).

Kiberg, D. and Strømsnes, K. (1992), *NSD-rapport nr.94 De norske valgundersøkelsene 1977, 1981, 1985 og 1989*, Norwegian Social Science Data Services.

Laugen, T. (1995), *Compliance with International Environmental Agreements. Norway and the Acid Rain Convention*, FNI Rapport 003:95, The Fridtjof Nansen Institute, Lysaker.

Roll-Hansen, N. (1986), *Sur nedbør - et storprosjekt i norsk miljøforskning*, NAVFs utredningsinstitutts Melding 1986:4, Oslo.

Rosencranz, A. and Wetstone, G.S. (1983), *Acid Rain in Europe and North America*, The Environmental Law Institute, Washington, D.C.

Underdal, A. (1993), 'Two models for predicting and explaining environmental policies', unpublished draft.

Governmental Papers and Reports

Forurensning i Norge 1994, The Norwegian Pollution Control Authority.
Lov om vern mot forurensninger og om avfall, 13.03.81, nr.6.
Miljøtilstanden i Norge 1996, The Norwegian Pollution Control Authority.
Miljøvernpolitisk redegjørelse 1994, The Norwegian Pollution Control Authority.
NOU 1976:47, *Miljøvernpolitikken. Funksjonsfordeling og administrasjonsordninger.*

NOU 1977:11, *Tiltak mot forurensninger.*

NOU 1982:23, *Utslippsavgifter.*NOU 1992:3, *Mot en mer kostnadseffektiv miljøpolitikk i 1990-årene.*

OECD Environmental Performance Review - Norway, OECD 1993.

Stortingsmelding nr.44 (1975-76), *Tiltak mot forurensninger.*

Stortingsmelding nr.51 (1984-85), *Tiltak mot forurensning.*

Stortingsmelding nr.47 (1987-88), *Om reduksjon av nitrogenutslippene i Norge.*

Sur Nedbør, 1995, Norwegain Ministry of Environment.

Virkemiddelbruk overfor utslipp av svoveldioksyd Rapport nr. ECON 45/93 for the Ministry of the Environment, 1993.

6 Reversing (Inter)National Policy - Germany's Response to Transboundary Air Pollution

DETLEF F. SPRINZ AND ANDREAS WAHL

Germany and Acid Rain

Perhaps like no other country, Germany has radically changed its policies towards regulating air pollution in the European context. Due to doubts about the relationship between causes and effects, it acted in the 1970s originally as a dragger with regard to the regulation of transboundary air pollutants. In the early 1980s Germany responded very decisively to its own damage assessment. In particular the adverse effects to forests ('Waldsterben' or forest die-back) led to the formulation of strict air pollution regulations in the domestic context, efforts to spread their regulatory system within the European Union, and activities within the United Nations Economic Commission for Europe (UNECE) to foster stronger, continent-wide emission reductions. Thus, Germany turned from a dragger state of the 1970s to an international leader on air pollution regulations in the mid-1980s (Sprinz and Vaahtoranta, 1994).

This chapter summarises the domestic foundations of Germany's air pollution policy in the national, European, and UNECE context and explains why Germany switched from dragger to leader by building on the conceptual lenses advanced in chapter three.

The Problem as Seen by Germany

The damaging effects of air pollution in general have been known in Germany for several centuries. Already in 1341, blacksmiths were forbidden to use anthracite whose smoke damaged vegetation (Klein 1984, p.32). And in 1883, the first book about environmental damage to forests by air pollutants - the main concern in the contemporary acid rain debate - was published in Germany (Schröder and Reuss, 1883). After World War II, air pollution was mainly an issue of the heavily industrialised areas, especially in North Rhine-Westphalia, and was seen primarily as a health problem, although already in the late 1960s, the problem of long-range transport of air pollutants and resulting damages to forests was, at least in principle, known in Germany.

For Germany, the problem of transboundary air pollutants is largely an issue of protecting German forests and advancing the state of technology (best available technology) to reduce emissions. Thus, protecting a vital part of Germany's (managed) nature and building on long-standing engineering traditions go hand in hand. While the economic damages are hard to quantify exactly (see below), the abatement costs have normally been comparatively low. Once massive damages were systematically documented in the forest sector in the early 1980s in Germany and elsewhere, Germany was able to respond to the challenges by building on its regulatory experience in dealing with air pollution (see below). However, as compared to many other countries, domestic effects of air pollution first had to enter the domestic political process, and subsequent international policies largely served to influence other countries to follow its lead. The German situation shows a slight reversal of the three-stage policy cycle used in this study. Domestic damages initiated substantial changes in domestic laws, aspects of domestic laws influence the position in international regulations, and domestic implementation and compliance is largely determined by the original German domestic laws – rather than being a response to international negotiations and the obligations stemming from international environmental agreements.

Domestic Policy and Regulations Prior to International Negotiations[1]

Since the 1950s, regulation of air pollutants has periodically appeared on the German policy agenda and led to a series of public statements and actual regulations. By 1957, the Federal Parliament (Bundestag) had received a report from the federal government concerning the problems originating from air pollution, and in 1960, the Association of German

Foresters passed a resolution at a meeting in Stuttgart targeted at those responsible for combating air pollution. However, the problem was considered, in principle, to be of a *localised* nature. The Technical Instruction for Air ('TA-Luft') entered into force in September 1964 and assisted the executive branch in implementing legislation. It specified the best available technology (BAT) for granting authorisation to build an industrial installation or plant. The first German flue gas desulfurisation plant followed in 1971. Already by 1967, the Federal Health Ministry pointed to the problem of the *long-range* transport of air pollution with emphasis on SO_2 - while also mentioning that the forests may be in danger. Early criticism of the policy of high smokestacks - which were used to solve the problem of *local* air pollution - was already expressed in October 1967 at a general meeting of the Federation of European Agriculture by Dr. Wenzel, who was responsible for Nature Conservation and Protection of the Countryside of the State (or Land) Government of Hesse. Referring to the adverse effects of SO_2, the Federal Interior Minister Höcherl pointed out in April 1968 that 0.8 per cent of German forests were already harmed. In the middle of the following year, Federal Minister of Health Strobel also drew attention to the dangers of air pollution.

Parliamentary responses followed. In June 1971, a hearing on the 'Preservation of Clean Air' took place in the Federal Parliament. By amending the Basic Law ('Grundgesetz' which serves as the German equivalent of a federal constitution) in 1972, the responsibilities for the well-established divisions of air pollution, noise pollution, and waste disposal were put under the prerogative of *federal* regulation (as opposed to the Länder) (Article 74, No. 24 Grundgesetz). Since then, the foundation of air pollution control measures lies with the 1974 Federal Clean Air Act (Bundesimmissionsschutz-Gesetz). More specific regulations can be grouped into plant-related regulations (TA-Luft, Ordinance on Large Combustion Plants, etc.), area-related provisions to control local problems (clean air plans), and product-related regulations (which specify quality standards for substances and products such as the Ordinance on the Sulfur Content of Fuel Oil and Diesel Engine Fuels or the Ordinance on Fuel Labeling and Fuel Quality). Originally, only *new* plants were affected by these regulations. With the amendment of the Technical Instructions Air in the 1980s and the Ordinance on Large Combustion Plants, regulations were also applied to *existing* plants.

Furthermore, in 1976, more incidents of dying pine trees in Bavaria were reported, and this led to increased scientific research in 1977. In the following year, the Federal Environmental Agency arranged a hearing on the damage to forests and long-range transport of air pollutants in preparation for the negotiations which ultimately led to the UN ECE Convention on Long-Range Transboundary Air Pollution in 1979.

Despite the suggested threats posed by air pollutants to forests, air pollution regulations were mainly geared towards the protection of human health. Particular emphasis was placed on large combustion plants and the domestic energy sector. The oil price shock of the 1970s gave rise to a domestic coal and nuclear energy strategy in order to further national independence of the energy sector. In 1975, Chancellor Schmidt met leading members of industry and trade unions at Schloß Gymnich. Because Germany was experiencing a decline in economic growth and a subsequent rise in unemployment (resulting from the oil price shock), the focus was put on 'pro-industry' environmental policies. In addition, the trade unions agreed on giving preference to economic issues over ecological ones. Therefore, no further environmental legislative initiative was taken until 1978. Even the first meeting of the Chancellor with representatives of the environmental NGOs in 1979 led to no tangible results regarding environmental policy. As a result of these developments, there was no national target on reducing sulfur emissions - the major air pollutant considered by the end of the 1970s. Instead a policy of 'more research' was pursued.

Until the early 1980s, the Federal Republic was a dragger in international negotiations on emission reductions. Especially the *transboundary* aspect of air pollution was not much attended to. Consequently, until the end of the 1970s, the regulations in the field of air pollution led only to higher smoke stacks to solve local and regional air pollution problems by way of dispersion (Hartkopf, 1984; Schärer, 1992, p.192).

Forest death ('Waldsterben') is not new to Germany, but forest decline in so-called 'clean air areas' (Reinluftgebiete) posed a new challenge to research, because only flue gas-related damage in the vicinity of sulfur emitters were well-known until the early 1980s in Germany. Most importantly in 1979, Prof. Ulrich (a soil scientist) and colleagues published their research results which suggested strong links between the dying of forests and SO_2 emissions (Ulrich et al., 1979).

Much of the policy process to follow was characterized by the interplay of science, public attitudes, and the mass media. The Council on Environmental Quality (Rat von Sachverständigen für Umweltfragen) published a special report on 'Energy and the Environment' in 1981 and concluded that even minor concentrations of SO_2 could damage vegetation. In the same year, the Federal Ministry of Agriculture and Forestry began its annual forest surveys. In November 1981, a very influential three-part series about the dying of forests and its causes appeared in 'Der Spiegel' (Der Spiegel 1981, No. 47 - 49) and made forest death a major issue for the mass public and policy makers alike - in Germany *and* beyond. By autumn of 1982, the Federal Government and the 'Committee for Protection Against Immissions [i.e. Air Quality] or Depositions' of the states (Länder) presented a report which concluded that 7.7 per cent of the national forest area was damaged. Air pollutants were highlighted as a major cause, especially SO_2, NO_X, heavy metals, photooxidants, as well as natural causes such as drought, frost, disease, etc. Scientifically conclusive proof could not be produced, but the report focused political attention on the issue. In March 1983, the Council on Environmental Quality presented an extensive, special report on 'Forest Damage and Air Pollution' (Sachverständigenrat für Umweltfragen 1983). This report contained recommendations which were similar to the draft regulations of the Ordinance on Large Combustion Plants. The results of the second survey of forest damage published in 1983 concluded that 10 per cent of the forest area of the Federal Republic should be classified as 'medium ill' to 'seriously ill' (forest damage classification categories 2–4). By the third forest survey in 1984, 17 per cent of the forests were classified as damaged (categories 2–4) (Bundesministerium für Ernährung, Landwirtschaft und Forsten 1985, p.12).

Around 1983, a large research program was launched on forest death. It was co-ordinated by the Inter-Ministerial Working Group 'Damage to Forests/Air Pollution' and convened by the federal government in the same year. As the name indicates, it focused on the link between acidifying pollutants and forest decline. Together with its scientific advisory committee and its office (which was established by the Federal Environmental Office, UBA), the working group spent about 250 million DM on ca. 660 different research projects between 1982 and 1988 (Gregor 1990, p.139). The fundamental conclusion that forest death is caused by a multitude of factors rather than by air pollution alone has remained unchanged.

In response to the environmental (and political) problems caused by forest death (see also below), Germany began to upgrade its air pollution regulations over time. For example, the second amendment of the Federal Clean Air Act of 1985 dropped the clauses on economic viability (which influenced the definition of BAT until then). In May 1990, a third amendment of the Federal Clean Air Act was enacted. And on 1 July 1990, the environmental regulations of the FRG were extended to the (then barely existing) GDR, the latter being a major emitter of sulfur to the European airshed. By way of the unification treaty of 23 September 1990 it was finally stipulated for the Eastern Länder of the FRG that

- all installations and plants had to comply with the best available technology standards and
- *existing* plants had to attain the standards of new ones within five years - or otherwise be shut down (see Art. 67A Federal Clean air Act, in Hansmann, 1995).

The deadline for implementing the regulations for reaching the targets of the Ordinance on Large Combustion Plants on the territory of the former GDR was one to three years longer than the one set for Western plants.

German policy on transboundary air pollution changed considerably over time. Beginning with a long history of scientific research into the causes and effects in the vicinity of emitters (in the past century), Germany began to develop air quality regulations which were health-based and technology-driven. However, with the advent of the international dimension of long-range transboundary transport of air pollutants and focused research on its vulnerable forests, Germany shifted from an international dragger in the late 1970s to become a persistent pusher since the early 1980s. Thus, the basic framework of air pollution policies existed long before acidification and forest death became major issues. By radically strengthening an existing regulatory framework, Germany was able to respond decisively to the new problem (Schärer 1992, p.192). In effect, German domestic regulations have been leading international regulations, both in terms of timing and in terms of stringency.

Role of Germany in LRTAP and EU Negotiations

Since the late 1970s, negotiations took place at the UNECE (United Nations Economic Commission for Europe). The Federal Republic of Germany has been actively involved in these negotiations since the beginning. Until 1982, the FRG was one of the most forceful *draggers* regarding SO_2-emission reductions as a result of assuming low damages to its ecosystems (see above as well as Sprinz and Vaahtoranta, 1994). This changed in the early 1980s when news of the dying forests reached the public and the Green Party began enjoying electoral success at the state and federal levels. Since 1982, the Federal Republic has become a *pusher* for strict international environmental agreements in alliance with the Nordic countries. In parallel to the UNECE negotiations, Germany has also used the institutions of the European Union to reduce emissions of air pollutants.

All major political programs on transboundary air pollution have been initiated and decided upon by the German authorities *prior to* signing the respective protocols - with the exception of the LRTAP Convention of 1979. Originally, the Federal Republic was one of the major *opponents* of agreements to reduce acidifying pollutants. The warnings and calls for help from the Scandinavian countries as well as their demand for a 30 per cent reduction of SO_2 emissions had no impact on the policy of the Federal Republic. Given the resistance against any specific pollution reduction by the U.K. and the FRG, the LRTAP Convention (1979) contained no binding obligations on emissions limitations. The text stated only, inter alia, that the signatory countries should 'endeavor to limit and, as far as possible, gradually reduce and prevent air pollution, including long-range transboundary air pollution', and therefore 'use the best available technology that is economically feasible' (United Nations Economic Commission for Europe 1979, Articles 2 and 6). Although this early international environmental agreement contained no binding obligations on emissions reductions, it still formed the starting point for subsequent negotiations.

The international negotiations on reducing sulfur emission ultimately resulted in two international environmental agreements, namely the Helsinki (1985) and Oslo (1994) Protocols (United Nations Economic Commission for Europe 1985 and 1994). At the first session of the Interim Executive Body for the Convention in June 1983, the Nordic countries renewed their demands for a 30 per cent reduction of SO_2 emissions. Because of the forest decline and domestic political attention, Germany now actively supported this proposal. The proposal of the Federal Republic, Austria, and Switzerland demanded additional regulations, namely to limit the content of sulfur in diesel and light fuel oil to 0.3 per cent.

At the International Ministerial Conference in Ottawa in March 1984, the Federal Republic signed a declaration which required signatories to reduce their SO_2 emissions between 1980 and 1993 by 30 per cent. For this group, consisting of the Nordic countries, Switzerland, Austria, Canada, France, and the FRG, the term the '30 per cent-Club' was coined. Later that year, the 'Multilateral Environmental Conference on the Causes and Prevention of Damage to Forests and Waters by Air Pollution in Europe' was hosted by the German federal government in Munich. On this occasion, many Eastern European countries declared their intention to reduce sulfur emissions or their transboundary fluxes by 30 per cent. This enlarged coalition of like-minded countries was needed to conclude the first Sulfur Protocol which required signatories to reduce national emissions of sulfur or their transboundary fluxes by 30 per cent from 1980 levels until 1993 (United Nations Economic Commission for Europe 1985, Art. 2).

When the deadline for achieving the reductions targets of the Helsinki Protocol began to approach in the early 1990s and in view of the substantial emission reductions under way, many countries wished to go beyond the objectives of this protocol. The international negotiations began in February 1990, when the UNECE Working Group on Strategies prepared documents which shaped the international negotiations on a second sulfur protocol. The Federal Republic supported (i) 60 per cent emission reductions of SO_2 based on BAT (best available technology) and (ii) the critical loads 'gap closure' approach. It was one of the states which initiated the negotiations, in alliance with the Nordic countries, the Netherlands, and Austria. In November 1993, Germany announced an 84 per cent emission reduction goal to be accomplished by the year 2000. At the end of the negotiations, the Federal Republic agreed to the target of an 83 per cent reduction by the year 2000 and an 87 per cent reduction by the year 2005 (United Nations Economic Commission for Europe 1994, Annex II). In parallel to its policy within the UNECE, the Federal Republic of Germany also pushed the issue of sulfur emissions by way of the EC's

Large Combustion Plant Directive - thereby narrowing the competitive disadvantage of Germany with respect to its most important trading partners.

Equally determined as in the case of sulfur emission reductions, Germany pursued reductions of nitrogen oxide (NO_X) emissions since the mid-1980s. Already on the occasion of the Conference in Ottawa (1984), the Federal Republic stated its intention to reduce NO_X emissions substantially. By June 1984, at the Munich Conference, the Federal Republic supported the proposal that the Executive Body should include the reduction of NO_X emissions in its working program. Perhaps not surprisingly, the UNECE Task Force on NO_X was chaired by the Federal Republic.

In addition to the UNECE forum, the NO_X issue was also pursued within the European Union (EU). After the domestic policy process led to the introduction of passenger cars with catalytic converters, Germany used the European Community (EC) as a forum to spread its regulations on a European scale. These efforts resulted, inter alia, in the Luxembourg compromise of 1985 – which regulated NO_X and hydrocarbon emissions for larger passenger cars. Since then, the scope of additional emissions regulations has been broadened to include all types of passenger cars and, ultimately, trucks. Within the UNECE domain, the Sofia NO_X Protocol was concluded in 1988 (United Nations Economic Commission for Europe 1988), and the Federal Republic of Germany committed itself not to exceed its 1987 emissions by the year 1994 and to introduce unleaded petrol. The latter was already mandated by the Luxembourg compromise within the European Union as a result of German pressure. The Sofia NO_X Protocol was signed by 26 countries and came into effect on 14 February 1991.

Given its ambitions to go further than a freeze, Germany played an active role as part of a group of twelve countries willing to reduce substantially NO_X emissions. A Sofia Declaration was signed by those countries to reduce their NO_X emissions by about 30 per cent by 1998 at the latest. Any year between 1980 and 1986 could be chosen as the reference year and Germany selected 1986 - close to its maximum emissions in this period.[2]

Considerably less is known about the German participation in the negotiations on non-methane volatile organic compounds (VOCs). In accordance with the work plan of the Executive Body for the Convention on Long-Range Transboundary Air Pollution, the UNECE Working Party on Air Pollution Problems established a task force on 'emissions of volatile organic compounds from stationary sources and possibilities of their control' under the leadership of the Federal Republic and France (Umweltbundesamt 1991, p.2). The national position of the Federal Republic in the international negotiations of the Federal Republic was a 30 per cent reduction (base year 1988) of the annual VOC emissions, which was to be accomplished by 1999 and it signed the Geneva Protocol on VOCs (United Nations Economic Commission for Europe 1991). As a result, the Federal Republic is obliged to cuts its emissions by 30 per cent between 1988 and 1999 once the Protocol enters into force - an obligation which is fully in line with the national position.

After a comparatively slow start due to questioning the scientific evidence presented in the 1970s, the Federal Republic embarked on ambitious national and international environmental policies after discovering substantial environmental (and national political) damage in the early 1980s. Since then, Germany has acted as a lead country across a range of air pollutants by pushing international environmental negotiations, by trying to convince other countries to subscribe to more stringent rule on emissions of pollutants, and by sponsoring relevant research and (co-) heading relevant working groups within the UNECE as well as within the parallel structure of the European Union.

Implementation and Compliance Record

Given its change in perspective on the LRTAP regime, Germany has been a country which undertook comparatively swift implementation of its obligations under the various LRTAP Protocols - because most of the substantive implications of these agreements were already embedded in national German air pollution laws by the mid to late 1980s (see above). The transfer of international obligations normally took 2-3 years (see Table 6.1). Furthermore, from the latest available data, it seems reasonable to conclude that the Federal Republic, despite its enlargement of jurisdiction in 1990, either fully complies with its international obligations, or, where the target year has not yet been reached, shows good progress in meeting these objectives (see Table 6.2).

Table 6.1   International signatures and national ratification

	Date of signature*	Date of national ratification*	Entry into force (inter-nationally)
LRTAP Convention	13 November 1979	15 August 1982	16 March 1983
Helsinki (First) Sulfur Protocol (1985)	9 July 1985	3 March 1987	2 September 1987
Sofia NO$_X$ Protocol (1988)	1 November 1988	16 November 1990	14 February 1991
Geneva VOC Protocol (1991)	19 November 1991	8 December 1994	not yet in force
Oslo (Second) Sulfur Protocol (1994)	14 June 1994	not yet ratified	not yet in force

* Refers to the Federal Republic in its respective borders at each point in time.
Source: (United Nations Economic Commission for Europe/Convention on Long-Range Transboundary Air Pollution 1995, 135).

Table 6.2   German emission reductions

	Sulfur emissions (1,000 t SO$_2$)	NO$_X$ emissions (1,000 t NO$_2$)	VOC emissions (1,000 t hydrocarbons)
Base year (Sulfur: 1980 NO$_X$: 1987 VOCs: 1988)	7,517	3,598 (1986: 3,683)	3,167
Target year (Sulfur: 1993 NO$_X$: 1994, VOCs: 1999*)	3,156	2,872	2,405*
Change in % of base year	-58%	-20% (-22% compared to 1986)	-24%*

* Data are for the year 1994.
Note: All data for the Federal Republic also contain the territory of the former German Democratic Republic.
Source: United Nations Economic Commission for Europe (personal communication, 02 June 1997).

In most cases, national implementation did not mandate substantive upgrading of national pollution laws. The LRTAP Convention itself did not require more than national ratification. For the Helsinki Sulfur Protocol, the Federal Republic could rely on the Federal Clean Air Act as amended on 4 October 1985 and the Technical Instructions Air (TA-Luft) as amended on 1 March 1986. The German Ordinance on Large Combustion Plants acted as the major regulative instrument for combating air pollution and reaching the national targets. As a result of the national regulations, the sulfur emissions had decreased by *more than* 30 per cent by 1986 as compared to the base year 1980. And by 1993, Germany had already reduced 58 per cent of its 1980 sulfur emissions - exceeding the requirements of the Helsinki Protocol within the enlarged territory. While the Oslo Sulfur Protocol has not yet been transposed into national law, available data indicate that Germany has also made reasonable progress towards compliance with its obligations under this international environmental agreement.

Reducing NO_x emissions is likely to be substantially more difficult since the transport sector has to be regulated. While the obligations of the Sofia Protocol could be easily accomplished, it will be more difficult to reach the 30 per cent reduction goal by 1998. Already by way of the Lead in Petrol Act of 18 December 1987, lead in regular gasoline was forbidden. On 1 January 1990, the financial incentives for low emission vehicles entered into force, and the introduction of catalytic converters has been subsidised for several years so as to ensure emission reductions. Furthermore, leaded gasoline has been taxed more heavily as compared to lead-free gasoline.

As a result of the measures taken, NO_x emissions decreased by 20 per cent between 1987 and 1994 (Sofia Protocol) and by 22 per cent between 1986 and 1994. Thus, the Federal Republic is well on the way towards reaching the stricter goal of 30 per cent of NO_x emission reductions, although the final increments may be harder to achieve than the initial gains in emission reductions which had been realised by creating incentives to switch to lead-free gasoline.

In order to achieve the goals set out in the Geneva VOC Protocol, the Ordinance on Surface Treatment and Dry Cleaning (2nd Federal Clean Air Ordinance, last amended on 10 December 1990) and the Ordinance on the Filling and Storage of Otto-Engine Fuel (20th Federal Clean Air Ordinance) and the Ordinance on the Refueling of Motor Vehicles (21st

Federal Clean Air Ordinance) were all passed on 07 October 1992. In view of a 30 per cent reduction goal for VOCs between 1988 and 1999, the Federal Republic has made progress by already achieving a reduction of 24 per cent by 1994.

Overall, the Federal Republic of Germany created its major legal framework for air pollution regulations in the mid-1980s and has been quite successful in implementing and complying with the provisions of international environmental agreements so far.

Explaining National Policy Formation and Implementation

Three models were introduced to explain national policy formation and implementation, namely the rational actor model, the domestic politics models, and the social learning model. We will turn to each of these perspectives in this Section.

The Rational Actor Model

The rational actor model relies on the logic of marginal abatement and damage costs being equal for determining optimal emission reduction levels. Thus, both types of data are needed. Regrettably, it appears impossible for existing studies to differentiate between the damage costs by type of pollutant (SO_2, NO_X, and VOC). Furthermore, it is not at all clear which fraction of damage costs is to be attributed to *other* causes. In particular, damage to forests has multiple causes. Estimates of damage costs of several billion DM per year caused by air pollution have been given. For example, Wicke estimated in the mid-1980s that the overall damage of air pollution in Germany was 48 billion DM per year (Wicke 1986, p.56). Given the methodological limitations of most studies, this estimate has to be regarded as a rather crude indicator of the damage costs due to acid rain. Reliable estimates exist for damage to materials which has been assessed as amounting to 3-4 billion DM annually since the 1980s (Heinz, 1985; Bundesministerium des Inneren, 1984; Isecke et al., 1991; Bundesministerium für Umwelt, Naturschutz und Reaktorsicherheit, 1992). While these figures have been characterized by the respective authors as lower bounds of actual damage, they are thought to be rather crude 'guesstimates'. Apart from any quantifiable costs, between ca. 15 per cent and 20 per cent of West German forests were assessed as being seriously ill (categories 2-4 for forest damages) in the second half of the 1980s, whereas this figure rose to ca. 25 per cent *after* the enlargement of the Federal Republic in 1990.

With respect to abatement costs for the 30 per cent reduction of SO_2 emissions demanded by the first Sulfur Protocol, estimates based on the RAINS model give the costs for Western Germany as 1.4 billion DM (equivalent to ca. 0.05 per cent of GDP) per year until the year 2000 (Amann and Kornai, 1987; Sprinz, 1993). For the Oslo Sulfur Protocol, RAINS 6.0 puts the costs of compliance for the enlarged Germany by 2000 at 4.81 billion DM (equivalent to ca. 0.14 per cent of 1994 GDP) annually. By modernising the East German electricity generation sector and enhancing energy efficiency, SO_2 emissions will be reduced as a side-effect at no extra cost. This may account for the fact that less than one third of the costs of compliance with the Oslo Protocol is expected to arise in Eastern Germany despite its relatively high emission level. Furthermore, in order to reduce NO_X by 30 per cent (base year: 1980), RAINS 6.0 computes abatement costs of 0.525 billion DM (equivalent to ca. 0.02 per cent of GDP) per year in the Western part of Germany (Amann, 1989; Sprinz, 1993), while compliance with the obligation under the Sofia Protocol to freeze NO_X emissions at 1987 levels by 1994 should cost the enlarged Federal Republic 0.1 billion DM. The abatement costs have been influenced in two different ways by the enlargement of the Federal Republic. To the extent that NO_X emissions stem from stationary sources (electricity generation), it has to be emphasised that as in the case of SO_2, general technological modernisation led to inexpensive pollution reduction. Consequently, less than 10 per cent of the costs of NO_X emissions stabilisation is expected to occur in the Eastern part of the Germany. However, a substantial increase in individual mobility has occurred in East Germany since 1990. Therefore, abatement costs for major emission reductions may become quite high because the sector consists of many independent actors.

The decision to accept the above-mentioned abatement costs had essentially been taken *before* the international protocols were concluded, most importantly by passing the Ordinance on Large Combustion Plants in 1983. Thus, the German Federal Government was able to state that it expected to comply with its international obligations under the Oslo Sulfur Protocol because of prior domestic legislation - without new, additional measures (Umwelt No. 7-8/1994, p.281).

In conclusion, taking the German case by itself, the rational actor model does not provide a precise explanation of the extent of German regulatory ambitions. Nevertheless, the rational actor model provides a reasonable directional prediction: high emission reduction goals by international comparison.

The Domestic Politics Model

The domestic politics model rests both on the demand for and supply of policies to improve environmental quality. Thus, this model extends the rational actor model outlined before by going beyond purely economic factors to be included in decision-making. Environmental protection has been perceived as important or very important by 80 to 90 per cent of the West German population since the early 1970s (Sachverständigenrat für Umweltfragen 1978, p.440-455; Eurobarometer, various surveys since 1981). At the end of the 1980s/beginning of the 1990s, 74 per cent of the German population regarded the environment as being in deep crisis, and more than 70 per cent responded that environmental protection should be a very important political task. More specifically, in 1980-1982 air pollution was regarded as one of the most serious environmental problems (Kessel/Tischler, 1984). Taking the number of media reports as an indicator, public concern about acid rain rose during 1982, at a time of economic recession, and peaked in 1983-1985 (Cavender Bares et al. 1995). While media attention declined afterwards, Eurobarometer surveys still showed strong public concern about acid rain, with 26 per cent (1986) and 29 per cent (1988) of the population selecting it as one of the top three environmental issues (Sprinz, 1992). In 1989 and 1991, nearly 75 per cent of the population of the Federal Republic mentioned acid rain as a very serious problem, while more than 20 per cent still believed it to be a fairly serious problem. At the end of the 1980s/beginning of the 1990s, a great majority of Germans in the Western as well as in the Eastern parts of Germany expressed great concern about air pollution (81-95 per cent). When the German population was asked about the most important environmental problems of the future in 1992, air pollution ranked second with 36 per cent of respondents mentioning this problem. In 1988, 24 per cent of the West German population complained about air pollution, while in 1993 this figure had declined to 21 per cent. In the Eastern parts, 28 per cent complained about air pollution in 1993, a pronounced decline from 56 per cent in 1990 (probably due to advances in combating local air pollution) (Statistisches Bundesamt, 1995, pp.568-569). In conclusion, while no time series data are available, public concern about acid rain appears to have risen sharply in the early 1980s despite the economic recession. It has declined somewhat since 1986, but, to the extent that it can be equated with concern about air pollution in general, it has stayed high on the public agenda of environmental problems.

In this context, the special attitude of many Germans towards their forests has to be taken into account. Although Germany had been largely a deforested area some 400 years ago and many of today's forests are monocultures planted within the last one hundred years, there is a strong tradition in Germany to regard forests as an important part of German culture, history, and identity. Thus, when the link between forest death and acid rain was established and made public in the early 1980s by the media (Cavender Bares et al., 1995), the issue easily achieved major public attention (on the special importance of forests to Germans see, inter alia, Cavender Bares et al., 1995; Boehmer-Christiansen and Skea, 1991, p.61;191).

On the industrial side, potential 'negative' demand for, i.e. opposition to strict air pollution control by the three most important German industries in terms of contributions to GDP ('fabricated metal products, machinery and equipment', 'wholesale and retail trade', and 'real estate and business services') or employment (the same as above, but 'real estate and business services' is replaced by 'construction') has to be considered as basically non-existent. Although the top three industries accounted for about 30-35 per cent of overall GDP or employment, none of them belongs to the sectors of the economy which are usually considered the main victims of acid rain (forest-related industries and inland fisheries) or the main polluters (electricity generation, cars, solvent industry). In general, the degree of exposure to damage as well as abatement costs of the top three industries has to be considered very low.

Only to the degree that the automobile industry is considered part of the category of 'metal products' is one of the top industries partially affected by abatement measures. However, it has to be noted that compensation was provided by the federal government to car owners/ purchasers by granting special tax exemptions and subsidies for low-emissions passenger cars between 1985 and 1991. At the same time, taxes on vehicles without catalytic converters were increased. Also, unleaded gasoline was granted a tax reduction in 1985, while taxes on leaded fuel were increased. Moreover, subsidies of 13.5 million DM were given to independent petrol stations to support introduction of unleaded gasoline during 1985-1987 (Bundesregierung, 1985; 1987; Bundesministerium für Umwelt, Naturschutz und Reaktorsicherheit, 1992). Otherwise, the government supported abatement measures by granting subsidies to those actors of the economy which have had to bear significant parts of the burden (but did not belong to the top three industries). Until the end of 1990, the government granted particularly favourable conditions for writing off environmental protection investments in general. Furthermore, industry could apply for low-interest credits for the support of measures taken to maintain clean air (total volume 1983-1990: 1.65 billion DM). Direct subsidies for investment in environment-friendly production amounted to 690 million DM until 1990, only part of which went to clean air investments.

In addition, technological development facilitated the implementation of far-reaching abatement measures. Industry made decisive progress in adapting foreign (in particular Japanese) technology to European and German conditions and in developing its own abatement technologies (Weidner, 1986; Prittwitz, 1984). This technological progress comprised desulfurization, denitrification, and catalytic converters. Germany is believed to have positioned itself ahead of most other nations, thus giving German industry special advantages as regards the export of abatement technology and catalytic converters (especially the lambda measuring device used in conjunction with catalytic converters). Consequently, Germany was able to reap a special benefit from international regulation of air pollutants (Sprinz 1992, p.121).

Industries which could be particularly affected by acid rain, namely forestry and inland fishery, do not play any significant role in the German economy. The costs of damage to materials correspond to a share of GDP of ca. 0.1-0.2 per cent per year.

Five possible (groups of) actors relevant for articulating societal demand for stringent acid rain policies have been identified:

(i)   environmental NGOs,
(ii)   Green Parties,
(iii)   representatives of regions particularly affected by acid rain,
(iv)   industries producing abatement technologies, and
(v)   industries selling environment-friendly production methods.

To the extent that these actors have influenced German policy-making on acid rain, they have mainly concentrated on *national* legislation and regulation. Thus, international obligations agreed upon later have (to a great extent) been internally 'agreed' upon and partially implemented beforehand. Therefore, supporting relatively strong international policies and accepting strict obligations posed few problems for German negotiators and made the international process itself relatively irrelevant to national environmental NGOs and the other groups.

Under these circumstances, it appears to be most appropriate to focus on the national political process regarding national regulations in order to account for the influence of relevant political actors on German acid rain policies. The most important of the national measures were the Ordinance on Large Combustion Plants of 1983, the amendment to the TA-Luft of 1986, the Ordinance on Small Combustion Plants of 1988 (all relevant to SO_2 and NO_x), the introduction of the catalytic converter (relevant to NO_x and VOCs), and the 20th (Ordinance on the Filling and Storage of Engine Fuel) and 21st (Ordinance on the Refueling of Motor Vehicles) Federal Clean Air Ordinances in 1992 (both relevant to VOCs).

The most important regulation of SO_2 is the Large Combustion Plant Ordinance of 1983 by which essentially even the emissions reductions of the Oslo Protocol will be achieved. In order to reduce NO_x emissions, the introduction of catalytic converters was the most relevant policy measure besides the Ordinance on Large Combustion Plants. In both cases, environmental NGOs have been active and can be assessed as influential. By climbing smoke stacks, collecting signatures in support of stringent regulation, organising information campaigns, distributing leaflets etc., environmental NGOs were able to generate considerable political pressure on decision-makers. However, NGOs were neither the first ones to put the issue on the public policy agenda (because this was done by the media; see Cavender Bares et al., 1995) nor were they the only ones articulating societal demand: NGO influence can be assessed as high, but not as extremely important. Their activism peaked in 1983-1984, but declined thereafter (Cavender Bares et al. 1995).

On the Länder level, the Bavarian state government, representing a region particularly affected by forest damage, was quite active in pursuing acid rain policies in the early 1980s. It influenced the process in two ways. On the one hand, it used its representation in the Bundesrat, which had to agree to the laws and regulations passed, to form coalitions which led to a tightening of the draft regulations. On the other hand, the conservative Christian Social Union (CSU), as the ruling party in Bavaria, became part of the federal government in 1982. A member of the CSU, Friedrich Zimmermann, became Minister of the Interior and was responsible for environmental protection at that time. He fought very hard for the Ordinance on Large Combustion Plants and, in particular, for the mandatory introduction of the catalytic converter. However, since other members of the ruling coalition, in particular officials of the liberal Free Democratic Party (FDP), were also committed to stringent clean air policy, the influence of the CSU/Bavaria can be assessed as high, but not extraordinarily high.

In contrast, the influence of the Green Party, according to most accounts, has to be given an extremely high score. In the early 1980s, when the Green Party organised itself at the federal level, it constituted a threat not only to the Social Democratic Party (SPD) but also to the Christian Democratic Union (CDU/CSU) because, in these times, it also included conservative factions. Thus, all parties felt the need to amend their party and election programs, with some changes already dating from the late 1970s, in order to take account of the environmental challenge. When the Green Party managed to win seats in the German Bundestag in early 1983 for the first time, the threat to the established parties became even more obvious. In the Bundestag, the Green Party fought hard to strengthen acid rain legislation and fulfilled the role of the voice for people concerned about environmental issues (Boehmer-Christiansen and Skea 1991, pp.90-91 and 198).

While the availability of abatement technology and, to a lesser extent, environment-friendly production methods certainly played a role in political decision-making (Sprinz, 1992; 1993; Sprinz and Vaahtoranta, 1994), little is known about the actual participation of the relevant industries in the decision-making process. The car industry was of some importance insofar as it ultimately did *not* oppose the introduction of the catalytic converter on the European Community level. The EC-wide introduction of catalytic converters held the promise for the German car manufacturers of gaining a competitive advantage because it had developed effective converter technology (Boehmer-Christiansen and Weidner, 1992; Sprinz, 1992, p.121).

By the time policy-making turned to combating VOC emissions in the second half of the 1980s, the attention of political parties, NGOs and others had declined dramatically. Interviews revealed that environmental NGOs did not influence the decision-making process to a great extent. Very little is known about the actual participation of the other relevant actors in the related political decision-making process. If existent at all, their influence on the political output and outcome has to be assessed as small.

On the institutional supply side of environmental policy, the Federal Republic has benefited from high political continuity of chancellors and cabinets. There was only one change of government during the period under investigation (1975-1994). In the 1970s, the government was formed by the Social Democratic Party of Germany and the FDP under former Chancellor Helmut Schmidt (SPD). He was succeeded by Chancellor Helmut Kohl (CDU) who has lead a coalition of the sister parties CDU and CSU with the FDP since October 1982.

With respect to the institutional capacity of the environmental branch of the federal government, few changes can be observed until the Ministry of the Environment (BMU) was established in 1986. Until then, obtainable data on the budgetary share of the environmental branch of the Ministry show figures fluctuating between 0.09 and 0.3 per cent, while the share of the expenses for personnel increased steadily from 0.05 to 0.08 per cent between 1975 and 1986.[3] In light of their relatively low level, these figures point to a rather weak environmental branch of government until 1986, especially vis-à-vis other ministries. The budgetary situation has improved somewhat after 1986 with the share of the Federal Ministry of Environment of the federal budget rising from 0.17 per cent in 1987 to 0.29 per cent in 1995 and its share of expenses for personnel even increasing from 0.19 per cent to 0.44 per cent during the same period. This development certainly strengthened the position of the Federal Ministry of Environment within the cabinet as it took over the role of the leading actor in drafting legislation and in interdepartmental decision-making from the environment department of the Federal Ministry of the Interior. However, the overall judgement is not as clear-cut as it may seem. While the Minister of the Interior had a senior standing in the cabinet that he could bring to bear for the environment, the Minister of the Environment had a junior institutional standing within the government. This may have been, at best, compensated to a certain degree by the personal skills of the former Ministers of the Environment Klaus Töpfer (1987-1994) - who has become the Under-Secretary Administrator of the United Nations Environment Programme in 1998.

Decision-making authority on issues of air pollution control in Germany is neither highly centralised nor very decentralised. It rests mainly with the federal government, albeit subject to relatively strong influence of the states as exercised mainly through the Bundesrat. By a constitutional amendment passed in 1972, air pollution control became part of the section on 'competitive legislation' in the constitution. Thus, the federal government received the authority to pass laws and regulations, while the states retained the power to pass legislation only if federal law did not exist. However, decisions of the Bundestag have to gain majority support in the Bundesrat for passing laws and issuing regulations in this field. Furthermore, implementation and enforcement of laws and regulations on air pollution control is conducted by the states *on behalf of* the federal government (Weidner, 1986).

The strength of the federal government is not enhanced by any direct financial dependence of the main target groups of acid rain policy on the executive. The main target groups in Germany are the electricity generating sector (SO_2 and NO_X), the transport sector (NO_X and VOCs), and the solvent industry (VOCs). The combined share of these groups amounts to 75 per cent or more of the overall emissions of any of the pollutants regulated internationally (United Nations Economic Commission for Europe/Convention on Long-Range Transboundary Air Pollution 1995, pp.74, 82, and 96). However, none of these actors has received subsidies from the government on a regular basis such that it would have made them dependent on the executive to any substantial degree (Bundesregierung 1985; 1987).

In terms of the domestic political process of curbing air pollutants, there was some disagreement within the government on air pollution control in the late 1970s and early 1980s which is difficult to attribute to *party* positions. Instead, disagreement appears to have been dependent on organisational roles. This is supported by the fact that the main conflict within the SPD/FDP government (until fall 1982) occurred between the Ministry of Economics, which acted as a dragger, and the environmental branch of the Federal Ministry of the Interior - with the latter supported by the Ministry of Agriculture and Forestry (Cavender Bares et al., 1995). This conflict, though, cannot be linked to specific parties, since at the time all three ministries were led by members of the FDP. After 1983, no information is available regarding major differences between ministries or governing parties. Some conflicts arose on transport policies in the 1980s and 1990s with regard to cars, the prescription of catalytic converters, a general speed limit on German highways, and alternative transport concepts. As mentioned before, however, air pollution control policy, in general, has been characterised by a high level of consensus among the relevant parties and actors in Germany since the second half of the 1980s.

This has also been the case with regard to the major opposition parties. At best, the opposition parties have even requested tighter air pollution controls. For example, the Green Party and parts of the SPD have advocated a general speed limit on German highways and a faster introduction of the catalytic converter technology. To conclude, political consensus on governmental policies and measures, seen as the lowest common denominator, can be considered to be very strong. This correlates well with the constant implementation progress regarding air pollution regulations since the early 1980s.

Germany experienced an era of relative political stability since the 1970s, measured in terms of the number of governments in power. The respective ruling parties commanded a quite comfortable majority between 1980 and 1994. Governing parties held between 54.2 per cent and 56.2 per cent of the seats in the Bundestag from 1980 to 1990. In 1990, the CDU/CSU/FDP government even won 60 per cent of the seats, while the same government can only count on the support of 50.7 per cent of the members of the German Bundestag since the fall of 1994. In terms of their power basis in the German Bundestag, the respective governments should have enjoyed fairly favourable conditions in the 1980s, whereas the situation became even more favourable in the early 1990s.

The political situation does not look as comfortable for the respective government, if public support for the ruling parties (as expressed in public surveys) is taken as an indicator. Before negotiations on the Geneva Convention started in July 1978, the ruling SPD/FDP coalition was supported only by about 48-49 per cent of the electorate. Due to the German political system, ca. 48 per cent of the votes is needed to win federal elections; hence, the federal government was politically vulnerable in 1978. The situation was better for the CDU/CSU/FDP government when negotiations on the Helsinki and Oslo Sulfur Protocols started in September 1984 and February 1990 respectively, because 52.2 and 53.4 per cent of the electorate had expressed their support only a few months earlier. During the preparation of negotiations on the NO_X and VOC Protocols in 1985 and late 1988, the position of the government apparently was weaker, since only about 48 per cent of the voters expressed their intention to vote for its parties.

Taking the German case study by itself, it appears that the explanatory power of the factors investigated on the 'supply-side' of formulating and implementing acid rain policies in Germany is rather limited. The pusher position taken by Germany in the international negotiations since the early 1980s as well as the drastic reductions of SO_2 emissions and comparatively high reductions of NO_X and VOC emissions which have been accomplished are in line with the governmental ideologically moderate position. The relatively high consensus that was reached during the 1980s in the ruling coalition as well as within the political system at large, the relatively stable legislative support of governments as well as the duration of the governing coalitions could be expected to have worked in favour of effective acid rain policies. Finally, times of relative political vulnerability immediately prior to international negotiations on the NO_X and the VOC Protocols should have supported a progressive stance in the international arena.

However, some of the variables investigated appear to be unfavourable for stringent German acid rain policies. The institutional capacity of the environmental branch of government, if measured in terms of its share of the federal budget, can hardly be said to be impressive. Also, the medium degree of centralisation of decision-making on matters of air pollution control in Germany is not believed to work towards stringent acid rain policies. Finally, even the most important target groups with respect to the control of SO_2, NO_X and VOC emissions are not dependent financially to any significant degree on the federal government.

The Social Learning Model

As noted above, the damaging effects of air pollution in general have been known in Germany for several centuries. After World War II, air pollution arose as a political issue mainly in the heavily industrialised areas as a response to health problems. Despite the accumulation of knowledge over time, the German Federal Cabinet initially decided not to support Swedish claims of transboundary effects, and the transboundary issue was kept off the political agenda (Cavender Bares et al., 1995).

Ministerial experts are reported to have been aware of the link between tall smokestacks and long-distance transport of pollution as well as soil acidification by the late 1970s (Müller, 1986). However, until the early 1980s, no specialised research efforts were undertaken on acid rain. This changed, however, when in the 1980s, the Council on Environmental Quality published several reports on acid rain and forest death, and in 1982-1983, a large research program on forest death invested 250 million in ca. 660 different research projects between 1982 and 1988. Intensive research and policy efforts were pursued in parallel during this period.

A special contribution to knowledge and learning in the German context has been made by the German Society of Engineers (Verein Deutscher Ingenieure, abbreviated VDI). The VDI has played an important role in defining best available technology (BAT), which is one of the most important legal norms in German air pollution control law (Boehmer-Christiansen and Skea, 1991, pp.169-170) and which influenced the technical annexes to several international environmental agreements on transboundary air pollution.

While international conferences and the activities of international organisations did not seem to play a very significant role in the scientific process, a consensus among German scientists developed after the publication of the research results of Ulrich et al. in the 1980s that SO_2, NO_X as well as ground-level ozone were major *contributors* to the problem of 'Waldsterben'. However, agreement could not be reached within the scientific community as to the *relative* importance of any of the pollutants. It was generally acknowledged that forest damage was a complex phenomenon which might have multiple causes, including anthropogenic emissions of air pollutants, but also pests, storms and climatic conditions (Boehmer-Christiansen and Skea 1991, pp.189-191; 199).

Conclusions

Germany has become a push country in terms of acid rain policies. This has meant that with regard to both national policy-making as well as partial implementation Germany was well on the way towards compliance even before it entered international negotiations on substantive protocols. Therefore, one may conclude from this country study that push countries may use the results of their national policy processes to influence the policy of *other* countries. Given the transboundary nature of acid rain policies, this lets them reap additional benefits from international policies to lessen their ecological vulnerability.

Three explanatory routes were assessed. The *economic* rational actor model pointed in the correct direction, but the missing damage costs of acid rain do not permit a narrowing of the predictive range of the model. Focusing purely on German forests and associated industries, the study shows that they do not constitute a major share of German GDP, and it seems unlikely that damage costs are extremely high for these industries. In addition, the abatement costs for the international environmental agreements (excl. VOCs) are of medium magnitude compared with other domestic social programs. Given the lack of specificity of economic damage cost data for each of the various pollutants involved, the economic rational actor model can only hint towards a generally ambitious policy of Germany. In the political actor model, we turn to a *political* demand and supply model for IEAs. On the demand side, environmental NGOs, the emergence of the Green Party, as well as media attention on environmental damages led to a strong domestic push in favour of demanding regulations, whereas the governmental supply side shows a mixture of enhancing factors (e.g., long duration of cabinets, partial electoral vulnerability) and unfavourable factors (e.g., low budgetary shares of the Federal Ministry of the Environment, junior standing of this ministry within the cabinet, little public sector control of emitters). Therefore, the domestic political actor model sheds partial light on the strong emission reduction program which Germany planned and has been implementing. Knowledge and learning explanations are particularly hampered by the (scientific) inability of attributing air pollution to be *the* major cause of forest dieback - the major environmental aspect in this policy domain. However, Germany has had a long history of regulating *local* air pollution which may have been a particular advantage once these policies were put into the international context.

Overall, Germany has been a committed pusher internationally, and it had the domestic backing needed for ambitious national and international policies. It embarked on strict emission reduction strategies domestically and endeavoured to make others join it or at least tried to induce other countries to upgrade their ambitions. Taken by itself, the domestic incentive structure made the execution of Germany's international obligations self-enforcing which accounts for full or overcompliance with the obligations accepted under international environmental law.

Notes

1. For an assessment of the impact of political actors, see below.
2. The NO_X Declaration was only signed by the Federal Republic of Germany, not the German Democratic Republic. Nevertheless, the obligation to reduce its NO_X emissions by 30 per cent now applies to all the Länder (states) of the Federal Republic.
3. In the latter figures only the items 'expenses for environmental protection' (including nuclear safety) and the budget of the UBA are included.

References

Amann, M. (1989), *Potential and Costs for Control of NO_X Emissions in Europe*, IIASA, Laxenburg, mimeo.

Amann, M. and Kornai, G. (1987), *Cost Functions for Controlling CO_2 Emissions in Europe*, IIASA, Laxenburg (IIASA Working Paper WP-87-065), mimeo.

Boehmer-Christiansen, S. and Skea, J. (1991), *Acid Politics*, Belhaven, London.

Boehmer-Christiansen, S. and Weidner, H. (1992), *Catalyst versus Lean Burn. A Comparative Analysis of Environmental Policy of Germany and Great Britain with Reference to Exhaust Emission Policy for Passenger Cars 1970 - 1990*, WZB für Sozialforschung, Berlin.

Bundesminister des Innern (1984), *Aktueller Bericht Waldschäden 1984 zum Aktionsprogramm 'Rettet den Wald'*, Bundesminister des Innern, Bonn.

Bundesminister für Ernährung, Landwirtschaft und Forsten (1985), Waldschäden in der Bundesrepublik Deutschland, *Schriftenreihe des Bundesministers für Ernährung, Landwirtschaft und Forsten*, No. 309, Bundesminister für Ernährung, Landwirtschaft und Forsten, Bonn.

Bundesministerium für Umwelt, Naturschutz und Reaktorsicherheit (1992), Fünfter Immissionsschutzbericht der Bundesregierung, *Umweltpolitik*, Bundesminister für Umwelt, Naturschutz und Reaktorsicherheit, Bonn.

Bundesregierung (1985), *Bericht der Bundesregierung über die Entwicklung der Finanzhilfen des Bundes und der Steuervergünstigungen für die Jahre 1983 bis 1986 gemäß §12 des Gesetzes zur Förderung der Stabilität und des Wachstums (StWG) vom 8. Juni 1967 (Zehnter Subventionsbericht)*, Drucksache 10/3821, 12. September 1985, Deutscher Bundestag, Bonn.

Bundesregierung (1987), *Bericht der Bundesregierung über die Entwicklung der Finanzhilfen des Bundes und der Steuervergünstigungen für die Jahre 1985 bis 1988 (Elfter Subventionsbericht)*, Bundesministerium der Finanzen, Bonn.

Cavender Bares, J., Jäger, J. and Ell, R. (1995), *From Laggard to Leader: Global Environmental Risk Management in Germany*, mimeo.

Gregor, H. D. (1990), Acidification Research in the Federal Republic of Germany, in A.H.M Brassier and W. Salomons (eds.), *International Overview and Assessment*, Springer, New York, pp. 139-58.

Hansmann, K. (1995), *Bundes-Immissionsschutzgesetz*, 14th edition with comments, Nomos, Baden-Baden.

Hartkopf, G. (1984), Emissionsbegrenzung bei Feuerungsanlagen - Bilanz und Perspektiven, *Natur + Recht*, vol. 1984 (4), pp. 128-31.

Heinz, I. (1985), Zur ökonomischen Bewertung von Materialschäden durch Luftverschmutzung, in *Kosten der Umweltverschmutzung*, Tagungsband zum Symposium im Bundesministerium des Innern am 12. und 13. September 1985, Umweltbundesamt, Berlin.

Isecke, B., Weltchev, M. and Heinz, I. (1991), *Volkswirtschaftliche Verluste durch umweltverschmutzungsbedingte Materialschäden in der Bundesrepublik Deutschland*, Umweltbundesamt, Berlin, mimeo.

Jäger, J., Cavendar Bares, J. and Blümhuber, C. et al. (1993), *The Issue of Acid Rain and Long-Distance Transport of Pollution in Germany*, Draft paper to the project on Social Learning in the Management of Global Environmental Risks, Wuppertal Institute, Wuppertal, mimeo.

Kessel, H. and Tischler, W. (1984), *Umweltbewußtsein. Ökologische Wertvorstellungen in westlichen Industrienationen*, Edition Sigma, Berlin.

Klein, H. (1984), Von Strabo bis Strauss. Kleine Geschichte der Luftverschmutzung, in B. Grill, and M. Kriener, (ed.), *Es war einmal. Der deutsche Abschied vom Wald?*, Gießen: Focus, Gießen, pp. 31-59.

Müller, E. (1986), *Innenwelt der Umweltpolitik. Sozial-liberale Umweltpolitik - (Ohn)Macht durch Organisation?* Westdeutscher Verlag, Opladen.

Organisation for Economic Cooperation and Development (1979), *The OECD Programme on Long Range Transport of Air Pollutants: Measurements and Findings*, Organization for Economic Cooperation and Development, Paris.

Prittwitz, V. v. (1984), *Umweltaußenpolitik. Grenzüberschreitende Luftverschmutzung in Europa*, Campus, Frankfurt a. M./New York.

Sachverständigenrat für Umweltfragen (1978), Der Rat von Sachverständigen für Umweltfragen, *Umweltgutachten 1978*, Kohlhammer, Stuttgart/Mainz.

Sachverständigenrat für Umweltfragen (1983), Der Rat von Sachverständigen für Umweltfragen, Waldschäden und Luftverschmutzung, Stuttgart, Mainz.

Schärer, B. (1992), Acidification Policy - Control of Acidifying Emissions in Germany, in T. Schneider (ed.), *Acidification Research: Evaluation and Policy Applications*, Studies in Environmental Science No. 50, Elsevier, Amsterdam et al., pp. 191-201.

Schröder, J. v. and Reuss, C. (1883), *Die Beschädigung der Vegetation durch Rauch und die Oberharzer Hüttenrauchschäden*, Parey, Berlin.

Sprinz, D. (1992), *Why Countries Support International Environmental Agreements: The Regulation of Acid Rain in Europe*, Ph. D. dissertation, The University of Michigan, Ann Arbor.

Sprinz, D. (1993), The Impact of International and Domestic Factors on the Regulation of Acid Rain in Europe: Preliminary Findings, *Journal of Environment and Development*, vol. 2, pp. 37-61.

Sprinz, D. and Vaahtoranta, T. (1994), The Interest-Based Explanation of International Environmental Policy, *International Organization*, vol. 48 (1), pp. 77-105.

Sprinz, D., Oberthür, S. and Wahl, A. (1995), *The Domestic Basis of International Environmental Agreements: Modeling National-International Linkages* (German Country Study) (EU Contract EV5V-CT92-0185), Potsdam Institute for Climate Impact Research, Potsdam.

Statistisches Bundesamt (1995), *Statistisches Jahrbuch 1995 für die Bundesrepublik Deutschland*, Statistisches Bundesamt, Wiesbaden.

Ulrich, B., Mayer, R., Khanna, P.K. (1979), *Depositionen von Luftverunreinigungen und ihre Auswirkungen im Waldökosystem im Solling*, Sauerländer, Frankfurt/Main.

Umweltbundesamt (1991), *ECE Task Force VOC - Emissions of Volatile Organic Compounds (VOC) from Stationary Sources and Possibilities of their Control*, in UBA-Texte 11/91, Umweltbundesamt, Berlin.

United Nations Economic Commission for Europe (1979), *Convention on Long-Range Transboundary Air Pollution*, United Nations Economic Commission for Europe, Geneva.

United Nations Economic Commission for Europe (1985), *Protocol to the 1979 Convention on Long-Range Transboundary Air Pollution on the Reduction of Sulfur Emissions or Their Transboundary Fluxes by at Least 30 Percent*, United Nations Economic Commission for Europe, Geneva.

United Nations Economic Commission for Europe (1988), *Protocol to the 1979 Convention on Long-Range Transboundary Air Pollution Concerning the Control of Emission of Nitrogene Oxides or Their Transboundary Fluxes*, United Nations Economic Commission for Europe, Geneva.

United Nations Economic Commission for Europe (1991), *Protocol to the 1979 Convention on Long-Range Transboundary Air Pollution Concerning the Control of Emission of Volatile Organic Compounds or Their Transboundary Fluxes*, United Nations Economic Commission for Europe, Geneva.

United Nations Economic Commissions for Europe (1994), *Protocol to the 1979 Convention on Long-Range Transboundary Air Pollution on Further Reduction of Sulfur Emissions*, United Nations Economic Commission for Europe, Geneva.

United Nations Economic Commission for Europe/Convention on Long-Range Transboundary Air Pollution (1995), *Strategies and Policies for Air Pollution Abatement: 1994 Major Review Prepared Under the Convention on Long-Range Transboundary Air Pollution*, United Nations, New York.

Weidner, H. (1986), *Air Pollution Control Strategies and Policies in the Federal Republic of Germany. Laws, Regulations, Implementation and Principle Shortcomings*, Edition Sigma, Berlin.

Wicke, L. (1986), *Die ökologischen Milliarden. Das kostet die zerstörte Umwelt - so können wir sie retten*, Goldhammer, München.

7 The Finnish Fight for the *Green Gold*

PETTERI HIIENKOSKI

The Problem of Air Pollution

Sulphur

Attention was first paid to acidification in Finland at the end of the 1960s, after noticeable damage had taken place in the lakes of Sweden and Norway. The acidifying sulphur deposits discovered there were mainly of Central European and British origin. As far as Finland was concerned, at the beginning of the 1970s there was no reliable information available regarding the damage caused by the sulphur deposits but it was assumed that acidification was as bad as in Sweden and Norway. However, it had been noted that air pollutants coming from West Europe entered Finland only under exceptional conditions. Stimulated by these initial observations, systematic research on acidification was started.

At the end of the decade it was determined that the prevailing winds carried a significant amount of sulphur dioxide (SO_2) from the industrial areas of Central Europe to Finland, while only a small part of the deposit came from the East. The size of the depositions of sulphur were perceived as presenting a threat to the valuable forests, the '*Green Gold*' of Finland, as well as to fishing waters. In this sense, acidification represented an enormous risk to the Finnish national economy because the forest industry and forestry formed its very core. Later, in 1983, it was reported that most of the sulphur of foreign origin was from the USSR and other East European countries in contrast to the initial assumptions regarding the role played by Central European countries. On the whole, more than two thirds of the total deposit originated abroad. The first estimates of the abatement costs also came in 1983. The annual costs of measures required by the Nordic proposal of a 30 per cent reduction by 1993 (see sections below)

167

were expected to reach several hundred million FIMs (Task Force on Sulphur, 1984).

The first scientific results concerning acidification became available at the beginning of 1984. The lakes in Southern Finland were already suffering from it, and it was predicted that the growth of trees would slow down within ten to twenty years. The Sulphur Commission, set up by the Minister of the Environment to plan a reduction programme, confirmed in May of 1985 that in Southern Finland, and partly even in the north, depositions of sulphur clearly exceeded the national target value that had been set based on ecological vulnerability. The Commission (1986) reported that the costs of reducing Finnish emission by 50 per cent would be high, in the order of 1,550 - 1,800 million FIMs by 1993. Later, the Sulphur Committee II (1993) reported that the costs for the 80 per cent reduction planned by the government would be even higher, running at about 4,000 - 4,600 million FIMs by the year 2000.

The Acidification Research Project, called HAPRO, indicated in the summer of 1987 that in Southern Finland the deposit was five to ten times higher than the natural sustainable level and in the north about twice as high. According to the project's final report published three years later, the acid deposit still exceeded critical loads nearly all over the country. It was causing slowly progressing damages to forests and water systems. In Southern Finland and Northern Lapland it was estimated that the deposit would have to be reduced by about 90 per cent from the 1987 level in order to avoid any acidification of the forest land, and by about 75 per cent if the land were to be allowed to acidify slightly but without danger to tree stand. Given Finland's location and its vulnerability to SO_2 imported from abroad, it became clear to the Finnish government that the sulphur deposits could not be reduced significantly by action taken by Finland alone.

Nitrogen

Reducing nitrogen oxides (NO_X) was considered 'important' but the primary concern was sulphur. Nevertheless, at the conference organised by the Nordic Council in 1986, a group of UNECE scientists reported that NO_X emissions should be reduced by 50 per cent within ten years in order to avoid exceeding the ecological tolerance for this substance. In addition, the Nordic Council of Ministers published the results of research which even proposed a 75 per cent reduction of the 1983 level by 1995.

According to HAPRO (1990) nitrogen did not cause immediate acidification. Instead, it increased the growth of trees. However, the deposit formed a threat in the future. HAPRO reported that a slight decrease of the nitrogen deposit, combined with a significant reduction of the sulphur deposit, would postpone damage into the more distant future. But here, again, the nitrogen deposit could not be significantly reduced by national means alone. Only approximately 16 per cent of the deposit were of domestic origin in 1985.[1]

The abatement costs for NO_x were much higher than those for SO_2. Measures designed to achieve a 15 per cent reduction from the 1980 level by the year 2000, as proposed by the Nitrogen Oxides Commission (1990), would have cost around 4,000 million FIMs. Together with the decision already taken on traffic emissions, the costs for reducing NO_x emissions were expected to be even greater, approximately 9,000 million FIMs by 2000.

Volatile Organic Compounds

Volatile organic compounds (VOCs) were not considered a significant threat in Finland. HAPRO (1990) indicated that VOCs together with NO_x formed ozone, which was poisonous to vegetation. The natural levels of ozone for the most sensitive plants were exceeded by as much as a factor of two. However, HAPRO could not confirm that ozone caused significant damages. According to the Reduction Strategy for VOCs (1992), research on the effects of VOCs or ozone damages did not provide a basis for evaluating possible ecological damage. Neither were the health effects of VOCs well enough known, even though high levels of ozone were considered to cause respiratory injuries.

Data regarding the size and effects of domestic emissions were uncertain. Transmissions from abroad and background levels of emissions from undefined sources were the most important causes for the high ozone levels observed when the winds blew from Central Europe. Domestic emissions affected the growth of the ozone levels of the background air only in part.

It was not expensive to reduce VOCs. Initial reductions were achieved by controlling nitrogen. A considerable part of VOCs and hydrocarbons (HC) originated from the same sources that produce NO_x, i.e. in transportation and energy production. The costs of reaching a 30-40 per cent reduction from stationary sources (see sections below) were estimated to be only about 400 - 550 million FIMs between 1988-1999. Expensive decisions concerning mobile sources had already been taken.

Beginning of Air Pollution Control: The Geneva Convention (1979)

Domestic Policy Prior to Geneva

Already at the beginning of the 1970s, the National Delegation on Environment Protection[2] considered Nordic co-operation important. It stressed the need to reduce sulphur and recommended some specific means for emission reduction. However, these proposals did not lead to any action.

Conflicting bureaucratic and political interests led to procrastination in preparing air pollution control legislation during the 1970s. Indeed, some Government organs even opposed the proposed Air Pollution Control Act submitted in the middle of the decade. Environmental NGOs demanded that the law be passed immediately. These demands were supported later by a few parliamentarians representing various parties. Even industry favoured new legislation in order to be able to include the necessary investments for the implementation of this law in its economic planning. The first proposals to establish a special ministry for environmental affairs were also made at the beginning of the decade. Disagreements continued until the beginning of the 1980s.[3]

Up until that time air pollution control functioned basically on a voluntary basis (Sarkkinen, 1993). However, already in 1977, the government had passed a law which made available low interest loans for investments by industry in air pollution control.

The Role of Finland in the Negotiations on the Convention

At the beginning of the 1970s Finland favoured Nordic co-operation and preferred international action to domestic measures. However, international co-operation on the control of acidification was sought with other, foreign policy objectives in mind. The primary goals of President Kekkonen's (1956-81) policy were to avoid conflict with the USSR and to support international détente. Environmental concerns and general foreign policy goals converged in the Final Act of the Conference on Security and Co-operation in Europe (CSCE). Signed in 1975 in Helsinki, this agreement promoted the control of sulphur and, later, other substances in the ECE area as was of the issues to be pursued within the broader context of détente. After that, Finland supported the idea of organising an environmental conference.[4]

From 1977 on, Finland participated actively in the preparation of a conference on transboundary air pollution. Here the Finnish goal was a European-wide binding treaty. In June 1978, the Nordic countries, with Finland among them, presented a draft for such a convention. The primary objective was to stabilise sulphur emissions and later gradually to reduce them. The final modified text was approved in the spring 1979 and in November Finland signed the Convention in Geneva. The provisions of the Convention were, however, more moderate than those proposed by the Nordic countries (Nurmi, 1990).

Implementation and Compliance with the Convention

The Government submitted the Convention to parliamentary proceedings[5] in November 1980 and it was adopted by Parliament in March 1981. Finally, the President, as the institutional figure ultimately responsible for foreign policy,[6] ratified it in April.

Although the Convention did not require specific action by national governments, it did encourage air pollution control within some signatory states. Despite its international activity, Finland still did not possess the systematic legislation and efficient administration necessary to implement the accord or to control domestic activities. The Convention hastened the passing of national legislation concerning air pollution control and the establishment of an environmental administration. Parliament started proceedings on the Air Pollution Control Act in 1981 and the act was adopted in December of that year. The Act, together with its related Decree,[7] came into force in October 1982. In the autumn of that year the Parliament also decided to establish the Ministry of the Environment. The new ministry commenced its work in October 1983. In March the President had already issued a decree by which the Convention took effect in Finland – this time as a part of domestic legislation.

First Phase of the Sulphur Policy: The Helsinki Sulphur Protocol (1985)

Domestic Policy Prior to the Helsinki Protocol

Both action by the Nordic countries and the appearance of the 'Greens' on the domestic political scene shaped Finnish sulphur policy. Around the year 1982 news about 'acid rain' and 'forest deaths' in Central Europe led to increased public concern about the Finnish forests. In the summer of 1982 the Stockholm Conference called for emission reductions and recommended an annual limit for sulphur deposit. At the same time, environmentalists in Finland demanded that domestic emissions be reduced by as much as two thirds.

At the beginning of 1983, a few weeks before the general elections, the Cabinet adopted in its energy programme a target of a 30 per cent reduction of SO_2 by 1993 from the 1980 level. This decision was part of a Nordic initiative: Finland and the other Nordic countries proposed that emissions be reduced in Europe and North America by the same amount and according to the same time schedule. In the elections the Greens won their first seats in the Parliament, which meant that environmental values had to be taken more seriously. The new Cabinet promised in its programme (1983) – which for the first time ever included a chapter on environmental protection – to 'pay attention' to the goals of the new Air Pollution Control Act. The focus was on sulphur. At about the same time the National Delegation on Air Pollution Control (a new government organ with advisory status) proposed that a research and evaluation programme on SO_2 be started 'immediately'. As a result, a task force was appointed which quickly drew up a plan for dealing with the sulphur problem. The Task Force on Sulphur (1984) reported that a 30 per cent target would be easy to reach because emissions, since the base year, had already fallen without any major investments. This was due to structural changes in industry and energy production. Nevertheless, research on the possibilities for further reductions was carried out in 1984-1985.

Alarming data on acidification were reported at the beginning of 1984. In response, the Government decided to start a national research project on acidification (HAPRO). Later, the Government set guidelines on air quality, including a target value for the reduction of sulphur deposits which was the same as that suggested by the Stockholm Conference. In

September of that year the Government informed the Parliament that, in addition to the 30 per cent reduction target, the second objective was to reduce emissions by 50 per cent during the 1990s. A special commission, consisting of representatives from various branches of the national administration and interest groups, began working in February 1985 on a programme aimed primarily at the 30 per cent reduction. The Sulphur Commission (1985) soon reported that even before the Nordic initiative emissions had decreased by 30 per cent. Therefore, the Commission decided to prepare a programme based on the idea of a 50 per cent reduction by 1993.

The Role of Finland in the Negotiations on the Helsinki Protocol

Finland continued to co-ordinate its efforts with those of the other Nordic states in order to conclude a protocol on concrete sulphur reductions. It was essential that other countries also should meet fixed emission reductions. The risk of being saddled with difficult and expensive reduction obligations forced Finland to take an active role in the negotiations. Prior to the negotiations Finland supported the principle of an equal percentage reduction for all states. Any other basis (e.g. GDP/GNP or per capita) would have required a larger reduction from Finland, which up until then had done 'very little' to cut its emissions. The Nordic proposal, made in March 1983, of a 30 per cent reduction by 1993 was based on the favourable 1980 level. In 1980 the level of Finnish emissions had been the second highest in history while subsequent reductions were due to reforms in industry and energy production. The Nordic countries officially presented their initiative in June. They received the support of some important states such as Canada and West Germany. A common Nordic concern was the opposition of the UK because a large part of the SO_2 'received' in the area was of British origin. Nevertheless, the '30 per cent Club' gained increased support at the Ottawa Conference in March 1984. Finally, at the Munich Conference in June of 1984, most of the countries supported an agreement based on the Nordic proposal. Negotiations on a protocol commenced in September and the Sulphur Protocol was signed in Helsinki in July 1985.

Finland had already 'met' the goal of a 30 per cent reduction in 1983, more than two years before the signing of the Protocol. The President ratified[8] the Protocol in May 1986. In August 1987 the President adopted a decree by which the Protocol came into force domestically as of September of that year.

Second Phase of the Sulphur Policy: The Oslo Sulphur Protocol (1994)

Domestic Policy Prior to Oslo

The Finnish sulphur policy went far ahead to what was agreed internationally. The Sulphur Commission, that had started to prepare a 50 per cent reduction, presented in 1986 the programme, which included specific means for reaching that goal. During that year the parties forming the Cabinet had turned to support reductions which exceeded the target of the Helsinki Protocol. However, industry still opposed the 50 per cent reduction by 1993 because those countries actually causing the deposits in Finland did not have the same target. Environmentalists, on the other hand, suggested more state subsidies for industry in order to 'buy' its support for the new target. In February 1987, one month prior to the general elections, the Cabinet made 'the most important environmental decision of the decade'. The Cabinet officially adopted the 50 per cent target and made the first decisions (regarding oil fuels, coal-using power plants, sulphate cellulose and sulphur acid factories) on the programme developed to achieve these objectives. In the elections, the Greens increased their support, gaining 4 per cent of the vote. Fear that the support for the Greens might grow even stronger induced the traditional parties to make their own programmes 'greener' (Välimäki and Brax 1991). This response was also reflected in the programme (1987) of a new 'conservative-socialist' Cabinet, which promised to strengthen the measures aimed at reducing acidifying substances, especially sulphur. The 50 per cent reduction programme was continued by the Cabinet, which made the last decisions (dealing with coal, oil refineries, and oil-using power plants) on the plan in November.

The interim report of HAPRO (1987) indicated a need for a new reduction target. In May 1988, the Cabinet promised to strive for a target which was 'much stricter than the international level'. The Minister of the Environment, Kaj Bärlund (SD), supported a 70-80 per cent reduction by the year 2000, but the Cabinet refused to accept any specific target.[9] In the summer of 1989 Parliament decided to promote the use of economic instruments for environment protection (Lommi-Kippola 1991). In this connection the Cabinet introduced environmental taxes but rejected a tax on sulphur suggested by environmentalists.

Already in 1989 emissions had decreased by almost 60 per cent from the 1980 level. Pressure for a new reduction target increased. In February 1990 the Greens, backed by the whole of the opposition, called for a vote of censure for reasons of inadequate action to reduce emissions. The Cabinet won the vote of confidence but all parties admitted that there was reason for concern. The Ministry of the Environment announced that it was preparing a proposal for a 70-80 per cent reduction target. The opposition groups, the Greens and the Centre Party, in turn, demanded even higher reductions.

According to the final report of HAPRO (1990) the domestic contribution to acid depositions had decreased during the 1980s from 35 to 25 per cent. The import of sulphur was now clearly higher than the export. The largest known source of imports was still the USSR with 26 per cent of the total deposition.[10] Nevertheless, the report supported further reductions because deposits still exceeded critical loads in nearly all parts of the country.

Finally, in August 1990, Prime Minister Holkeri (Cons.) promised that the Government would reduce emissions by 80 per cent by 2000.[11] Environmentalists and the forest industry supported the new target, while industry as a whole criticised it. Just before the general elections in 1992 the Sulphur Committee II was appointed to prepare the programme. At the elections the Greens once again increased their support, although the Centre Party ended up as the winner. The programme of the new 'centre-right' Cabinet paid special attention to the aim of diminishing emissions from neighbouring areas. However, the environmental policy of the Government suffered drastic changes during its term in office because Finland faced a national economic crisis and entry into the European Union. Despite all of this, in 1992 the Government took the first decisions (dealing with industrial fuel-oil) based on the proposals of the Sulphur Committee II. The following year (1993) the Committee proposed a ten-year-programme together with measures designed to achieve a reduction of emissions of 80 per cent by the year 2000 or so. That target was achieved already in 1994. Since then, no new targets have been set.

The Role of Finland in Negotiations on the Oslo Protocol

During this time, Finland continued to pursue a policy of international co-operation in order to reduce emissions further. A conference organised in September 1986 by the Nordic Council concluded that the Helsinki Protocol was insufficient from an ecological point of view. When the negotiations on a new sulphur protocol began in the autumn 1992, Finland favoured emissions reductions based on critical loads and country-specific reduction schedules for achieving these objectives by the year 2000 or 2005. The following spring Finland supported, in addition to on-going national programmes, a country-specific target, called the '60 per cent Gap-Closure', according to which the difference between the 1990 deposit and the critical load was to be narrowed by 60 per cent. In June 1994, the Oslo Protocol was signed. The parties committed themselves to reduce their deposits below critical loads, but without a common schedule. Finland also agreed to decrease emissions by 80 per cent in accordance with its domestic programme. At the same time, Finland criticised the other countries because they did not agree to tighten their low reduction targets.

The 80 per cent target, stipulated for Finland in the Oslo Protocol, had already been reached in the year of its signing. Implementation of the measures leading to this result had begun almost four years earlier. The economic depression at the beginning of the 1990s also contributed to the decrease of emissions. Consequently, the government did not need to take many explicit actions to meet its international obligations under the new protocol. Nevertheless, the implementation 'still continues' in case the emissions begin to rise. Although Finland had not, as of the summer of 1997, ratified the Protocol, a 'quick ratification' was planned.

Nitrogen Policy: The Sofia Nitrogen Protocol and Declaration (1988)

Domestic Policy Prior to Sofia

Traffic emissions   Curbing NO_x began with the controlling of emissions from automobiles. Already in 1984 the Cabinet had decided on air quality guidelines which also included target values for nitrogen. In the same year the Cabinet promised to tackle pollution from motor vehicles 'more effectively'. Nevertheless, in contrast to the other Nordic countries and some other Western states, Finland refused[12] to sign the Stockholm Declaration in July 1985. The objective of the Declaration was to harmonise European auto emission standards with the strict United States standards of 1983 (US-83). However, the Declaration did affect the control of traffic emissions in Finland by providing an example to follow (Nurmi 1990).

In December 1986 the Greens presented a parliamentary question on the Government's nitrogen policy. In his answer, two months before the general elections, the Minister of the Environment, Matti Ahde (SD), promised that Finland 'would follow the international development' in reducing exhaust emissions. As mentioned above, the growing political support of the Greens influenced the programme of the new Cabinet. In this case it promised 'more action' in order to reduce NO_X and car emissions. Partly as a result, in the spring of 1987, the Ministry of the Environment, together with the Ministry of Transport and Communications, presented a draft proposal according to which car exhaust emissions would be limited in a first phase in line with the EEC norms and then later (by 1990) they were to be brought in line with the US-83. Meeting the US standards would require the use of catalytic converters.[13]

Among interest groups there was disagreement on the proposed reductions. Representatives of the auto dealers and the oil industry criticised the compulsory use of catalytic converters. For their part, the environmental NGOs called the plan inadequate and demanded a 75 per cent reduction of the 1980 level of NO_X caused by traffic by 1992. This demand was based on a report of the Nordic Council of Ministers (1986). Moreover, environmentalists proposed that the US-83 standard be put into effect immediately.

Finally, in April 1988, at the same time that negotiations on the Sofia Protocol were being concluded, the Government[14] decided that auto emission regulations similar to the US-83 norms would come into effect for new cars, although gradually and only in 1990 and 1992 (Salo-Asikainen 1988). In order to 'buy' the support of the auto dealers, the Government also provided for a transition period with tax breaks for unleaded petrol (required for catalytic converters) and lower taxes on cars using such control devices.

Emissions from energy production The impetus to cut NO_x from energy generation came from the Greens. Their parliamentary question on nitrogen policy was focused on controlling emissions from power plants. The Minister of the Environment announced that emission reductions 'were already being studied' but made no promises concerning energy production.

Not until the summer of 1988 did a working group propose standards for energy production. The reason for establishing the Task Force on Nitrogen Emissions from Energy Production was the Finnish decision to sign the NO_x Protocol. However, the proposals concerned only new plants. Industry as a whole supported the flexible implementation of the proposed standards due to the need to proceed in a cost-efficient manner. However, the environmentalists, who had been excluded from the Task Force, criticised a reduction strategy which contained no specific targets and, furthermore, considered earlier proposals for reductions of traffic emissions inadequate. They, in turn, demanded a 50 per cent reduction of total NO_x emission by 1992.

The Role of Finland in the Negotiations on the NO_x Protocol and Declaration

In September 1986 a conference organised by the Nordic Council recommended that a protocol on NO_x be prepared. As a result, Finland, together with the other Nordic countries and a group of like-minded states, soon proposed that work on such a protocol be started. By the spring of 1987, however, the positions of these countries on reductions had grown more cautious. Co-operation among the Nordic countries was weakening. In May Finland and Sweden considered a freeze of the emissions to be a minimum target. However, Norway announced that even a freeze might be impossible. In September Sweden, in addition to few other Western states, suggested a 30 per cent reduction as a first step. This time Finland joined with Norway and proposed, as a compromise, that emissions be frozen to a freely chosen level after 1990. The Finnish goal was to achieve a protocol with 'as wide support as possible'. Consequently, Finland favoured flexible reductions in the first phase, because there was no need for 'emergency solutions', and a second phase based on critical loads.

Agreement on a protocol derived from the joint proposals of Finland and Norway was reached in April 1988. In October, Finland signed the

Sofia Protocol which stipulated that emissions were to be reduced to the 1987 level by the end of 1994 and, after that, below the critical loads, which still needed to be determined. In Sofia Finland, surprisingly, also signed a declaration according to which it would try to cut its emissions by about 30 per cent from the 1980 level before the end of 1998.

Implementation and Compliance with the NO_X Protocol and Declaration

At the time the Sofia Protocol and Declaration were signed, domestic policy on this problem was still incomplete. An initial decision on traffic emissions had been made but nothing had been decided on the emissions from energy production.

The Task Force on Nitrogen Emissions from Traffic reported that the provisions of the Protocol could be achieved by rationalising transportation, decreasing the volume of car traffic, and promoting the use of catalytic converter for new cars. These proposals once again elicited conflicting reactions from the interest groups. On the basis of its earlier decision, the Government banned low octane leaded petrol and decided that unleaded petrol should be available at almost all service stations beginning in September 1989. In addition, the purchase of new catalyst equipped cars was to be supported by gradually reducing the tax on such cars during 1989-1991. However, the lower car tax did not decrease the purchase prices and did not, therefore, increase the number of cars sold (see Ruokoranta 1989). The first regulations for heavy diesel vehicles were also issued in 1989. Another task force, preparing norms for old power plants – the Task Force on Nitrogen Emissions from Energy Production, which also included a representative of environmentalist groups – reported that emissions could be reduced by 30 per cent either by tighter standards or by structural means, such as taxation.

During the spring of 1989 there was a public debate on environmental taxes. As a result, the Government imposed, in addition to a normal tax on petrol, a new tax on fossil fuels (based on their coal content). These decisions affected both traffic and energy production. However, the Government rejected a special nitrogen tax proposed by the environmentalists. Not until September 1989 was a commission appointed, representing various interest groups,[15] to prepare a reduction programme for NO_X based on the targets of the Sofia Protocol and the accompanying Declaration. The Greens demanded 'more regulations' in order to reduce

NO_x emissions. Moreover, in a parliamentary interpellation on the condition of Finnish forests they attacked the Government's proposals for cutting emissions as being inadequate.

The President ratified the Sofia Protocol in January 1990. In May the NO_x Commission, which disagreed with the reduction targets, proposed 'specific technical means' by which emissions would only be reduced by 15 per cent compared with the 1980 level by 2000. The Commission reported that the freezing of the emissions, in accordance with the Protocol, would be possible at a reasonable cost by 'purely technical means'. However, the reduction required by the Declaration would not be possible without 'new technical or specific structural means'. Two opposing statements were included in the report of the Commission: one representing the interests of industry and labour unions, the other the environmental point of view. In addition, industry criticised the signing of the Declaration because it was done without 'adequate information' regarding the possibilities of implementation. It was of the opinion that it would be unreasonable to attempt to exceed the 15 per cent target by technical measures. Furthermore, industry opposed structural means. For their part environmentalists considered the proposals inadequate because ecological considerations required reductions of even 70-80 per cent (Laurila 1990).

Among opposition parties the feeling grew that a much higher reduction target was required This was especially the case with regard to the Greens and the Centre Party. HAPRO (1990), however, reported that nitrogen surprisingly increased the growth of trees even though it contributed to the long-term threat of acidification. HAPRO suggested slight reductions of NO_x emissions. Nevertheless, in the autumn 1990 the Cabinet[16] promised to keep the 30 per cent reduction target despite the difficulties.

In March 1991, just before the elections, the Cabinet decided to cut the NO_x emissions in energy production, in accordance with the proposals of the NO_x Commission. The new norms came into effect already in April for new plants but not until 1995 for old plants. The implementation of the NO_x reductions was continued by the new Cabinet. On the basis of the proposals of the NO_x Commission, the Cabinet decided on new limits for traffic emissions in the autumn 1991. These were aimed especially at diesel vehicles. Some of them were, however, suppose to be implemented by the next Cabinet. Nevertheless, a new, low-sulphur diesel-fuel which allowed the use of catalyst converters in diesel vehicles, was introduced in 1993. This fuel had been developed with state support.[17]

The Sofia Protocol came into effect in February 1991. The freezing target of the Protocol was achieved already in 1992: emissions had declined by more than 6 per cent from the 1987 level. In 1994 total emissions were still at about the same level. Although a new reduction programme was 'on the boards' at that time, it was obvious that the target of the Declaration would not be achieved. Nevertheless, preliminary data suggest that there was in fact a clear decline in emissions. The economic crisis at the beginning of the 1990s, which diminished industrial activity and traffic volume, contributed significantly to this reduction.

VOC Policy: The Geneva VOC Protocol (1991)

Domestic Policy Prior to the VOC Protocol

No specific policy on VOCs existed in Finland prior to the international negotiations. The reduction of VOCs began with the imposition of limits on auto emissions. The decisions made were mainly based on the Finnish nitrogen policy. An exception to this was the introduction, in 1991, of a new type of petrol[18] that had been developed with state support. VOCs themselves did not evoke any political interest since they did not threaten the forests as acidifying substances did. Even reliable data on domestic emissions and their effects was unavailable.

In January 1991 the Ministry of the Environment appointed a one-person task force to prepare a reduction strategy for VOCs; this work went on simultaneously with the on-going international negotiations. A steering group overseeing the preparation of the strategy consisted mainly of civil servants drawn from the environmental administration. The results of the task force were used by the Finnish representatives to the negotiations on the international protocol. At the same time, these negotiations affected the formation of the national reduction strategy.

The Role of Finland in the Negotiations for the VOC Protocol

At the beginning of the negotiations in 1988 Finland had not yet defined a reduction target. There existed only a general goal concerning the curbing of exhaust emissions and hydrocarbons (HCs), which are the most important group of VOCs. The general objective was to achieve the broadest possible co-operation at the international level for taking action on this problem.

In August 1989 the question of a reduction of up to 30-50 per cent came up in the negotiations. The Finnish position on this issue remained open until March 1990 when Finland adopted the target of the Nordic Programme on Air Pollution Control (1990). The goal of this programme was to reduce HCs and other ozone forming compounds by about 50 per cent from the 1988 level before the year 2005 by means of the best available technology. Nevertheless, the target proposed by the Nordic group was given up during the negotiations. Denmark, for example, In November 1990, accepted a 30 per cent reduction, as proposed by Germany, to be a 'good intermediate target' although it still considered a 50 per cent reduction as the main goal. Norway, in turn, adopted a new policy which meant that it was unwilling to join in on emission reductions. At the beginning of 1991 the negotiations led to a draft which contemplated an initial 30 per cent reduction of emissions according to a time schedule which was, for the time being, to be left open. Consensus on the main content of the protocol was reached in June. Finland confirmed that it could meet a 30 per cent reduction target although it was not in favour of meeting this goal by 1997. Finland, with Norway, Denmark and some other countries, preferred 2000 as the target year. Nevertheless, in November Finland signed the Protocol in Geneva, which specified that emissions were to be reduced by 30 per cent from the 1988 level by 1999.

Implementation and Compliance with the VOC Protocol

According to the Reduction Strategy for VOCs, which was finished in January 1992, the goals of the VOC Protocol should have been easy to achieve. The emissions would most probably have declined by 30-40 per cent by 2000 even without any specific action to this end, due to other decisions which had already been taken, or were likely to be taken, by the government and business.[19] Administrative guidance and supporting action were considered necessary, however, to cut emissions by 50 per cent by 2005, in accordance with the Nordic Programme. It was felt that emissions could be reduced by 40-50 per cent even before 2000 if a number of economically feasible technical measures were taken. Industry supported the implementation of the Protocol but disagreed on a reduction target higher than 40-50 per cent. Industry suggested that the reductions be sought through co-operation between governmental authorities and industry, and welcomed financial support. Environmentalists claimed that the Reduction Strategy was ecologically poorly reasoned and demanded special action in order to reach the goals of the Protocol.

By 1990 emissions had increased to more than 10 per cent above the 1988 level. In 1993 the Ministry of the Environment began to prepare a reduction programme of VOCs. However, in light of Finland's pending entrance into the European Union, it was finally decided that the implementation would be carried out by the EU directives already in place or being contemplated. The President ratified the VOC Protocol in December 1993.

Explaining the Finnish Policy

Finland as a Rational Actor

Sulphur policy The main reason why Finland wanted to reduce sulphur emissions through a treaty was simply that depositions caused by SO_2 represented a threat to the forests which were vital to the national economy. A decline in forest assets would have decreased Finland's international competitiveness, export incomes, and economic welfare. Prior to taking domestic action Finland favoured stipulating emission reductions by means of a binding protocol for at least two reasons. Firstly, domestic reductions alone would have been insufficient since less than one third of the deposit was from domestic sources. It was, therefore, important that the countries causing the deposit should reduce their emissions as well. Secondly, Finland's domestic emissions were relatively high and reductions without a treaty would have been very expensive. It was important to Finland that its main trading partners were also reducing their emissions. The increased costs of Finnish products would otherwise have led to a loss of competitiveness in the market. It was particularly important to Finland that its main competitors in the export of forest products reduced their emissions.

Finland actively sought to obtain reductions in foreign emissions in accordance with the Protocol, even though it was a net exporter of SO_2 at the beginning of the 1980s. Pushing international co-operation gave Finland the chance to play an active role in the negotiations and to push for reduction targets that were most favourable to its own interests. In order to underline its 'leading role', Finland even adopted a reduction target that was higher than required by the provisions of the Protocol. The provisions were not very demanding for Finland since its emissions had already decreased more than required before the Protocol was signed. In addition, most of the important countries also reduced their emissions more than required.

Prior to the Oslo Protocol domestic emissions were reduced even further. In the late 1980s Finland had been a net importer. These new reductions were based on the ecological considerations and country-specific targets which were proposed in the international negotiations. Finland supported the critical loads approach as a basis for calculating the reductions for the new protocol because its main concern was, quite simply, the condition of its forests. The problem in the international negotiations was that later many countries were not willing to commit themselves to a protocol with emission reductions. This left Finland, together with a number of other countries, alone with its high reduction targets.

Nitrogen policy Finland preferred controlling NO_X emissions on the basis of an international agreement rather than by domestic action because it was felt that a widely supported treaty was what it needed in order to diminish significantly deposits within Finland. Here, too, domestic reductions alone would have been insufficient since less than one-fifth of the total deposit was of domestic origin. Still, domestic emissions were also relatively high. In fact, Finland was a net exporter of nitrogen oxides. In the beginning Finland supported a reduction of NO_X emissions, but later it pushed only for a freeze rather than a reduction because new information had become available on the environmental effects and reduction possibilities of NO_X. Basically, nitrogen was causing acidification and was, like sulphur, a threat to the forests. However, for the moment nitrogen was found to be contributing to increased growth of trees and thus formed only a long-term threat. It was, therefore, not reasonable to resort to any 'emergency solutions'. In addition, the reduction of domestic emissions appeared to be expensive and difficult to achieve. Most of them were from transportation, but a significant part of them was from energy generation. A feasible and inexpensive reduction technology was not available. Consequently, the reductions would be a burden to industry and would have had consequences for Finland's international competitiveness. Furthermore, regulations on car emissions were politically difficult because they were partly directed at the private car owners. Finland signed the Sofia Declaration, even though it lacked sufficient information about the possibilities of implementation, only because it was a political statement without any legal status.

VOC policy The main reason Finland was not in favour of strict reductions on VOCs was that they were not considered a serious threat to the forests as were acidifying substances. There was no incentive for strict reductions. However, reductions undertaken as part of an internationally agreed control program would be relatively inexpensive for the national economy when compared to reductions based on unilateral measures. A widely supported treaty would not adversely affect Finland's competitive position internationally. Irrespective of these considerations, Finland adopted the Nordic Programme because it was without any legal status and could, therefore, be ignored. Later in the negotiations on the VOC protocol, Finland's supported only a 30 per cent reduction and fought against proposals for a strict reduction schedule. The reason for the lower target was that research on the environmental impacts of VOCs failed to indicate that strict reductions were required. In addition, the domestic reduction strategy was based on an easier reduction schedule. Finland ultimately signed the VOC Protocol primarily because VOCs were expected to decrease significantly in any case as a result of the reductions of NO_x emissions.

Conclusions on rational action From the rational action point of view there are three main factors that explain Finland's action. Firstly, the ecological impact of emissions on national forest assets was the factor that essentially determined Finland's position on emission reductions. The greater the threat to the health of the forests, the greater the reductions Finland was ready to carry out. The possibility of other material damages, such as erosion of buildings and monuments or a threat to human health, did not play a significant role in shaping Finnish policy. Secondly, the contribution of foreign emissions to deposits of acidifying substances in Finland also influenced significantly the country's response to the call for international agreements on emission reductions. The larger the contribution of foreign emissions, the more important it was considered that especially those countries causing the domestic deposit should commit themselves by treaty to emission reduction targets. The net balance of pollution exchange did not, however, substantially affect either Finnish domestic policy or its position in international negotiations. Thirdly, the costs of reducing domestic emissions also played a role in determining the level of emission reductions that Finland was ready to accept. The greater the abatement costs, the less willing Finland was to commit itself to strict emission reductions. It is interesting to note that Finland was not willing to commit itself to strict reductions of VOCs, even though these reductions would have been inexpensive. As noted above, this was due to the fact that VOCs were not considered to be a threat to Finnish forests.

Finnish policy can largely be explained by these three factors. However, in order to see how these factors emerged and how they worked to shape Finnish policy, we have to take a closer look at the domestic politics of Finland during this period.

Domestic Politics

Affected values   Acidifying substances, especially sulphur, directly threatened the Finnish forest industry. From the 1970s to the 1990s the forest industry, measured in terms of gross value of production, was the largest, and, measured in terms of the number of employees, the second largest sector of industry in Finland. The total economic value of the forest industry remained considerable, although its relative significance decreased in the 1980s.[20] However, its position was controversial because the forest products industry also was one of the chief sources of polluting emissions. In addition to emissions from pulp and paper processes, the industry consumed tremendous amounts of energy which also caused emissions. Consequently, emission reductions entailed significant abatement costs for the forest industry. On the other hand, nitrogen also had a positive effect on the growth of timber, which encouraged the forest industry to support reducing sulphur emissions instead. Other important sectors of industry were only indirectly affected by acidification. Nevertheless, they directly bore a share of the abatement costs. Especially the metal industry suffered high abatement costs.

The need to maintain the international competitiveness of Finnish industry was an important factor determining Finland's policy. The international competitiveness was vital to the national welfare because about one fourth of the national income was from export.[21] The forest industry dominated export with its share of 56 per cent in 1970. At the end of the 1980s its contribution was still around 40 per cent, the same as that of metal industry.

Public Opinion Finns can be regarded as people who appreciate environmental values very much. Even the serious economic recession after 1990, when the country experienced a negative annual growth of the GDP for the first time since 1976, did not diminish the popular concern for the environment.[22] Public support remained an important motivation for the Government's environment-friendly policy. Due to the deep concern for the forests, several groups in Finnish society consistently demanded reductions of acidifying emissions.

Intermediate Actors Domestic factors have been mainly taken into consideration by appointing national commissions to prepare reduction programmes. Important interest groups were represented in these commissions, which have included representatives from environmental and other administrative branches, various sectors of industry, abatement technology, environmental research and NGOs. Such a procedure is quite usual in neo-corporatist systems, such as Finland, where participation of interest groups is often channelled through special commissions (Nousiainen 1985). The integration of target groups in this way into the implementation process has improved compliance with the programmes. Its influence on policy formation was much smaller because the work of the commissions was based on targets set by the Government. Only rarely did the commissions deviate from the tasks they had been given to provide an independent input to policy formulation.

The Finnish scientific community played an important role in decision-making because the Government based its policy on – or at least justified it in terms of – the results of environmental research. The scientists studying ecological vulnerability took part in the processes of policy formation indirectly, especially via HAPRO (1985-1990). In spite of some disagreements, the studies tended to encourage high reduction targets for SO_2 with less stringent targets proposed for NO_X. Ecologists were also represented in both sulphur commissions.

The most important environmental NGO in Finland was the Finnish Association for Nature Conservation. Although it had little official power, the Association enjoyed much prestige and its opinions carried a certain amount of political weight. Environmental NGOs took part in air pollution control policy mainly by issuing statements and by participating in the commissions. The environmental NGOs had more success in influencing the official sulphur policy than in the case of nitrogen policy.

Despite their limited parliamentary power, the Greens influenced official policy by the mere fact of their existence.[23] The Green movement participated in the elections from the beginning of the 1980s but it did not organise itself under the name of the Green League until 1987. It acquired the status of a party only in April 1988. Since the mid-1980s, other parties as well have acquired a certain amount of 'greenness'. The traditional parties set high reduction targets for sulphur and nitrogen partly in order to counter the growth of the electoral support of the Greens.

The media played an important role as intermediary by publicising the results of foreign and domestic research on the ecological vulnerability of forest and lakes to acidification. A strong public debate on acid rain developed in 1982-1983, peaking around 1985. The issue stayed in the headlines through the 1980s, although environmental issues lost much of their public attention in the beginning of the 1990s. The public debate in the media, especially in the press, which was associated with the products of forest industry, pushed for high reduction targets, especially for sulphur.

The providers of environmentally friendly technology had contradicting interests in the political processes because they were also among the worst environmental offenders. On the one hand, the abatement costs resulting from the new standards had an immediate impact on them, while, at the same time, they were able to profit by selling to others the new technology by which those standards could be met. Various sectors of industry and commerce strongly influenced the reduction targets. The Government co-operated with industry throughout the processes of policy formation and implementation. The interests of industry were not only represented in the commissions but also in the task forces which preceded them. In some cases, the development of environment-friendly technology was even supported by the Government.

Government ideology Finland's activity in environmental policy did not derive directly from a leftist or rightist tendency of the Cabinets. In general, the Cabinets were supported by majority coalitions in which the leftist parties favoured higher reduction targets than the parties on the political right and in the centre. The conservative-socialist Cabinet (1987-1991) was more active in environment policy than its socialist-agrarian predecessor (1982-1987), or the subsequent centre-rightist Cabinet (1991-1995). Yet they all emphasised environmental issues more than the earlier popular front Cabinets (1977-1982).

At the beginning of air pollution control policy and in the first phase of the sulphur policy there was a relatively high consensus between the Cabinet and the opposition on the nature of the problem and the general line to be followed in dealing with it. The consensus facilitated the making of decisions of specific programs and measures. However, when the parties in opposition demanded higher reduction targets (especially for NO_x) decisions on the reduction levels were difficult to reach. The Cabinets have often adopted new reduction targets immediately prior to general elections, usually in response to political demands. Despite the conflicts between the old Cabinet and the opposition on reduction targets, newly-installed Cabinets, which included members of the former opposition parties, have generally simply implemented the earlier targets.

Institutional capacity Up until the beginning of the 1980s limited administrative capacity prevented effective implementation of air pollution control policy. It was only after the Air Pollution Control Act and Decree had entered into force and the Ministry of the Environment had been set up that resources started to grow significantly. This growth continued until the early 1990s when it slowed down. Nevertheless, from the end of the 1980s onward the improved institutional capacity facilitated an active environmental policy.

Until 1983, when environmental affairs were still the responsibility of the Ministry of the Interior, the political priority given to these issues within the Government was quite low. After extensive political debate and bargaining, the Ministry of the Environment was given relatively wide powers in the field.[24] However, the status of the ministry remained low because it was so new. Since the minister for the environment came to be included in the group of ministers deciding on economic affairs in 1987-1991, the political weight of the environmental branch has grown.

The state of the economy indirectly affected the Government's ability to adopt and implement high reduction targets. At the beginning of the 1990s the state increased its revenues by borrowing money. Overall, however, the condition of the national economy as a whole, which depended on the competitiveness of Finnish industry, was more important than state finances. Air pollution control was officially based on the polluter pays principle, according to which the state should not directly finance abatement measures of industry. Especially in the case of NO_X, high reduction targets were easier to adopt when the national economy was flourishing; during the economic depression the realisation of these objectives was more uncertain, even though the depression itself diminished emissions. Industry was not, however, financially dependent on the Government for its abatement measures. No specific state support was provided as part of the reduction programmes for SO_2. Investments in air pollution control by industry were, however, supported by low interest state loans. Nevertheless, the Government controlled parts of industry because some of the most important single polluters, especially sources of SO_2, were state-owned companies. NO_X and, partly, VOCs as well, were more difficult to control because most of the emissions were from dispersed sources, mainly from traffic. Therefore, the Government used economic instruments, such as taxes, to achieve its auto emissions reduction targets.

Conclusions on domestic politics From the domestic politics point of view the main forces driving Finnish policy were public environmental awareness and the need to protect the competitiveness of industry and commerce. While the former contributed to the demand for air pollution control, the latter sought to block expensive emission reductions. The stance taken by the forest industry was decisive and its position on the issue varied with the pollutants under consideration.

The international competitiveness of Finnish industry and commerce, especially of the forest industry, essentially influenced how the Government dealt with air pollution control. The larger the burden on the export industries, the lower the emission reductions it was willing to accept. On the other hand, public concern with the quality of the forests, fed by the results of environmental research, also affected significantly the Government's behaviour on air pollution control. Public concern was channelled through media, environmental NGOs and political parties, especially the Greens. The greater the domestic political pressure for emission reductions, the higher the reduction targets the Government was ready to adopt. Public concern on forests even led to competition between political parties. The Greens demanded that the other parties and the Cabinets support strict air pollution control. The more other parties feared losing electoral support, the greater their willingness to accept higher targets for emission reductions. Outside of parliament, especially in the commissions planning the reduction programmes, public concern was represented by the environmental NGOs. Stories about the threats to the forests, derived from environmental research, were especially prominent in the printed media. The more unsettling the news was reported, the higher the reduction targets the Government was willing to support. Without the media the results of research and the demands of the environmental movement would not have attracted as much public attention. Nevertheless, without HAPRO there would not have been domestic results to publish. The growing public environmental awareness contributed also to the establishment of the environmental administration and the passing of legislation, which in turn formed the legal and organisational basis for achieving emission reductions.

The analysis of domestic politics provides a detailed overview of national factors and actors, that played a role in the formation and implementation of Finnish policy. By looking at the domestic processes, this analytical focus adds important insights to the picture drawn by the rational action approach. Neither approach, however, explicitly considers the contextual factors affecting the positions taken by the Finnish government in international negotiations and decisions on acidification.

Contextual Factors

From the contextual point of view there are various foreign policy goals as well as industrial and economic forces that help explain Finland's action. Besides concern for the environment, and the repercussions of acidification for the economy, international protection of the environment served as a tool for reaching more general goals in Finnish foreign policy. Finland favoured international action during the Cold War era because co-operation in 'non-political' areas, like environmental protection, would encourage détente. Through its international activity Finland asserted its neutral position between the East and the West. The strategy was aimed, primarily, at preventing potential Soviet pressure. Finland's early unwillingness, in the beginning of the 1970s, to support international action on air pollution control can partly be explained by the general attempt to avoid a conflict with the USSR, even though this Eastern neighbour was responsible for transmitting significant amounts of pollutants to Finland. Finally, in 1989, when the Cold War was drawing to a close, Finland and the USSR were 'ready' to sign a bilateral treaty on emission reductions.

Since the late 1980s, Finland's international activity in environmental affairs was also used to reinforce its status as a Western democracy. Finland even signed the Sofia NO_x Declaration partly because it wanted to behave like a 'good European' country. By signing, Finland attempted to separate itself from the East European socialist countries which had the reputation of being heavy polluters. Important Western countries, by contrast, supported high reductions of NO_x. Especially the signing of the VOC Protocol was used to foster new goals of Finnish foreign policy. After the collapse of the USSR, Finland, wanted more than ever to be accepted as a Western country. Again, important West European states favoured strict reductions. By demonstrating that it, too, supported progressive environmental policies, Finland hoped to obtain important support from the leading EC countries, especially Germany, for its application for membership in West European institutions. In turn, one reason behind Finland's interest in becoming member of the EC was its fear of being exposed to pressures from its Eastern neighbour in future.

Nordic co-operation provided a traditional framework for Finnish foreign policy. It also enhanced Finland's status as a Western democracy, and later its public image of as a leading country on environmental issues. The Nordic framework was also an important arena for social learning and diffusion of ideas. As is well known, the problem of acidification was originally perceived and defined by Sweden and Norway.

Finland actively supported the target of the Helsinki Protocol because its domestic emissions had already decreased from the base year, in large part due to economic reasons. On the other hand, however, such reductions may also have started because it was already expected that they would be included in the Geneva Convention. Therefore, as a result of domestic developments, Finland was later easily able to accept and achieve higher reduction targets. Likewise, the latest target for SO_2 contained in the Oslo Protocol and the target of the NO_x Protocol have been achieved partly because of an economic recession in the beginning of the 1990s which diminished industrial activity and traffic volumes, and, subsequently, emissions as well.

Contextual factors provide an overview of the background for the Finnish air pollution control policy. While, as such, they suggest important 'side-motivations' explaining the outcome, they cannot alone capture the full rationale behind Finland's action.

Conclusions

Formally, Finland has a good record of compliance with the protocols. After the LRTAP agreements had been ratified and entered into force, they were also incorporated into national legislation.

In an important sense, Finland actually was a pusher in reducing sulphur. Its domestic reduction targets were higher than those agreed to at the international level, and the obligations of the treaties were achieved long before the stipulated deadlines. The goals of the nitrogen policy were, however, lower. The domestic reduction plans for NO_X were still unfinished when the international agreement was signed. The target of the Sofia NO_X Protocol was achieved ahead of the schedule, but fulfilling the obligation of the Declaration was later considered 'uncertain'. In the negotiations Finland took an ambivalent position, shifting from a laggard to a pusher and back. In the VOC policy Finland can be regarded more as a laggard than a pusher, although it initially adopted the high Nordic target, and finally even signed the VOC Protocol. Reduction programmes for SO_2 and NO_X were prepared by national commissions in which important interest groups were represented. In the case of VOCs, national reduction measures are to be derived from the requirements of the pending EU directives.

For Finland, forest health was a 'question of fate'. To understand the rationality of the policy it is essential to appreciate that, given the vulnerability of the forests, the perceived impact of the pollutants on the quality of the forests was the principal concern throughout the entire process. The contribution of foreign emissions to the domestic deposits was a more important motivation for Finland's international activity than the *net balance* of pollution exchange. The concern with forest damage was, however, tempered by a concern with abatement costs.

It was crucial to maintain the international competitiveness of Finnish industry. Reducing acidifying emissions was motivated by the need to conserve the supply of raw materials for the forest industry which was the most important export sector. However, the abatement costs were a burden to all sectors of the economy. The fundamental basis for the Government's policy was the environmental awareness of public. Public concern for the health of the forests, awakened primarily by the results of research programmes such as HAPRO, played an important role. Furthermore, the Government based its decisions on scientific data. Without news, provided especially by the printed media, alarming results would not have mobilised public concern. The Greens influenced the policy by their mere existence. Other parties set relatively high reduction targets partly in order to counteract the growth of the electoral support of the Greens.

Finland's international activity in environmental affairs was also a means for pursuing other goals in foreign policy. During the Cold War era it promoted détente and used environmental issues to strengthen Finland's position as a neutral state. Co-operation with other Nordic countries was a useful framework for Finnish policy. Later, the need to be identified as an 'environment-friendly' actor provided an additional reason for following the Nordic examples. After the collapse of the USSR the image of a 'good European' state – defined, in part, as one that takes care of the environment – was also a means of getting support in order to further Finland's integration into Western institutions.

Notes

1. In addition to domestic sources, the other Nordic countries, the Western part of the European continent, the USSR and the rest of East Europe as well as undefined sources, each caused about one-sixth of the total deposit (HAPRO 1990, pp.19-20).
2. The first government organ on environmental affairs, established in 1971, had only advisory status (Markkula 1993, pp.552-3).
3. The Centre Party favoured narrow while the leftist parties favoured wide powers for the new ministry. The final compromise was reached in early fall 1982. The Conservatives, which were in opposition, surprisingly supported the leftist parties (Ministry of the Environment 1988, pp.138-9), probably for tactical reasons: less than five years later the Social Democrats formed the Cabinet together with the Conservatives.
4. Cooperation in the field of environment was originally one of the three issues suggested by the Soviet leader Brezhnev as a continuation of the CSCE process.
5. The Constitution Act would not have required an adoption by the Parliament but it was

motivated mainly due to lack of domestic air pollution control legislation (Government Proposal 1981, p.282).

6. According to the Constitutional Act (§ 33) the President decides on relations with foreign states. Parliament's adoption is required only for treaties which include regulations belonging to legislature (e.g. long-term budgeting) or which according to the Act itself requires such an adoption (see Nousiainen 1985, p.339-40).

7. A decree is a procedure by which the President, on the basis of a Cabinet proposal, may give legislative rules specifying how to interpret existing law. The decrees have lower status than the parliamentary acts but are binding legislation (see Nousiainen 1985, p.310).

8. Parliament's adoption was unnecessary since the Helsinki Protocol was 'only' a protocol added to a convention that had already been adopted (see notes 5 and 6).

9. Of the Cabinet parties the Social Democrats (1987) supported 'further reductions' and international treaties on an ecologically safe emission limit by 2000. However, the Conservatives (National Coalition Party), still in the spring 1988, supported only a 50 per cent reduction by the year 2000, although the previous Cabinet had decided on a tighter schedule. Their target was consistent with the views of industry.

10. In order to diminish the Soviet origin deposit, Finland signed a treaty in October 1989. It stated that areas on the Finnish-Russian border should reduce their emissions by 50 per cent by 1996 (see e.g. Vaahtoranta 1989). The agreement required practically nothing of Finland.

11. The Government made an official decision about the 80 per cent reduction programme in January 1991 (Sulphur Committee II 1993, pp.32-3).

12. Of the Cabinet parties, the Swedish People's Party urged that Finland join the 'leading countries' by sharpening exhaust norms from 1989 onwards (Stenlund and Granfors 1986, p.6), but the Centre Party (1986) argued for a more gradual implementation.

13. The norms of the first phase were the same as those contained in the 'Luxembourg compromise' of 1985. However, by norms of the second phase, the emissions from new cars would be ca. 80-90 per cent lower than before (Sarkkinen 1987, p.26).

14. Of the leading Cabinet parties the Social Democrats (1987) had emphasised the need to conclude a NO_x protocol and also the need to reduce traffic emissions by new technology, but the National Coalition Party (1988) favoured a 'slower' schedule.

15. A forest researcher Prof. Satu Huttunen, also a known environmentalist, did not participate in the work of the Commission. This predicted 'less advanced' results.

16. The Conservatives and the Social Democrats confirmed the 30 per cent target in their programmes later in 1991.

17. The new 'city diesel' was developed by the Ministry of Trade and Industry in co-operation with the state-owned oil-company and catalyst-manufacturer, Technical Research Centre of Finland, and Helsinki City Transport (Mikkonen 1993, p.17).

18. The new 'city gasoline' could be used in old cars. It reduced HCs by 5-10 per cent and also VOCs by 13-17 per cent as compared to old fuels. The fuel was developed by the state's oil-company and alcohol monopoly, and by the Technical Research Centre of Finland (Mikkonen 1993, p.20).

19. The measures were the emission reductions of traffic, already in process, and, regarding stable sources, administrative decisions, economic guidance, and agreements between authorities and industry. (See Reduction Strategy of VOCs 1992, p.155-7.)

20. The forest industry's share of the Finnish GDP was over 20 per cent in 1980 but at the beginning of the 1990s only about 12 per cent.
21. The most important single country in terms of total export until the end of the 1980s was the USSR (contributing about 20 per cent). After its collapse, Germany (about 14 per cent) became the most important market. Sweden (about 13 per cent) has remained the second most important and the UK (about 11 per cent) the third most important market.
22. During the economic downturn people were even more willing to lower their living standards for the benefit of the environmental protection. (See Centre for Finnish Business and Political Studies, 1993.)
23. Support for the Greens has remained constant at around 5-7 per cent in the general elections, while the number of MPs it has held reached ten (out of 200) at the most.
24. Apart from 'pure' environmental issues, the Ministry's responsibilities also covered housing, physical planning, and building.

References

Centre for Finnish Business and Political Studies (1993), *Kansa tienhaarassa - Raportti suomalaisten asenteista 1993*, Elinkeinoelämän valtuuskunta EVA, Helsinki.

Centre Party (1986), *Suomalainen suunta - Keskustapuolueen tavoiteohjelma 1980-luvun loppupuoliskolle*, puoluekokous, Lappeeranta 13-15.6.1986, Etelä-Saimaan Kustannus Oy, Lappeenranta.

Geneva LRTAP Convention (1979), 'Asetus valtiosta toiseen tapahtuvan ilman epäpuhtauksien kaukokulkeutumista koskevan yleissopimuksen voimaansaattamisesta', nr. 15/1983 (incl. 'Convention on Long-Range Transboundary Air Pollution'), *Treaty Series of the Statute Book*, SopS nr. 265/1983, VAPK, Helsinki.

Geneva VOC Protocol (1991), 'Protocol to the 1979 Convention on Long-Range Transboundary Air Pollution Concerning the Control of Emissions of Volatile Organic Compounds or Their Transboundary Fluxes', unpublished version, pp. 49-94.

Government Proposal (1981), 'Hallituksen esitys Eduskunnalle valtiosta toiseen tapahtuvaa ilmanepäpuhtauksien kaukokulkeutumista koskevan yleissopimuksen hyväksymisestä', nr. 208, 21.11.1980, *Vuoden 1980 valtiopäivät, Asiakirjat A3: Hallituksen esitykset 141-233*, VAPK, Helsinki.

Green League (1990), *Energialaskun aika - Vihreän Liiton energiaohjelma*, Vihreän liiton syyskokous marrask. 1989, Helsinki 15.1.1990.

HAPRO (1987), *Happamoituminen Suomessa; HAPROn väliraportti (Acidification in Finland; Interim Report of the HAPRO Programme)*, P. Kauppi, P. Anttila, L. Karjalainen et al. (eds), Ministry of the Environment, Department on Environment Protection, Sarja A 57/1987, VAPK, Helsinki.

HAPRO (1990), *Happamoituminen Suomessa; ·HAPROn loppuraportti (Acidification in Finland; Report of the HAPRO Programme)*, P. Kauppi, P. Anttila, L. Karjalainen et al. (eds), Ministry of the Environment, Department on Environment Protection, Sarja A 89/1990, VAPK, Helsinki.

Helsinki Sulphur Protocol (1985), 'Asetus vuoden 1979 valtiosta toiseen tapahtuvaa ilman epäpuhtauksien kaukokulkeutumista koskevaan yleissopimukseen liittyvän, rikkipäästöjen tai valtiosta toiseen kulkeutuvan rikkivuon vähentämisestä vähintään 30 prosenttia tehdyn pöytäkirjan voimaansaattamisesta', nr. 50/1987 (incl. 'Protocol to the 1979 Convention on Long-range Transboundary Air Pollution on the Reduction of Sulphur Emissions or Their Transboundary Fluxes by at least 30 Per Cent'), *Treaty Series of the Statute Book*, SopS nr. 727/1987, VAPK, Helsinki.

Hiienkoski, P. (1995), *Air Pollution Control Policy in Finland: Basis of the Finnish Air Pollution Control Policy within the Framework of the Geneva Convention on Longe-range Transboundary Air Pollution, and its Protocols*, October 1995, The Finnish Institute of International Affairs.

Laurila, J. (1990), 'Typpitoimikunnan ehdotus jää puolitiehen', *Suomen Luonto* 4/1990, 31. toukok. 1990, vol. 49, pp. 2-3.

Lommi-Kippola, C. (1991), 'Ympäristöverot ja –maksut', *Ympäristökatsaus* 4/1991, p.7.

Markkula, M. (1993), 'Ympäristöministeriön ja luonnonsuojeluliiton välimaastosta', *Ympäristö- ja terveys* nr. 9/1993, vol. 24, pp. 525-556.

Mikkonen, S. (1993): 'Reformulated Fuels Reduce Automotive Emissions, Air pollution control in Finland', *Ilmansuojelu-uutiset/FAPP-News* 5-6/1993, pp. 17-21.

Ministry of the Environment (1988), *Environmental Protection in Finland, National Report 1987*, Environmental Protection Department, Series A 67/1988, VAPK, Helsinki.

National Coalition Party (1988), *Energiapoliittinen ohjelma 1988*, Kansallinen kokoomus, kevät 1988, Helsinki.

Nitrogen Oxides Commission (1990), *Typenoksiditoimikunnan mietintö* (Report of the Nitrogen Oxides Commission), Komiteanmietintö 1990:11, Ministry of the Environment, VAPK, Helsinki.

Nordic Council of Ministers (1986), *Euroopan ympäristö - Euroopan ilma*, Pohjoismaiden ministerineuvoston raportti Pohjoismaiden neuvoston ilmansaasteita käsittelevään kansainväliseen konferenssiin Tukholmassa 1986, PMN:n ympäristökysymysten virkamieskomitean toimeksiannosta, Br. Aniansson (ed.), Tukholma: Norstedts Tryckeri.

Nordic Program on Air Pollution Control (1991), *Pohjoismainen ilmansuojelun toimintasuunnitelma* (Original: Nordiska Ministerrådet, Nordisk handlingsplan mot luftföroreningar, NU 1990:3), Pohjoismaiden Neuvosto Reykjavik 2.3.1990, O. Järvinen (ed.), Ympäristöministeriön ympäristönsuojeluosaston muistio 2/1990, VAPK, Helsinki.

Nousiainen, J. (1985), *Suomen poliittinen järjestelmä*, WSOY, Juva.

Nurmi, S. (1990), 'Ilmansuojelun kansainväliset sopimukset', *Ympäristö- ja terveys* no. 9-10/1990, vol. 21, pp. 606-610.

Oslo Sulphur Protocol (1994), 'Protocol to the 1979 Convention on long-range transboundary air pollution on further reduction of sulphur emissions', ECE EB.AIR/R.84 (rev. version 14 June 1994), United Nations, pp. 1-30.

Program of the Cabinet (1983), 'Pääministeri Kalevi Sorsan 4. hallituksen ohjelma', 6.5.1983, *Virallinen lehti* 52/1983.

Program of the Cabinet (1987), *Pääministeri Harri Holkerin hallituksen ohjelma*, 30.4.1987, VAPK, Helsinki.

Reduction Strategy of VOCs (1992), *Haihtuvien orgaanisten yhdisteiden (VOC) vähentämisstrategia* (Reduction Strategy of VOCs), Ministry of the Environment: Department on Environment Protection 14.2.1992, Muistio 6/1992, VAPK, Helsinki.

Ruokoranta, M. (1990), 'Liikenneministeriö korostaa kansainvälisyyttä', *Ympäristönsuojelu* 1/1990, pp. 15-16.

Salo-Asikainen, S. (1988), 'Typenoksidipäästöjen rajoittaminen alkusuoralla', *Ympäristön-ja luonnonsuojelu* 3/1988, pp. 7-9.

Sarkkinen, S. (1987), 'Ilmansuojelu: Henkilöautoliikenteen päästöjen vähentäminen', *Ympäristön- ja luonnonsuojelu*, 2/1987, p. 26.

Sarkkinen, S. (1993), 'Air Pollution Control Policies in Finland, Air Pollution Control in Finland', *Ilmansuojelu-uutiset/FAPP-News*, 5-6/1993, pp. 6-7.

Social Democratic Party in Finland (1987), *Ympäristöpoliittinen ohjelma*, SDP:n XXXIV puoluekokous kesäk. 1987, unpublished.

Sofia Nitrogen Protocol (1988), 'Protocol to the 1979 Convention on Long-range Transboundary Air Pollution Concerning the Control of Emissions of Nitrogen Oxides or Their Transboundary Fluxes', ECE/EB.AIR/21, United Nations, pp. 1-19.

Stenlund, Peter and Granfors, Stig (1986), *Renare avgaser, PM om avgasrening för bilar*, Rapport från SFP om lufvårds samband med bil- och bränslebeskattningen, Januari 1986, unpublished, pp. 1-7.

Sulphur Commission (1985), *Rikkitoimikunnan välimietintö I* (Interim-Report I of the Sulphur Commission), Komiteanmietintö 1985:35, Ministry of the Environment, VAPK, Helsinki.

Sulphur Commission (1986), *Rikkitoimikunnan mietintö* (Report of the Sulphur Commission), Komiteanmietintö 1986:33, Ministry of the Environment, VAPK, Helsinki.

Sulphur Committee II (1993), *Rikkitoimikunta II:n mietintö* (Report of the Sulphur Committee II), liitteenä rikkitoimikunnan välimietintö (incl. the Interim-Report of the Sulphur Committee), Komiteanmietintö 1993/6, Ministry of the Environment, ensipainos, Ympäristöministeriö, Helsinki.

Task Force on Sulphur (1984), *Ilmansuojelun rikkiselvitysten suunnittelutyöryhmän mietintö*, Ympäristöministeriö: Ympäristön- ja luonnonsuojeluosaston julkaisu C:2, VAPK (Annankadun monistamo), Helsinki.

Vaahtoranta, T. (1989), 'Suomen ja Neuvostoliiton kahdenvälinen ilmansuojeluyhteistyö', *Ulkopolitiikka*, 4/1989, pp.39-40.

Välimäki, P. and Brax, T. (1991), *Vihreä ABC-kirja*, Vihreän liiton julkaisuja 4, Arpatehtaan kirjapaino, Tampere.

8 Switzerland: From Pusher to Laggard in Clean Air Policy

STEPHAN KUX AND WALTER SCHENKEL

Acidification and the Vulnerability of the Alpine Eco-System

The Swiss Alpine region is particularly vulnerable to acidification. High NO_X and ozone loads are caused by long-range transboundary air pollution originating from the industrial zones of northern Italy and from urban zones of Switzerland itself. Trans-Alpine traffic (tourism and trucks) is also a major pollution source. The Alpine cantons are confronted with forest damages as a result of sulphur depositions and ozone concentrations. Air pollution does not primarily threaten vital economic interests, but could jeopardise the equilibrium of ecosystems and the security of Trans-Alpine traffic routes. Alpine forests are essential for the protection against soil erosion, falling rocks and avalanches. Furthermore, they provide resources for the local economy (timber) and recreation areas (tourism). During the 1970s, damages also became visible in urban areas in the form of air pollution from stationary sources and seasonal smog. Later on, high ozone concentrations shaped the public debate. Clean air discussions primarily focused on human health, but increasingly shifted to eco-system damages.

At the beginning, abatement costs were low, since binding federal laws were missing and the directives concerning sulphur contents did not take into account the highest possible technological standards. As a consequence of the 'acid rain' debate in the early 1980s, industries and car associations were increasingly willing to accept financial burdens and to invest in new technologies, but they were still opposed to market-oriented measures. After the introduction of binding federal legislation in the 1980s, a conflict emerged over the distribution of abatement costs between the national, cantonal and local governments emerged. Some cantons and communes started to complain that they were being forced to carry too

199

much responsibility and too high costs. As a consequence of the economic slowdown at the beginning of the 1990s, not only the cantons but also the polluter groups began again to oppose additional environmental measures. For this reason, the Federal Council paid more attention to market-oriented programs, namely taxes (SO_2, VOC) and self-commitments (CO_2) by industries. Yet those measures which are supposed to be most effective have not yet reached a consensus between environmental and economic interests.

The 'willingness-to-pay' was great in the case of SO_2 and NO_X, focused mainly on national abatement costs. By advocating stricter reductions at the international level, Switzerland tried to ensure that other countries were also sharing the burden. During the NO_X negotiations and in view of the self-declared 30 per cent reduction target, Switzerland was even willing to pay much more than other European countries. Accordingly, the Federal Council pushed for the introduction of stricter vehicle emission regulations in Western Europe. Since the national reduction targets were stricter than the international commitments, abatement costs were calculated independently of the international agreements. Cost-damage calculations also influenced Switzerland's position regarding the critical loads approach. Since SO_2 abatement measures were taken at an early stage and most national legislation was already in force, Switzerland profited from the switch from a flat reduction rate to a more differentiated approach. Thus the second sulphur protocol had no direct cost implications, but the systems of measurement and data collection had to be revised. In the case of the VOC protocol, Switzerland had not yet developed a proper domestic abatement strategy and was thus opposed to the high international targets demanded by Scandinavian countries, Germany and the Netherlands.

An Early Starter

In the 1960s, environmental policy at the national level was dominated by the discussion on the introduction of a constitutional article on environmental protection. At first, the Federal Council was of the opinion that the existing national and cantonal laws provided a sufficient legal basis in order to initiate national environmental protection measures. Based on the Labour Law or the Road Traffic Law, for instance, the government could combat noise, dust, air pollution and other damages of stationary or mobile sources (Rausch, 1977, p.122ff.). Clean air policy rested mainly on non-binding guidelines regarding control and limitation of loads from combustion and industrial plants. The new constitutional provision met broad support. The two chambers of parliament voted unanimously in favour of it (APS, 1970, p. 124). In 1971, the Swiss population passed the constitutional article with a 93 per cent majority, an overwhelming result in the history of Swiss direct democracy.

The clear vote put pressure on the federal government to act. First, a Federal Office for Environmental Protection was created. Second, preparatory work on an Environmental Protection Law (USG) was initiated. Encouraged by the overwhelming popular support, an expert commission prepared a detailed and comprehensive draft taking into account demands by both environmental and economic organisations. This first draft was very progressive regarding both objectives and approach (environmental taxes, precautionary and polluter pays principles). It failed, however, because business circles were no longer prepared to give environmental protection high priority in view of the energy crisis and recession in the mid-1970s. The majority of the cantons also criticised the extensive transfer of competencies to the national authorities, which violated the principles of federalism (Nüssli, 1987, p. 31; Ammann, 1990, p. 93). Subsequently, the federal government initiated a broad dialogue with political parties, non-governmental organisations and interest groups in order to build a consensus. After substantial concessions, the watered-down framework law finally entered into force in 1985, at a time when concerns over 'acid rain' started to influence environmental considerations (Nüssli, 1987).

Third, the controversial debate on regulating motor vehicle emissions started. In 1970, the parliament demanded the introduction of UNECE Regulation No. 15. In 1973, doubts emerged regarding the effectiveness of these norms. The lawmakers referred to the stricter U.S. norms and charged the Federal Council to negotiate the tightening of the UNECE regulation. In 1974, federal clean air policy also came under strong pressure from environmental organisations. Two initiatives were submitted. The first, the so-called Albatros Initiative, called for a massive reduction of vehicle emissions by 80 per cent for new cars until 1977, and to the technically feasible minimum for old cars until 1978. The second pushed for twelve car-free Sundays. Both initiatives were based on first emission charts for stationary and mobile sources and on scientific studies which described air pollution and lead emissions as 'a time bomb with unknown threat potential' (APS 1974, p. 99; NZZ, 16. /17.8.1974).

The Federal Council reacted with a 'Report on Emissions and Noise of Motor Vehicles' (BBl 1975 I 25-41). It agreed with the overall objectives of the Albatros Initiative, rejected however the tight time frame as unrealistic for economic, political and technical reasons. The Federal Council promised to introduce by 1978, respectively 1982 more severe vehicle emission thresholds. It did not exclude unilateral steps by Switzerland and took up the argument of the environmental organisations that Switzerland – as a country without heavy or automobile industry – should play a pioneering role in environmental protection. The automobile and road traffic associations focused on fighting the two initiatives, yet did not oppose the Federal Council's more moderate strategy. In 1977 and 1978 respectively, the two initiatives were defeated by a clear majority. Yet the Federal Council failed to implement its strategy of tightening car emission standards over time and came under increasing pressure from parliament, environmental organisations and the public (Ammann, 1990). The two referenda resulted in a strengthened mobilisation and organisation of environmental interests (Giger, 1981). At the same time, the car lobbies gained substantial influence on federal policy, namely in alliance with the Federal Office for Police (BAP) and the Federal Office for Foreign Trade (BAWI). International influences also played an important role. Existing commitments in the framework of the UNECE and substantial pressure from multinational car manufacturers forced the Federal Council and parliament to postpone the unilateral introduction of stricter vehicle emission thresholds until the mid-1980s.

In addition to the drafting of the Environmental Protection Law (USG) and the vehicle emission regulation, the federal government also undertook less controversial measures to improve air quality. The Expert Commission for Air Hygiene (EKL), for instance, set non-binding guidelines for stationary plants and elaborated quality standards for combustibles and fuels. Based on these recommendations, the Federal Council - on the basis of the Law on Poisonous Materials - reduced the lead content of normal gasoline (Geschäftsbericht, 1971, p.77). Finally, the collection of scientific data on air pollution was improved. In 1975, a first SO_2 register for 1970 was published. In 1975, the Federal Council initiated the creation of the National Observation Network for Air Pollution

(NABEL). In 1978, the first installations were operational (Der Bund, 2.3.1981). Air pollution problems also started to influence other policy areas. Relatively early, experts working on a Comprehensive Transport Conception and a Comprehensive Energy Conception demanded that environmental objectives and targets should be taken into account. Traffic's share of energy consumption, for instance, should not grow beyond 25 per cent. In the case of the Transport Conception, however, the Federal Council was unwilling to confront the road traffic associations and pursued a conservative course (Klöti, 1993).

With the exception of traditional health policy, the cantons and communes had not yet to implement federal clean air measures. Cities and densely populated cantons were, however, increasingly confronted with air pollution problems and started to pass their own clean air legislation within their competencies. Thus they played a pace-setting role. Already in the 1960s, the City of Zurich introduced binding norms for stationary sources. In 1972, the Canton of Zurich introduced a decree regulating stationary sources. In 1978, it passed a Clean Air Program based on Germany's Technical Guidance on Clean Air (TA Luft). In 1973, Basel-Land was the first canton to pass a proper law on clean air. Both cantonal and communal guidelines primarily set emission standards; pollution threshold had no sufficient scientific basis yet.

To sum up, before negotiations started at the international level, Switzerland entered a controversial process of formulating a national position and building a consensus on environmental and clean air measures. After the broad support for a constitutional article on the environment, implementation proved to be much more difficult and protracted than anticipated. At the end of the 1970s, however, a political consensus emerged on the environmental protection law, clean air policy and vehicle emissions. The following account of the formulation, negotiation and implementation of Switzerland's national/international clean air policy helps to illustrate that domestic politics matter (Model II).

Switzerland's Role in the LRTAP Negotiations

These early discussions at the domestic level were also reflected in Switzerland's international activities. In the late 1960s and early 1970s, the Swiss federal government pushed for better co-ordination of environmental policy in the framework of the Council of Europe and the OECD. Switzerland was an active participant at the 1972 Stockholm conference on the environment and – as non-member of the UN – pledged a major financial contribution to the UN Environmental Program (UNEP). In 1972, the country joined the UN Economic Commission for Europe (UNECE) as full member and became an active participant in the newly created consultative body for environmental questions (APS, 1972:96). In 1973, Federal Council and parliament also approved UNECE Regulation No. 15 on the Equipment and Components of Light Motor Vehicles in order to establish a legal foundation for national vehicle emission regulations. Under pressure from environmental organisations, but also following positive experiences at the international level, the Federal Council pledged that Switzerland should play a leading role in international environmental protection, an aspiration which was facilitated by early national measures and by favourable economic-structural conditions (little heavy industry etc.) (APS, 1977, p.115). In the negotiations on the Geneva Convention on Long-range Transboundary Air Pollution (LRTAP), Switzerland sided with Austria, Germany and the Scandinavian countries pushing for binding reduction targets for SO_2. This demand was, however, dropped in order to win a majority of UNECE members to sign the framework convention (Böhlen and ClÈmenÁon, 1992).

For Switzerland, the Geneva convention came at a time when - after several expert committees had been formed and drafts of the Environmental Protection Law (USG) had been rejected - the political-administrative actors and the target groups started to find a consensus on clean air strategies. The corresponding programs and measures were not yet legally binding, but they were likely to be put into force soon. Ratification of the Geneva Convention was, however, postponed to 1983, because the Federal Council did not want to overburden the parliamentary debate on the USG by international obligations (BBl, 1982 III 336). At the first meeting of the executive organ, Switzerland continued to push for binding measures, namely the reduction of the sulphur content in fuel oil. The meeting produced no tangible results; the discussion of substantial issues was postponed. Nevertheless, the parties agreed to a joint action plan (NZZ, 15.6.1983).

In 1984, Switzerland joined the '30 per cent Club' demanding, together with Austria, Germany and the Scandinavian countries, a substantial reduction of SO_2 emissions by 1993. The federal government also advocated the introduction of low-lead gasoline and catalytic converters. In 1985 it signed the Helsinki Protocol on the Reduction of Sulphur Dioxide. Since its objectives and measures coincided with the Swiss national strategy, the protocol was ratified immediately and entered into force in 1987. The 'acid rain' debate thus not only accelerated legislation at home, but also resulted in a 'green diplomacy' of the Swiss government with the objective to push technical measures against air pollution in neighbouring countries as well. Together with eleven other UNECE members, the Federal Council supported the so-called Saas-Fee declaration, which pressed for negotiations on reducing nitrogen oxide and hydrocarbon emissions (NZZ, 13.2.1986; Geschäftsbericht, 1987, pp. 34; 112). In 1988 Switzerland signed the UNECE Protocol on NO_X, which,

Table 8.1   National positions and international negotiations

Treaty	Outcome	Preferred regulation	Maximum concession
Geneva Convention 1979	No reduction targets	Substantial reduction targets	See: outcome
SO₂ Protocol 1985	30% reduction from 1980 to 1993	30% reduction from 1980 to the end of the 1980s	See: outcome
NOₓ Protocol 1988	Stabilisation from 1987 to 1994	30% reduction from 1984 to 1994	Official position combined with declaration for 30% reduction from 1984 to 1998
VOC Protocol 1991	30% reduction from 1984 to 1999	30% reduction from 1984 to 1999	30% reduction from 1988 to 1999
SO₂ Protocol 1994	Critical loads Approach with individual targets (52 %)	52% reduction From 1980 to 2000	See: outcome

Source: Kux, Schenkel, and Zeyen, 1997.

however, did not specify reduction targets. In a joint declaration, twelve countries - under the leadership of the Sofia conference's chairman, Swiss federal counsellor, Flavio Cotti - committed themselves to reduce nitrogen oxides by 30 per cent within the next ten years. Protocol and declaration entered into force in 1990.

The Swiss government thus successfully pursued its objective to act as a leading force at the international level. The outcome of the Geneva Convention, the SO_2 and the NO_X protocols did not correspond with the more ambitious objectives of Swiss clean air policy (Table 8.1). In the subsequent negotiations, Switzerland turned however from a pusher into a laggard. In view of growing domestic conflicts over clean air policy and the obvious implementation deficits of national clean air legislation, the Swiss delegation exerted restraint in the negotiations on the VOC protocol. Switzerland signed the protocol in 1990, ratified with a delay in 1993 and started to implement it in the same year.

The switch from the flat reduction rate to the critical loads approach offered Switzerland a new opportunity to play an active role, since it clearly favours progressive countries. The basic principles were defined at a UNECE workshop in Bad Harzburg in 1988. The new approach provided the basis for negotiations on a SO_2 follow-up protocol. None of the parties to the Oslo meeting opposed the critical loads approach, mainly because the cost-effective regional approach takes account of the variations in effects and abatement costs between countries. Switzerland, together with Austria, France, Germany and the Netherlands, pushed for a gradual reduction to the critical load level based on best available technology (BAT), including reductions in emissions, improved energy efficiency, use of renewable energy and reduction of sulphur content in fuel. These countries were willing and able to introduce lower SO_2 and NO_X emission levels for new large combustion plants within one year, and for existing ones within five years of ratification. Given the opposition of Great Britain, Norway, Spain and the USA, the thresholds for existing plants were declared as non-binding guidelines (WWF/Greenpeace, 1993; Fuhrer/Achermann, 1994; Thiemann, 1994).

Domestic Implementation and Compliance Records

In the early 1980s, concerns over 'acid rain' started to mobilise public opinion and to dominate the domestic debate. First newspaper articles appeared in 1980 mobilising an increasingly concerned public (Weltwoche, 'Wer rettet unseren Wald?', 30.1.1980). Subsequent emission measurements and studies confirmed that forest damages also affected Switzerland and paid attention to the critical share of domestic pollution sources.[1] Nevertheless, it took another two years until 'acid rain' started to shape the political debate and resulted in tangible measures. The draft Environmental Protection Law and the follow-up directives took the parliamentary hurdles without any problem. Politicians, cantons, environmental and medical organisations demanded drastic additional steps in order to restore air quality at the level of the 1950s. Technical measures on stationary and mobile emission sources, the intensification of scientific research and data collection as well as international environmental activities (ratification of the Geneva Convention, EMEP protocol etc.) remained largely uncontested. The draft Environmental Protection Law covered most of the provisions of the Geneva Convention. The corresponding strategies and measures were contained in the government guidelines for 1979 to 1983 (BBl, 1980 I 588). The SO_2 protocol was also ratified without problems, at a time, when domestic legislation came into force facilitating compliance with international agreements.

In spring 1984, a large demonstration in favour of forests in Berne emphasised the importance of clean air policy. Reacting to public opinion and more than a hundred parliamentary motions, cantonal initiatives and interventions of environmental organisations, the Federal Council submitted an action plan to protect the forests and reduce air pollution (BBl, 1984 III 1129-1421). Key elements included the accelerated introduction of low-lead gasoline and catalytic converters and the passing of the Clean Air Directive (LRV). After lengthy consultations with political parties, target groups and environmental organisations, the LRV was adopted in 1986. Overall objective was to restore air quality by 1995 at the level of the 1950s. The LRV defined thresholds for 150 substances, most of them stricter than those agreed in the LRTAP protocols (see Table 8.2). Moreover, a series of national research programs on air and forest quality were launched in the early 1980s. Intensive conflicts emerged however over measures affecting road traffic. Environmental organisations demanded stricter traffic restrictions such as speed limits, car-free days, fuel rationing and taxation. Directives on car emissions were passed in 1982 and 1987 establishing the US-83 norms. Thus the emission thresholds promised by the Federal Council in 1974 were introduced with more than ten years delay (Ammann, 1990:123).

The Clean Air Directive (LRV) provided the legal basis for implementing the NO_X protocol. In the second half of the 1980s national clean air policy started to slow down. Scientific evidence was disputed; opposition against measures affecting road traffic increased and first deficiencies in the implementation of national clean air legislation became visible. While measures on stationary sources proceeded relatively well, the implementation of new regulations on road traffic proved to be difficult. Moreover, new scientific studies and evaluations of the Clean Air Directive indicated that the binding measures adopted were not sufficient to achieve the set pollution targets by 1995. In urban agglomerations, critical loads for NO_X and ozone were frequently trespassed. As a consequence, the Federal Council seriously considered taxation measures.

The cantons were unable to submit their action plans on time. In August 1990, with a delay of one year, the first cantonal clean air action plans were submitted, namely from Zurich, Basel-Stadt, Basel-Land, Schaffhausen, Zug and Uri. In September 1991, the WWF filed a complaint against all 26 cantons on the ground that only two-thirds had submitted action plans or parts of them. Some cantons extended the deadline for reductions from 1995 to 2005. In 1992, the Federal Council took position on the complaint and stated that there was no reason to take measures against the cantons. In the same year, a second complaint was launched, this time against the Federal Council itself. Some cities also complained about the slow implementation of the cantonal action plans and the lack of federal support. The cantons, in turn, complained about the lack of federal support, especially after the Federal Council rejected initiatives by the cantons to lower speed limits in the agglomerations (APS, 1988:175). Many measures, such as traffic, emission regulations, gasoline rationing, and financial incentives, were not in the competency of the cantons, but of the federal government (NZZ, 19.5.1990). In urban cantons, the flow of motorised commuters across communal, cantonal and national borders had increased significantly; i.e. the effects of many cantonal clean air measures were counteracted by traffic. Today, most of the nitrogen and ozone thresholds continue to be exceeded by huge margins.

While the Federal Council's incoherent clean air policy repeatedly made headlines, Switzerland tried to pursue a forward strategy at the international level, on the one hand in order to satisfy parliamentary demands, on the other in order to push compatibility of Swiss clean air policy with European measures. While the Swiss government emphasised that it would not make concessions in its national environmental policy, it increasingly adjusted its agenda to that of the EC. It preferred, for instance, the EC standards for measuring emissions for heavy vehicles to the US norms. And the planned taxes for CO_2, SO_2, and VOC was gradually adjusted to those of the EC. The adaptation to weaker EC standards and the growing opposition of domestic actors resulted in a levelling-down of Switzerland's efforts at the national and international level.

In the early 1990s, clean air policy was no longer at the centre of the national discussion on environmental policy, despite the fact that deficits in implementation had become obvious. The targets of the Clean Air Regulation (LRV) could not be achieved within the time set. The UNECE VOC Protocol was ratified with delay and the government was reluctant to press for stricter international agreements, at a time, when the possibilities for technical improvements were largely exhausted and market-oriented instruments had come under strong domestic criticism. In its opinion on the ratification of the VOC protocol, the Federal Council emphasised that critical ozone loads are frequently trespassed in Switzerland. The government also recalled that it played a decisive role in 1986 in Saas-Fee and in 1988 in Sofia. International commitments were thus instrumentalised to put pressure on domestic actors. Subsequently, the federal authorities engaged in an intensive dialogue with the target groups in industry and trade, also in order to evaluate the acceptance of a VOC tax (NZZ, 16.11.1991). In 1992, the revised Clean Air Directive (LRV) entered into force, and Switzerland introduced the most stringent standards for stationary plants in Europe (NZZ, 1.2.1992).

The VOC protocol required a reduction of 30 per cent - based on the year 1984 - by 1999. According to the Clean Air Directive, VOC emissions had to be reduced by 57 per cent by 1995 - based on the year 1984 - in order to reach the level of 1960. Without the revision of the Clean Air Directive, the introduction of a VOC tax and other measures, the national and international reduction targets could not be achieved on time (Botschaft, 1993:9). For the first time, Switzerland ran into serious difficulties with implementing national and international targets. In addition, a study by the Federal Commission for Air Hygiene indicated that NO_X and VOC emissions have to be reduced by 70 to 80 per cent in order to avoid ozone overloads in summer (BUWAL, 1989).

Table 8.2   State of implementation of UNECE and Swiss clean air targets

Pollutants	Swiss targets	UNECE targets
SO₂	Emission level 1950 (-57%): *Achieved* (1980-1995: -57%)	Protocol 1985 requires a reduction of 30% from 1980 to 1993: *achieved*
		Protocol 1994 requires a reduction of 52% from 1980 to 2000: *in progress* (ratified)
NO_X	Emission level 1960 (69%): *Not achieved* (1984-1995: -41%)	Protocol 1988 requires a stabilisation from 1987 to 1994: *achieved*
		Declaration 1988 requires a reduction of 30% from 1984 to 1998: *achieved*
VOC	Emission level 1960 (-57%): *not achieved* (1984-1995: -29%)	Protocol 1991 requires a reduction of 30% from 1984 to 1999: *almost achieved*
Ozone	Concentration limits: *Not achieved*	Combined protocol 1987 (VOCs, NO_X, NH₃): *in progress*

Source: Kux, Schenkel, and Zeyen, 1997.

Switzerland strongly supported research on the critical loads concept and was in favour of a binding agreement on corresponding reductions, since it was compatible with the national thresholds in the LRV. Data collection started early in 1990. The maps showed that 63 per cent of forest soil and 100 per cent of mountain lakes were surpassing critical loads, the Alpine region along the Gotthard route was particularly badly affected.

Switzerland seems to have few problems in implementing the SO_2 protocol, which requires a reduction of 52 per cent between 1980 and 2000, according to the RAINS scenario for a 60-per cent gap closure; i.e. depositions will still exceed the critical loads. The Clean Air Directive envisages 57 per cent, yet it is based on flat rates and does not take into account the sensitivity of ecosystems. In the revised Clean Air Directive (LRV) of 1992, all the necessary national measures have been introduced before ratification of the Oslo SO_2 Protocol. The introduction of a SO_2 tax in the revised Environmental Protection Law, energy legislation and saving programs and current cantonal implementation plans make a 52 per cent reduction of sulphur emissions by the year 2000 feasible (Botschaft, 1993).

To sum up, in Switzerland, clean air problems were defined and policy solutions developed largely independent of international agreements. Until recently, international negotiations did not have any significant impact on national strategies. Switzerland only signed the LRTAP protocols when the national legal basis and implementation programs were already introduced, or at least in preparation. The federal government thus pursued a bottom-up strategy of pre-implementation of international agreements. In the 1980s, technical measures were largely exhausted and legislation on road traffic met strong resistance. The domestic room for manoeuvre became narrower, the strategy of national pre-emption ran into difficulties, especially regarding NO_X and VOC. With the signature of the second sulphur protocol, national strategies, measurement systems and reduction targets also had to be adapted. International agreements started to put national clean air policy under pressure and to shape the domestic agenda.

Consequently, Swiss clean air policy reached a turning point in 1996. Signing and ratification of the Oslo SO_2 Protocol coincided with the domestic discussion on the introduction of taxes on VOC, SO_2 and CO_2. Due to the growing opposition, the general economic recession and budget cuts, the Federal Council changed to a 'new' approach including more self-responsibility, voluntary agreements and economic instruments. The 1992 revision of the Clean Air Directive did not yet reflect the 'new' approach, and has to be seen as a strengthening of the regulative approach. At that time, however, proposals on SO_2, VOC and CO_2 taxation were in preparation or in consultation already. Opposition to economic instruments was still strong. For tactical reasons, SO_2 and VOC taxation was given priority in order to establish the principle, not to internalise all environmental costs. While the

Federal Council emphasised that it would not be willing to level down national environmental policy, it nevertheless started to co-ordinate environmental taxes with the EU in order to avoid competitive disadvantages in production and trade. Accordingly, the introduction of a CO_2 tax was postponed in line with the EU (BBl, 1997 III 410). Thus, the 'Europeanisation' of Swiss clean air policy has had a retarding effect.

The revised Environmental Protection Law (USG) introduced in 1997 includes ecological taxes on VOC and fuel oil 'extra-light' starting in 1998 (NZZ, 13.11.1997). The new law establishes the co-operation principle in the relationship between federal authorities, polluter groups and cantons. Industries, for instance, are given the choice between voluntary efforts or paying ecological taxes. Voluntary agreements should – as far as possible – be transformed into general rules. These changes are seen as being in line with the concept of sustainable development, i.e. internalisation of environmental costs and integration of economic, environmental and social policies.

As Table 8.2 shows, Switzerland achieved both the UNECE and the stricter national targets of SO_2 emissions, i.e. the reduction to the level of 1950, on time. In the case of NO_X and VOCs, there were few difficulties in fulfilling the UNECE targets. Yet the federal government missed the stricter national NO_X and VOC targets by huge margins. Reductions between 50 per cent and 80 per cent would be required to reach the level of 1960. The progressive nature of the Swiss environmental standards is misleading. Implementation of non-technical and transport-based clean air regulations often failed in the face of internal conflicts within the Federal Council and the administration and strong opposition by influential interest groups.

Switzerland's main air pollution problems remain acidification and tropospheric ozone. Even if the national emission targets have been or will be reached, the Swiss ecosystems are still exposed to harmful effects of sulphur depositions. To get below the critical loads, Switzerland would have to introduce more substantial emission reductions, including market-oriented instruments. In this respect, the country could be a frontrunner in two ways: First, in view of its lack of heavy industries it could pursue the 100 per cent gap closure strategy instead of the 60 per cent strategy. Second, Switzerland has a long-standing tradition of including subnational actors such as cantons and organised interests in national decision-making. This experience in subsidiarity could contribute to a more open, transparent and interactive policy formulation and implementation process in environmental protection. The new approach could improve the political understanding of ecological problems and contribute to the attempt to find negotiated, cross-cutting solutions for intra-state transboundary air pollution. The critical loads/levels approach also provides a basis for a combined strategy of controlling more than one impact and pollutant at the same time within spatially limited areas. By fine-tuning this approach in collaboration with regional authorities and polluter groups, the potential for reducing emissions could be increased and the way for sustainable and less-expensive solutions opened. The scientific basis underlying the critical loads approach could thus help to untie the knots in the Swiss clean air policy. Switzerland thus turned from a pusher into a laggard in national and international clean air policy.

Affected Values and Public Opinion

In general, the Swiss public is highly aware of domestic environmental problems, but takes hardly notice of international activities. At the beginning of the 1970s, environmental issues had very high priority in public and politics. 'Acid rain' mobilised large segments of the population that so far paid little attention to environmental issues. The visibility and the symbolic nature of the 'Waldsterben' put clean air policy on top of the agenda. Only a few years later, the energy and economic crisis led to a slight decrease of environmental awareness. Nevertheless, environmental organisations gained more support from the public. The launching of popular initiatives against motorised traffic emissions was probably the most effective action by ecological groups. In the 1980s, with the growing concern over 'acid rain', there was a breakthrough of many controversial proposals made by environmental organisations years ago. Again, at the end of the 1980s, Swiss public opinion was losing interest in environmental issues. Reports on air pollution and dying forests became part of routinised processes and everyday life. There are few protests against high NO_X and ozone loads in the large agglomerations or along the trans-Alpine routes. Environmental concerns have been replaced by other topics such as economic recession, unemployment, or European integration. Resistance against 'over-regulation' increased. Thus, societal demand for national/-

international clean air policies is declining. The main focus in environmental affairs is on climate policy.

Capacities of Intermediate Agents

In the Swiss case, the relevant actor networks are almost exclusively dominated by federal agencies. The formal consultation process is limited to the federal administration. Until the 1990s, the cantonal authorities and other intermediate actors received no information about Switzerland's international activities. In the formulation of national policies and clean air strategies, the following intermediate agents play an important role: parties, cantonal authorities, expert committees, trade and industrial organisations, environmental organisations and automobile associations. Generally, the capability of NGOs depends much more on their capacity to mobilise members, their access to the federal decision-making process and their links to government agencies than on financial resources, manpower and membership size. Regarding the size of membership and staff, WWF Switzerland is the most powerful environmental group. On the other side, the trade and industrial association (Vorort) represents the most important business interests.

In the 1970s, the traditional interest groups and political parties could largely assert themselves. Yet environmental organisations started to become more influential - with the support of centre-left parties, other NGOs and broader circles of society. Among the political parties, a typical left-right-cleavage prevailed at first: The conservative parties emphasised competitiveness, the Social Democrats and the Independents adopted more or less the green positions of the environmental organisations. The results were typical consensus solutions. In 1977, parties with an explicit ecological program took part in local and cantonal elections for the first time in the French speaking part. A year later, a Green Party was created in the Canton of Zurich (NZZ, 14.10.1978). In 1979, the 'Groupement pour la protection de l'environnement' in the Canton Vaud entered the national parliament (APS, 1979, p.201). In general it was assumed that the Green Parties had little chances to succeed, since the environmental organisations still preferred candidates in the ranks of the established parties. Since mainstream politicians started to include ecological issues in their political agenda, the environment gained importance as an election issue (APS, 1978, pp.112, 177).

Among the non-governmental groups, mainly economic and environmental organisations became active in clean air policy. At the beginning of the 1970s, the WWF criticised the federal environmental protection efforts and published its own programs (APS, 1971:119). The professional environmental experts in administration, economy and political parties supported more or less the federal government's policy. With the creation of the Swiss Society for Environmental Protection (SGU), a group consisting of environmental, experts, scientists and organisations, the influence of environmental interests was strengthened substantially (Nüssli, 1987, p.16). In order to broaden public support and to gain influence in the Swiss debate, these organisations founded the Swiss Traffic Club (VCS) in 1979, a mass organisation apart from the traditional Automobile associations (Ammann, 1990, p.92). *The environmental organisations could strengthen their membership basis and their financial resources in the 1970s and 1980s. In the 1990s, they experience stagnation or a slight decline in membership and public attention.* Following the enlargement of the Federal Office for Environmental Protection, environmental organisations could also strengthen their informal influence in the federal administration (Kummer, 1997). *While they exerted substantial influence on the formulation of the national position on the SO₂ and the NOₓ Protocol, their impact was limited in the case of the VOC and the 1994 SO₂ Protocol.*

The economic, trade and automobile associations worked mainly in expert commissions or co-operated closely with federal authorities such as the Federal Office for Police (BAP) and the Federal Office for Foreign Trade (BAWI). In principle, they supported the Federal Council's policy, in public they mainly became active to combat people's initiatives launched by environmental organisations. *They increased their influence on the formulation of Switzerland's international environmental policy in the 1980s and 1990s.*

In the clean air networks, the cantons played a pivotal role. In the 1980s, most cantons acted like 'victims' and supported more or less the position of environmental organisations. In the 1990s, cantonal authorities tended to join business organisations to form coalitions opposing government regulation. To sum up, it can be observed that the influence of intermediate agents was strong and there was a margin of consensus during the preparation of the SO_2 and NO_X Protocols. The impact of environmental organisations declined in the late 1980s and early 1990s. First, following economic recession and organisational difficulties the size of membership and funding declined. Second, these organisations were split over the question whether European integration would jeopardise national clean air policy or not. And third, they began to shift their focus to climate change.

Governmental Capacities

The Swiss governing coalition is very stable due to the so-called 'magic formula', a power-sharing agreement among the four leading parties. The result is substantial continuity in federal policies. On the other side, government performance is based on consociationalism. New policies have to win the support of all federal counsellors; this facilitates compromises but limits innovation. As a consequence, government and parliament pursued more or less the same strategies. After the introduction of the constitutional article, they put considerable emphasis on environmental protection. In the light of the energy and economic crisis, however, the preparatory bodies had to make concessions to economic interests. In the federal administration, the newly-created Federal Office for Environmental Protection had little influence at first. Given their superior resources and political influence, the traditional agencies usually asserted themselves. Only after the reorganisation and enlargement of the environmental office and with increasing criticism of the federal government's clean air policy, could the agency increase its influence on the Environmental Protection Law and the vehicle emission regulations (Knoepfel and Zimmermann, 1991).

In the 1980s, the governmental and non-governmental actors took into account the international conditions much more frequently than in the 1970s. The Federal Council had the aspiration to play a leading role in environmental and vehicle emission regulation at the international level. Accordingly, its international activities increased substantially. At the same time, the environmental office was criticised because Switzerland's clean air policy was much more restrictive than that of other countries. The federal agency was, however, of the opinion that the Swiss measures were based on the experiences in other countries and were co-ordinated with the strategies of the neighbouring countries. Thus, also in this way, an internationalisation of Swiss clean air policy can be observed during this period (Epiney and Knoepfel, 1991).

Governmental actors frequently made reference to international clean air activities. Despite the rejection of the European Economic Area in 1992, trade and market oriented arguments remained of great importance: Business circles demanded a clean air policy in line with EC provisions, while the environmental organisations were split over the question whether European integration would jeopardise national clean air policies or not. There was broad agreement, however, that efforts at the international level should be intensified and that the various interest groups should further co-operate.

Despite the fact, that some of the relevant competencies fall in their realm, the cantons did not take any active part in the preparation of the SO_2, NO_X and VOC protocols. In Switzerland, the signing and ratification of international agreements are not accompanied by a formal consultation process. Ratification has only to be confirmed by the federal parliament. The UNECE protocols passed the democratic bodies without any significant opposition. Despite the fact that international clean air policy addresses not only transboundary air pollution, but also domestic sources, the LRTAP protocols were too weak to provoke any important veto group, including the cantons. The cantons mainly had to focus on the Clean Air Directive, since it contains binding limits for emission and pollution. In the 1980s, the cantonal experts did not have enough time and capacities to take notice of international agreements. They were fully occupied with implementing national and cantonal regulations. National clean air policy started to slacken, scientific evidence was disputed and opposition against restrictions on road traffic increased, especially at the cantonal level. On the one hand, some cantons complained that the national government did not take enough responsibilities itself. On the other hand, they became more and more aware of the necessity to solve clean air problems at the international level. The cantons agreed with the federal government that Switzerland's international activities had to complement and support national and regional measures.

Table 8.3   **Subsidiarity and clean air policy**

Treaty	Degree of centralisation	Victims/polluters within same region	Effectiveness of implementation
Conv.1979	*High*; no cantonal implementation	*Same*; SO_2 emissions in agglomerations	*Effective*
SO_2 1985	*Middle*; fuel oils in federal competence, stationary sources in cantonal competence	*Different*; acid rain in rural and alpine zones; sources in urban zones	*Effective*
NO_X 1988	*High*; traffic emissions in federal competence	*Same*; NO_X emissions in agglomerations	*Effective*
VOC 1991	*Low*; commerce and industrial emissions in cantonal competence	*Same/different*; ozone pollution in agglomerations and rural zones	*Ineffective*
SO_2 1994	*Middle*; sulphur taxation in federal competence, stationary sources in cantonal competence	*Different*; sulphur depositions in alpine zones	*Too early to assess*

Source   Kux, Schenkel, and Zeyen, 1997.

As illustrated, at the beginning progressive cantons such as Basel and Zurich played a pace-setting role. Switzerland's clean air policy was initially formulated and shaped bottom-up. Transboundary and inter-cantonal transmission of air pollution required, however, national and international co-ordination of abatement measures. A growing centralisation of clean air policy can be observed which varies depending on the division of competencies and the spatial distribution of sources and polluted areas (Table 8.3). Only the federal government could regulate, for instance, the lead-content of gasoline, the introduction of catalytic converters or vehicle emission standards. Legislation dealing with stationary sources largely remains in the realm of the cantons. Acid rain and ozone problems are caused by polluters located in other regions than the 'victims', such as alpine and transit zones. NO_x pollution is more or less local within urban regions. Polluters and victims tend to be part of same societal segments. Switzerland thus developed a multi-level approach, in which the federal government negotiated international agreements and passed framework legislation setting reduction targets, while the cantons were free to chose implementation strategies and measures in their action plans (executive federalism). As the difficulties in implementing the Clean Air Directive demonstrate, federal overlaps and conflicts over competencies blocked the process.

Specific problems emerged with VOC. Main polluters are stationary commercial and industrial installations. The regulation of emission standards is mainly delegated to the cantonal authorities. There are, however, significant differences between cantonal efforts. Some cantons are implementing VOC measures, since their clean air action plans have come into force. Others have not yet elaborated action plans or VOC measures have been added only recently. Industrial interest groups rightly point to these asymmetries in cantonal standards and economic attractiveness. The introduction of a uniform VOC tax provides a solution. Yet since taxation is a federal competency, the cantons are losing influence, as is the case with most economic measures.

With the second SO_2 protocol, the political-strategic character of the first protocols shifted to a scientific one. Hence, the critical loads approach demands more information about specific ecosystems and regional pollution situations. Up to now, federal data collection, measurement systems and data standards were compatible with cantonal activities. The development of the critical loads approach, i.e. the definition of new mapping and monitoring standards, could turn out to be as incompatible with the national and cantonal approaches. Cantonal administrations are already concerned about possible consequences. The new approach could result in a redefinition of the national and cantonal clean air targets and duties. The inter-cantonal working group on clean air problems has taken measures to get more information about the international activities. Thus the switch to the critical loads approach requires more co-ordination between the federal and cantonal level. Also the new approach in the revised Environmental Protection Law establishes the principle of partnership and co-operation among all actors concerned. Consequently, national and international clean air measures have significant impacts on the distribution of roles and may result in a new form of co-operative federalism in Switzerland. The future direction and effectiveness of clean air and climate policy will depend on whether a new bargaining system involving all the political actors and polluter groups emerges (Kux, 1995; Schenkel, Kux and Marek, 1997).

The Scientific Community and Learning Processes

Despite the lack of systematic surveys and research programs in the 1970s, science had a great influence on Swiss clean air policy. First, the experts in the preparatory commissions and the Federal Commission for Air Hygiene shaped the legislative process and the 'guideline policy' of the federal government, the latter in co-operation with the Swiss Norm Association (SNV). Second, university institutes, governmental research centres and cantonal and communal experts produced much noted studies on the extent of air pollution. While the programs suggested by these experts still had largely non-binding character, the number of cities, cantons, politicians and interest groups pressing for transferring guidelines into binding decrees increased. In the 1980s, the cross-cutting function of clean air policy gained in importance. 'Acid rain' forced the Federal Council and parliament to initiate comprehensive analyses, concepts and programs, which linked environmental protection policy with agriculture, energy, traffic, planning and other policies.

In the early years, Swiss clean air policy was a trial-and-error process. Each city, each canton formulated its own strategy. There was a competition between rules and regulatory approaches. An additional source of learning was the increasing contacts between experts and government representatives that facilitated the diffusion of new ideas and experiences among the participating countries.

The consensual knowledge and the influence of scientific circles increased, while the incorporation of suggested new rules and norms as well as the degree of compliance decreased. At first sight, this seems to be a contradiction. Yet there are two explanations: First, environmental policies in neighbouring states and at the international level have been pushed forward, while Swiss environment policy stagnated. Second, through the change of the decision basis from the flat rate reduction to the critical loads/levels approach, the international system of measurement and thresholds are not anymore adaptable to Switzerland to the same extent as in the case of the protocols of the first generation. Switzerland will have to revise its clean air strategies. Neither the revised Clean Air Directive nor the new Environmental Protection Law takes into account the sensitivity of the Alpine ecosystems.

Conclusions

Switzerland is neither a country with a very high vulnerability nor a country with an extreme negative (or positive) pollution exchange balance. However, the results of the country study tend to confirm the unitary rational actor model. Vulnerability was high in the case of SO_2, NO_X and VOC, earlier scientific calculations suggested negative pollution exchanges, and abatement costs tended to be low in relation to most other parties to the convention. Looking at the national level, the abatement costs expressed in quantitative terms have increased during the 1980s, while Switzerland's activities at the international level declined. Scientific studies also showed that air pollution is to a large extent home-made. Pushing forward international agreements, the Swiss could justify national measures and shift the burden also to its main competitors.

In the case of Switzerland, domestic politics (Model II) played an important role shaping national and international clean air policy. Societal demand varied substantially over time, while the government's strategy to supply clean air policies was based on a long-term approach solidly based on scientific evidence, consociationalism and cost-benefit-distribution. Given this cautious, long-term approach, in the context of the particularities of the Swiss political system, it is difficult to demonstrate a causal relationship between changing demand and supply. In the case of air pollution, the spatial and societal distribution of perceived damage and abatement costs is extremely complex. There is, however, a close relationship between changing public opinion, governmental capacities, and the emergence of intermediate actors and abatement measures at the national and international level. In this case, the robustness of the Swiss political system is an advantage, even if it proved to be very slow in adapting to new challenges and changing circumstances. The aspiration to build consensus and to find equitable solutions along horizontal and vertical lines renders spectacular moves and leadership at the international level difficult, as Switzerland's declining role in the LRTAP negotiations demonstrates. At the same time, it facilitates solid, sustainable and long-term strategies.

Social learning and policy diffusion (Model III) played an important role in Switzerland's clean air policy. The openness of the political system, the fragmentation of political organisation along horizontal and vertical lines, and societal mobilisation facilitate the diffusion of new ideas and scientific knowledge. Yet as illustrated, social learning, policy diffusion and behavioural change are slow, protracted processes in a system oriented toward consensus and stability. The growing understanding of the air pollution problem, the strategic use of scientific knowledge, and the slow emergence of a transnational epistemic community had a certain impact on Swiss abatement strategies. As demonstrated in the case of the switch to the critical loads approach and the new strategy in the revised Environmental Protection Law, adaptation takes time, especially if it touches standard operational procedures and the very bases of the established political system such as federalism and consociationalism.

Note

1. In 1984, 34 per cent of the forests showed substantial damage, 42 per cent in the Alpine region (NZZ, 14.1.1985).

References

Ammann, D. (1990), *Die Autoabgabe-Policy der Schweiz. Eine Vollzugsstudie*, Kleine Studien zur Politischen Wissenschaft, No 267-268, Zürich: Forschungsstelle für Politische Wissenschaft.

Annèe politique suisse (APS) (1960ff.), published by Forschungszentrum für Schweizerische Politik, Bern.

Böhlen, B. and ClÈmenÁon, R. (1992), 'Die internationale Umweltschutzpolitik der Schweiz', in: Riklin, Alois; Haug, Hans, and Probst, Raymond (eds.), *Neues*

Handbuch der schweizerischen Aussenpolitik. Bern: Haupt, pp.1015-1026.

Botschaft zu einer Änderung des Bundesgesetzes über den Umweltschutz (USG) vom 7, Juni 1993.

BUWAL (1989), *Ozon in der Schweiz. Status-Bericht der Eidg. Kommission für Lufthygiene* (Schriftenreihe Umwelt, No. 101), Bern: BUWAL.

Epiney, A. and Knoepfel, P. (1991), *Die schweizerische Umweltpolitik vor dem Hintergrund der europäischen Herausforderung* (Cahiers de l'IDHEAP, No.77).

Fuhrer, J. and Achermann, B. (1994), *Critical Levels for Ozone,* A UNECE Workshop Report (Schriftenreihe der FAC Liebefeld, No. 16), Bern: EDMZ.

Giger, A. (1981), Umweltschutzorganisationen und Umweltpolitik, in *Schweizerisches Jahrbuch für Politische Wissenschaft* 21/1981, pp.49-65.

Klöti, U. (1993), 'Kapitel III: Verkehr, Energie und Umwelt - Die Infrastruktur und ihre Begrenzung', I.

in *Handbuch Politisches System der Schweiz,* Band 4, Politikbereiche, pp.225-300.

Knoepfel, P. and Zimmermann, W. (1991), *Evaluation des BUWAL. Expertenbericht zur Evaluation der Luftreinhaltung, des ländlichen Gewässerschutzes und der UVP des Bundes,* Schlussbericht vom 18. November 1991.

Kummer, L. (1997), *Erfolgschancen der Umweltbewegung. Eine empirische Untersuchung anhand von kantonalen politischen Entscheidungsprozessen,* Bern/Stuttgart/Wien: Haupt.

Kux, S. (1995), 'Internationaler Rahmen und innerstaatliche Abst,tzung der schweizerischen Umweltaussenpolitik', in M. Jochimsen und G. Kirchgässer (Hrsg.), *Schweizerische Umweltpolitik im internationalen Kontext,* Basel: Birkhäuser, pp.140-172.

Kux, S; Schenkel, W. and Zeyen, P. (1997), Die Schweizerische Luftreinhaltepolitik im internationalen Umfeld, EC Contract EVSV-CT 930.85, Basel: mimeo.

Neue Zürcher Zeitung (NZZ).

Nüssli, K. (1987), Neokorporatismus in der Schweiz. Chancen und Grenzen organisierter Interessenvermittlung, Umweltpolitik. Zürich: FPW (Kleine Studie, No. 251).

Rausch, H. (1992), *Studien zum Umweltrecht,* Zürich: Schulthess.

Schenkel, W.; Kux, S. and Marek, D. (1997), 'Subsidiarität und Luftreinhaltung in der Schweiz – aus internationaler, nationaler und lokaler Perspektive', *Swiss Political Science Review,* 3, pp.229-256.

Schwager, S. (1988), 'Das schweizerische Umweltrecht im Lichte der Umweltschutzbestimmungen der Europäischen Gemeinschaften – ein Rechtsvergleich', in S. Schwager; P. Knoepfel, and H. Weidner, (1988): *Umweltrecht Schweiz - EG* (Ökologie und Gesellschaft, vol.1), Basel: Helbing and Lichtenhahn, pp.7-78.

Schweizerischer Bundesrat (Geschäftsbericht) (various years): *Bericht des Bundesrates an die Bundesversammlung über seine Geschäftsführung im Jahre ...* Bern: EDMZ.

Tages-Anzeiger (TA).

Thiemann, A. (1994). *Die Entwicklung der ECE und ihrer Prinzipien im Umweltbereich,* Münster/Hamburg: Lit.

Widmer, T. (1991), *Evaluation von Massnahmen zur Luftreinhaltepolitik in der Schweiz,* Zürich: Rüegger.

WWF/Greenpeace (1993), 'Meeting Critical Loads of Sulphur. A Study of Energy Use and Emission Reduction Strategies in Europe'.

9 What's this got to do with me? France and Transboundary Air Pollution

JIM SKEA AND CAROLINE DU MONTEUIL

Introduction

France has participated fully in all of the UNECE agreements on transboundary air pollution. However, it has not played a leading role. In the early days when the primary focus was on sulphur, the agreements appeared largely irrelevant to French interests. France participated as a 'good international citizen', but had little other motivation. The reasons for this lack of active engagement were threefold. First, there was a perception that France suffered little damage from transboundary air pollution. Second, the structure of French energy supply and, in particular, an increasing reliance on nuclear power during the period when the early agreements were being negotiated, meant that France had to take no special action to meet its international obligations. Third, public opinion in France was little engaged with the issue of transboundary air pollution. In contrast to other European countries, this dimension is almost completely absent from the French story. No environmental assets of cultural significance appeared to be threatened, there was comparatively little press attention and the French policy-making system was not transparent to the wider public. Transboundary pollution did not become to any degree a politicised topic. The tenor of French policy-making in this area was unaffected by significant political changes between 1980 and the late 1990s which encompassed socialism, conservatism and two different varieties of 'cohabitation'. To put it crudely, international agreements on sulphur were a non-event in France.

As the UNECE regime progressed, and as EC Directives began to come forward, French interests became more strongly engaged. Although it was easy to comply with the sulphur regimes, the NO_X and VOC agreements have proved more problematic. In this context, agreements on transboundary air pollution began to generate some domestic controversy. Tensions between international measures and domestic pollution control procedures on the one hand, and the economic impacts of certain international measures on key industrial interests tied closely to the French state on the other, have acted as domestic barriers to acceptance of the need for wider abatement policies. However, in one case, these same industrial interests prompted France to take a pro-active role in pushing for a more ambitious approach to sulphur control within the EC.

France has signed and approved each protocol agreement under the Geneva Convention. However, implementation in relation to the NO_X and VOC agreements has been problematic.

This chapter examines the relationship between French policy on acid rain and relevant international agreements. It focuses on the Geneva Long-Range Transboundary Air Pollution (LRTAP) Convention, associated protocols and various EC Directives. The chapter covers the period from the early 1970s to the early 1990s.

French Perceptions of Transboundary Air Pollution

Emissions and Deposition

French perceptions of transboundary air pollution are strongly conditioned by its unusual pattern of emissions. There were substantial changes during the 1980s. French SO_2 emissions rose from under 1,500 kilotonnes in 1960 to peak at nearly 4,000 kilotonnes in 1973. Then, emissions fell to 3,338 kilotonnes in 1980 and to only 989 kilotonnes in 1990 (Table 9.1). Unusually, electricity generation is not the largest source of SO_2 emissions in France. It accounted for only 26 per cent of SO_2 emissions in 1990 (Table 9.2), far less than in most other industrialised countries. The largest source of SO_2 emissions was industrial combustion (39 per cent). Oil, rather than coal, now makes the largest contribution to SO_2 emissions in France. The issue of SO_2 emissions from the petroleum chain has recently caused some problems for France and has influenced its international position.

Table 9.1 shows that NO$_X$ emissions have shown less of an obvious trend. Transport accounted for 65 per cent of NO$_X$ emissions in 1990 (Table 9.2) while electricity generation accounted for 7 per cent and industrial combustion 10 per cent. The increasing proportion of the vehicle stock fitted with catalytic converters tends to drive emissions down, while traffic growth is working in the opposite direction. Transport policy and technical measures to curb vehicle emissions are at the heart of France's policy on NO$_X$.

Estimates of emissions of non-methane volatile organic compounds (VOCs) are much less reliable. Table 9.2 shows that 41 per cent of VOCs emissions derive from transport, but 22 per cent comes from solvent use (for example paint application), 8 per cent from combustion in the residential and commercial sectors and 22 per cent from natural sources. VOC policy is targeted, as with NO$_X$, at road transport but also at industrial uses of solvents.

Table 9.1   Trends in French emissions of transboundary air pollutants (kilotonnes)

	1980	1985	1990	1995
Sulphur dioxide (as SO_2)	3338	1470	1298	989
Nitrogen oxides (as N_2O)	1823	1615	1585	1666
VOCs	2,150	2,874	2406	2620

Note: There have been methodological revisions to NO$_X$ and VOC data which means that 1995 data is provisional and not comparable to previous years

Source: UNECE EMEP Programme

Tables 9.3 and 9.4 show France's position with respect to 'trade' in transboundary air pollutants. Table 9.3 refers to SO_2 in 1993-1994 while Table 9.4 assesses the NO$_X$ situation. France is a net exporter of sulphur but about 28 per cent of emissions are deposited in the North Sea, the Atlantic and the Mediterranean. About 32 per cent of French emissions are deposited in France itself. Germany is the biggest single 'victim' of French sulphur exports, receiving about 11 per cent of French emissions. French emissions account for 37 per cent of domestic deposition - but Germany, Spain (being 'upwind'), the UK, Italy and Belgium are also significant sources of deposition. France is a significant net exporter of sulphur to Germany but 'imports' from Spain and the UK.

Table 9.4 shows a similar pattern for trade in nitrogen. However, in this case, Germany is, in relative terms, a more significant contributor to deposition in France, while a greater proportion of French emissions is exported.

Table 9.2   **Emissions of transboundary air pollutants by sector 1990** (kilotonnes)

	SO_2	NO_X	VOCs
Electricity	343.7	105.9	1
Residential, commercial combustion	116.2	88.6	214
Industrial combustion	514.2	165.0	7
Production processes	110.9	30.8	100
Extraction and distribution of fossil fuels	23.8	3.5	123
Solvent use	0.0	0.0	636
Road transport	145.3	1037.8	1170
Other mobile sources and machinery	24.6	129.1	122
Waste treatment and disposal	19.2	23.9	21
Agriculture	0.0	0.0	10
Natural Sources	2.5	5.5	462
TOTAL	1300.4	1590.1	2866

Source: CORINAIR

Table 9.3   **French exchanges of sulphur in 1993/94** (kilotonnes of sulphur)

	Deposited in France from:	Foreign depositions attributable to French emissions	Balance
France	133	133	0
Spain	45	7	+38
Germany	28	45	-17
UK	28	10	+18
Italy	17	14	+3
Belgium	13	11	+2
North Sea	4	33	-29
Atlantic	4	36	-32
Mediterranean	-	47	-47
SO_2, other countries	31	73	-42
Indeterminate sources	59	n/a	+59
TOTAL	362	409	-47

Source: UNECE EMEP Programme

Table 9.4   **French exchanges of nitrogen in 1993/94** (kilotonnes of nitrogen)

	Deposited in France from:	Foreign depositions attributable to French emissions	Balance
France	72	72	0
Spain	18	7	+11
Germany	27	37	-10
UK	20	8	+12
Italy	13	15	-2
Belgium	8	5	+3
North Sea	3	22	-19
Atlantic	4	34	-30
Mediterranean	-	21	-21
NO_X, other countries	19	96	-77
Indeterminate sources	26	n/a	+26
TOTAL	210	317	-107

Source: UNECE EMEP Programme

Environmental Damage

During the 1970s, understanding of the effects of long range pollution on forests and other ecosystems was confined to a small section of the academic community, and one with little prestige. In 1974, Professor François Ramade published a book showing that French forests were being damaged by acid rain. He was labelled a 'catastrophist' by the head of the Institut National de la Recherche Agronomique (INRA). INRA had a laboratory specialising in forest damage, but it examined only localised pollution effects such as caused by aluminium production in the Valle de la Maurienne.

The publication of a German article on forest death in Der Spiegel in 1981 was followed by several French articles on air pollution and forests. Nevertheless, the reaction in France to the news of forest damage in Germany in 1981-1982 was one of scepticism. The official body responsible for French forests, the Office National de la Forêt (ONF), gave no credibility to the reports and so no research was carried out. The geographical situation of the country and the direction of prevailing winds led people to believe that there was no risk. As a result of its nuclear programme, France was believed to be in a strong position regarding sulphur pollution.

Information about forest damage in France spread from Alsace via groups such as the Club Vosgean which organises walks through the forests. By 1983, the authorities themselves began to note forest damage in the Vosges. As part of the EMEP Programme, from the beginning of 1984, France began monitoring atmospheric pollution at 2000 sites throughout the country. The results revealed that air pollution occurred far from major sources. 1984 saw the height of media attention, with articles in the regional and national press, influenced by the *Der Spiegel* article on forest dieback in the Black Forest. In February 1984, the Forestry Department published a special acid rain issue of one of its journals. In April, members of Greenpeace climbed the chimney of a refinery at Le Havre and demanded more active measures on SO_2 reduction.

In 1984, a research programme, 'Forest damage attributed to atmospheric pollution' (DEFORPA) was set up. DEFORPA ran from 1984 to 1991. In February 1985, ONF reported that 21.5 per cent of coniferous trees and 4 per cent of deciduous trees in 300,000 hectares of public forest in Alsace and Lorraine had badly damaged leaves, and that 19.5 per cent of coniferous and 1.5 per cent of deciduous trees were damaged out of 100,000 hectares of forest in the Franche-Compté.

The final report on the DEFORPA research produced a more nuanced account of the causes of tree damage (Barthod et al., 1993). It concluded that ozone was the air-borne pollutant most likely to cause damage. The doubling of ozone concentrations over a century in the lower levels of the atmosphere, and in particular, the high concentrations to be found in the Vosges and the South-East of France, merited constant vigilance, the report concluded. Thus by the time the first sulphur protocol was signed, forest damage had been admitted, and some understanding of the causes was emerging.

The proportion of French conifers affected declined in the period 1986-1992 while the additional area of broadleaf trees affected grew, but to a lesser extent than elsewhere. Forest damage in France's cross-channel neighbour, the UK, has grown far more quickly. French forests are affected to a far greater degree by forest fires than are those of Germany, Sweden and the UK. These striking differences in patterns of forest damage provide a plausible partial explanation for a lower French concern, in relative terms, about ecological damage linked to transboundary air pollution.

Domestic Policy Prior to International Negotiations

The Policy Framework for Air Pollution

Air pollution in France is the responsibility of the national government. The lead responsibility lies with the Ministry of the Environment which was established in 1971. Environmental programmes on air are dealt with in the Bureau of Atmosphere and Energy, which comes under the Industrial Environment Service, which in turn is located in the Division of the Direction de l'Eau, de la Prévention des Pollutions et des Risques. The Ministry is supported by the Agence de l'Environnement et de la Maitrisse de l'Energie (ADEME) which assists the government in environmental policy formation and technological development. ADEME's work is guided by three Ministries: Industry; Environment; and Research. These arrangements indicate the extent to which air pollution is institutionally linked to industry, risk elimination and energy use.

Many ministries other than the Ministry of the Environment, notably the Ministry of Health, have some responsibilities for environmental matters. In order to deal with specific environmental matters, inter-ministerial committees are set up to co-ordinate action. In addition, other agencies, and members of special interest groups may be brought in to form *ad hoc* committees. These 'multipartite' committees draw in a variety of interests in an attempt to build consensus.

The Legal Framework

Legislation covering fixed sources   Legislative control on factory emissions has long existed in France because of concerns for public health at the local level. In 1917 a Law on Classified Installations was passed to control emissions from fixed facilities. This law controls the operation of all factories, workshops or other installations which might create danger or nuisance to health, safety, amenities, agriculture, the environment and sites and monuments. A Décret of 1953 defines these dangerous, noisy or noxious facilities. They are divided into three classes. Classes 1 and 2 require authorisation from the Préfet while Class 3 facilities are subject only to 'declarations'.

The 1917 Law was upgraded through the 1976 Environmental Protection Law. This was still motivated by concerns for public health and it made no provision for national emission targets. Transboundary air pollution was not a consideration. The 1976 Law incorporated the concept *of best available technology within reasonable economic costs*, thus enabling the 1984 EC Air Framework Directive to be transposed fairly readily into French law.

Authorisation for classified installations is given by the departmental Préfet. A request for authorisation must specify: the site; the nature and volume of activity; and the manufacturing processes and materials used. A detailed study of treatment of emissions and discharges, removal of waste, transport of materials and products and emergency plans must be provided. If granted by the Préfet, the authorisation states emission limit values for the most important pollutants. Maximum limit values are defined at the national level through technical instructions referring to specific certain types of installation, such as combustion plants and household waste incinerators.

Other legislation Air quality legislation was passed in 1961 and upgraded in 1974 through a Law which integrated legislation on air and water pollution and established the national framework for setting emission limits. The 1974 Law requires that emission controls on fixed and mobile sources should take account of air quality standards. This required the setting up air quality monitoring networks. It also extended the 1948 Law on energy efficiency and included the concept of pollution alerts.

Following the 1961 Law, Zones of Special Protection (ZPS) were designated in 1964. In such zones, the use of lower sulphur fuels is required. The City of Paris was declared a ZPS in 1964 and this Zone was enlarged in 1978 to include the outskirts. Lyon and Lille were designated ZPS in 1974 and Marseille in 1981. In addition, Zones of Alert (ZA) were created. In these, the main emitters of pollutants are obliged to reduce their emissions significantly the moment an alert on air quality is declared. They can do this either by reducing levels of activity or by using lower sulphur fuels.

France has also adopted a number of other measures relating to emission from fixed and mobile sources. A 1985 Decree, for example, controls the level of lead in petrol. Decrees have been used to implement other requirements originating in EC Directives. A programme to combat emissions of VOCs has been initiated. This aims to reverse the rise in emissions and to reduce them by at least 30 per cent over a period of fifteen years.

Although SO_2 emissions are decreasing, NO_x emissions are on the rise. Consequently, the Ministry of the Environment is carrying out research into traffic control systems which would limit NO_x emissions. The Minister of the Environment is currently developing plans to rewrite French air pollution laws through a mixture of statutory and fiscal incentives to curb the use of cars, improve public transport and raise health standards.[1] Frequent meetings between the Environment Minister and interested parties, including scientists, doctors, representatives from the car industry and cyclists, the air monitoring associations and the environmental NGOs, have been held in developing this law.

This process of wide public consultation goes against long standing traditions that public servants can act behind closed doors to take decisions which will promote the broader public interest. In 1993, for example, the Ministry of the Environment argued strongly against publishing a report (ERPURS, 1994) linking health problems to peaks of pollution in the Paris area.[2] This tradition partly explains the weakness of NGOs within the French system.

Fiscal Measures

Unlike most other EC countries, France has used fiscal instruments relating to atmospheric emissions for some time. The main role of these taxes is to raise revenue for pollution monitoring and prevention. The *Parafiscal Tax on Atmospheric Pollution* was set up in 1985 for a period of five years and was subsequently extended in 1991. This applied initially to 480 large combustion installations. The tax was fixed at FFr130 per tonne of SO_2 emitted, and initially raised about FFr100m a year. The tax has subsequently been extended to take into account additional pollutants including H_2S, NO_X and HCl. The tax now covers about 870 installations, including medium-sized combustion installations and household waste incinerators. The tax has now been raised to 150FFr per tonne of pollutant, raising about FFr 190m in revenue.

The tax is managed by an ad hoc inter-ministerial committee, consisting of representatives from the Ministry of the Environment, the Ministry of Energy (DGEMP), the Ministry of Industry, the Ministry for Research, the Ministry of the Interior (in its role of managing local authorities), the Ministry of the Budget and local authority representatives.[3] The revenue generated by the tax is used by ADEME for the development of pollution prevention techniques and monitoring. ADEME prepares documents for key committees which gives it some agenda-setting powers.

Implementing Air Quality Policy

Enforcement of pollution control regulations takes two forms. The licensing of Classified Installations under the 1976 Law is carried out at the local level under guidelines established in Ministerial Circulaires from central government. Once licensed, factories must continuously monitor their own emissions as required under the 1988 Circulaire. Compliance is checked through periodic monitoring carried out by organisations approved by the Inspectorate of Classified Installations. Contravention of such a licence can be dealt with by the courts. Sanctions can be penal – in the form of fines, imprisonment and closure – or administrative, for example consigning a sum of money for work which will bring the installation up to standard or the public authorities carrying out the work at the owner's expense.

The regulation of pollution from small sources generally takes the form of fuel restrictions or pollution control requirements on motor vehicles. The efficacy of these controls is monitored both to deal with emergency situations and as an input to policy analyses.

While the Departmental Préfets are formally required to implement air pollution legislation, day-to-day control is exercised by the Directions Régionales de l'Industrie, de la Recherche et de l'Environnement (DRIREs). These are regional outposts of the Ministry of Industry whose primary role is to implement the policy of regional industrial development. The DRIREs constitute a sort of 'technical magistrature' (Knoppfel and Larrue, 1985). They are part of the Police Commissariat which gives them the equivalent of police powers in matters of monitoring air pollution. The actual measurement of pollution is done by a central police laboratory which collaborates with the local authorities and the DRIRE.

There is, however, no system of monitoring the performance of these officials. If pollution levels exceed the norms, they negotiate with the owner of the offending factory. Owners often argue that, if they are penalised for pollution, they may have to close the factory down with the consequent loss of jobs. This form of pressure has been highly persuasive in France, not merely because of the levels of unemployment over the last decade, but because sympathy for the unemployed is a highly accepted social value. As a result of such pressures, a 'relaxed' regime operates in some areas. A further problem is quite simply the number of inspectors. There are currently only 700 covering 50,000 installations.[4]

Air Pollution Monitoring

Monitoring stations have existed since 1850 in France. Industrial operators are obliged to monitor their own emissions and local air quality. Since this was expensive, it often proved to be more economical to do so collaboratively. As a result of their industrial origins, many French monitoring networks were established in urban and industrial areas.

The monitoring networks are run by Associations regulated by a 1901 Law which treats them as non-profit making organisations. The system of governance of these monitoring stations is uniformly believed in France to ensure transparency and credibility. If central government agencies, departmental authorities or industry were to make measurements independently, it might be assumed that special interests could have distorted the information. At the national level the stations are associated with three agencies: the Ministry of the Environment, ADEME and the Laboratoire Central de Surveillance de la Qualité de l'Air (LCSQA).

Under the demands of EC legislation, the existing networks have been extended, co-ordinated and modernised, with the first automatic and computer controlled network set up in Paris in 1982. Networks are now being extended into the rural areas. During 1988 and 1989, new strategies and methods were developed for monitoring large towns with help from EC funds. A 1990 Decree allowed the parafiscal environmental taxes to be used to develop the networks from 1992 onwards. Between 1989 and 1993 the Paris network was modernised to monitor not only SO_2 from industry but also traffic.

The national system for monitoring background pollution in France, the MERA (MEsure des Retombées Atmosphériques) Programme was set up in 1990 in response to the Geneva Convention with the specific mission of understanding the phenomenon of acid rain. It consists of ten stations spread over France, collecting atmospheric samples. Three additional stations have been added from the international BAPMON (Background Air Pollution MOnitoring Network) managed by the Meteorological Office.

The many different monitoring stations, which were set up at different times for different purposes, often use different measurement techniques and protocols. ADEME has the task of harmonising methods and results to create a truly national system.

France and the LRTAP and EC Agreements

France participated in OECD work on transboundary air pollution in the 1970s but did not play a driving role. France took a relaxed position during the negotiations leading to the signature of the 1979 LRTAP Convention because of perceptions about the limited extent of ecological damage in France and the limited impact of control policies on the French economy.

As of the end of the 1970s, French administrators had little understanding of transboundary pollution. Even if it existed, France was not implicated either as a receiver or a sender. Geographically, France was felt to be well placed and relatively unaffected. France had no important forests at risk and forests were in any case not considered to be of great economic or social importance. As far as sulphur was concerned, French emissions were low compared to those in many other countries, a situation which would only improve as the nuclear programme built up. France also had a well-established system of control over air pollution and the beginnings of an air quality monitoring programme.

France therefore signed without difficulty the UNECE Convention on Long Range Transboundary Pollution in 1979 and ratified it in 1981.

The Sulphur Story

The sulphur story is structured round four themes: the first UNECE sulphur protocol; the negotiation of the EC Large Combustion Plant Directive (LCPD); growing French interest in petroleum-related emissions; and the second sulphur protocol. French participation in the UNECE sulphur protocols has been completely uncontroversial and signing up to aggregated national emission targets has had no domestic consequences. On the other hand, the more technically focused EC measures have raised questions about compatibility between Directives and French domestic procedures. Here, more discussion is warranted.

The first sulphur protocol 1985   French perceptions of ecological damage associated with acid rain were very limited at the beginning of the 1980s. At the time France began to consider the transportation of pollution across frontiers, sulphur was perceived to be the main problem. However, it was thought to come largely from heavy industry in Eastern Europe, especially Poland. France did not have any major areas in which bodies of water were threatened by acidification.

In spite of domestic reservations about the science of acid rain, France was well able to participate in international agreements relating to SO_2 emissions. During the period of negotiations of the first sulphur protocol, France felt so confident in her ability to reduce emissions that it not only agreed to reduce by the standard 30 per cent, but committed itself in 1984 to a 50 per cent reduction between 1980 and 1990. Emissions fell by 60 per cent from 3338 kilotonnes in 1980 to 1,298 kilotonnes in 1990. Emissions fell further to 989 kilotonnes in 1995. The protocol was simply of no consequence at the domestic level. France signed the First Sulphur Protocol in July 1985 and accepted it in April 1986.

The large combustion plant directive, 1988 The LCPD, proposed by the European Commission in 1983 and based on domestic German legislation, contained two components. The first contained emission limits for new combustion plants including conventional fossil fuel-fired power stations and larger industrial installations. The second component set national aggregate emission quotas for all existing large combustion plant. Unlike the UK, France had little difficulty with the second part of the package: investment in nuclear power had, by 1983, already brought emissions from existing power stations down dramatically. The first component, less controversial in most countries, caused France greater difficulties.

In negotiating the Large Combustion Plant Directive, France objected because it did not want combustion norms for new plants imposed by Europe since these are traditionally negotiated by regional officials. France had set up a system of regulation of emissions from large plants many years ago. The limits for these were negotiated locally under the authority of the Départmental Préfets as discussed above. French administrative practice was to allow local decision-making, under the guidance of 'Circulaires'. A Circulaire does not have the power of compulsion – it is up to a Préfet to decide whether to put it into practice, but no-one has the right to do things which contradict it.

In 1985, France was taken to task by the European Commission for refusing to enact legislation to transpose Directives, especially those containing guide and limit values for air quality. Circulaires rather than Décrets were used since, in the mind of the French administration, these had sufficient force and did not upset traditional practice. There was also a commonly held view that if specific numerical limits are given for pollutants, this has the force of giving permission to pollute – and people will emit up to these limits. It was not until 1991 that France capitulated and published a Décret which incorporated EC Directives.

Nevertheless, France consented to the LCPD when it was finally agreed on a unanimous basis by the Council of EC Environment Ministers in June 1988. Although France expected to meet the emission quotas for existing plants established in the LCPD without taking active abatement measures, some investment in flue gas desulphurisation has been required in order to compensate for the rising output from coal-fired plants used to substitute for nuclear plants affected by drought.

Sulphur and the petroleum chain One specific set of EC negotiations on the sulphur content of liquid fuels provides a unique example of France operating pro-actively in the international air pollution domain. Strikingly, the motivation was not to promote higher environmental standards but to protect the French petroleum refining industry against what it saw as imbalanced competition.

The problem for France lay in the specificities of French fuel consumption. A larger proportion of French oil refinery output goes to middle distillates, such as gas oil and diesel, compared to other European countries. This is partly caused by lower taxes on diesel relative to petrol and partly due to restricted markets for heavy fuel oil because of the heavy reliance on nuclear power in electricity generation.

France has been differentially affected by a series of EC Directives restricting the sulphur content of gas oil and diesel, and felt that these had put it at a competitive disadvantage with respect to countries such as the Netherlands. The Dutch can use international bunker (shipping) fuel as a dump for sulphur. The original Directive on gas oil (75/716) dates from 1975 and was motivated primarily by the desire to remove barriers to trade. It has been tightened on several occasions, most recently in 1993.

The 1993 Directive (93/12) specified a limit of 0.2 per cent on the sulphur content of gas oil and diesel beginning in October 1994. At the time, the average French sulphur content of gas oil was 0.34 per cent. Being aware of plans for this Directive and the adverse effects on its domestic refining industry, France proposed a broader Memorandum on sulphur to the Commission in 1990.[5] This Memorandum tackled two concerns: that existing measures to control SO_2 would only have a limited effect as the Large Combustion Plant Directive only applies to about half of total sulphur emissions; and that measures affecting only some fuels could cause major distortions in competition.

The Memorandum recommended lower sulphur limits for fuel oil and setting SO_2 emission limits for refineries and for combustion plants burning heavy fuel. By limiting sulphur content throughout the petroleum chain, France hoped to level up the playing field. This idea was said by the Directorate General de l'Energie et des Matières Premières (DGEMP) to have originated with them, but in discussion with the Ministry of the Environment, with the final decision in the hands of the Prime Minister. The Memorandum has influenced subsequent sulphur policy in the European Commission. French oil companies backed the 'French

Memorandum' as, initially, did the French subsidiaries of multinational oil companies.

The Netherlands strongly objected to the Memorandum as it would have severely altered the pattern of bunker fuel supply at Rotterdam. In 1992, the European Commission commissioned a report from Arthur D Little in order to investigate the economics of various options for restricting sulphur emissions from the European petroleum chain. As a result of this report, the EC has, after considerable debate, agreed a 'Liquid Fuels Sulphur Directive' which represents a compromise between the French and Dutch positions. This episode provides an interesting example of how France had come to recognise how its domestic interests were affected by international air pollution policies. However, the interests which it sought to protect were economic rather than environmental.

The second sulphur protocol 1994 Like the first sulphur protocol, the second UNECE sulphur protocol caused few problems for France. This Protocol, based in part on the idea of critical loads for deposition of sulphur, commits the 32 signatories to further reductions from a 1980 base by the years 2000, 2005 and 2010. France committed itself to reduce emissions by 78 per cent by 2010 from 1980 levels. Already by 1995, sulphur emissions had been reduced by 70 per cent.

Effectively, France has undertaken to reduce sulphur emissions by 26 per cent below 1995 levels under this Protocol. France signed the protocol in June 1994 and accepted it in June 1997. Action on SO_2 emissions from the petroleum chain will help France, and other countries, to meet the new emission reduction requirements.

The NO_X Story

The NO_X story primarily refers to the French vehicle industry. Proposals on vehicle exhaust emissions made by the EC during the latter part of the 1980s had a significant impact on costs and technology for French manufacturers. At the same time, there was a much higher recognition by that time that air pollution arising from vehicle emissions caused ecological damage, especially to forests. The NO_X story is built round two themes: EC motor vehicle emission controls; and the 1988 UNECE NO_X protocol.

EC motor vehicle emission controls France first became aware of acid rain because of its effects on forests. The causal links between this and transport came later, and were at first denied. The former Minister of Industry, Laurent Fabius, wrote to Mme Bouchardeau at the Ministry of the Environment, saying, 'I forbid you to make any link between cars and forests'.[6] Despite this, the Report of the Office Parliamentaire commissioned by Mme Bouchardeau, published in November 1984, concluded that there was such a link and its conclusions favoured clean technologies.

French car manufacturing is dominated by two major groups, Peugeot and Renault (Barrier-Lynn et al., 1988). From producing 10 per cent of the world private car production, they have been in continual decline, reaching a low point in 1985, where their market share was only 8.1 per cent. The rate of growth of exports fell from 2.6 per cent per year in the 1970s to 0.2 per cent in the 1980s.

Car manufacturers have traditionally had the ear of government not merely for reasons of the trade balance, but in their role as major employers. In addition, Renault is owned by the French state. These close ties to the state partly help to explain the initial reaction of the industry when the 'acid rain' problem emerged at the beginning of the 1980s, and for the first time, pollution was understood to emanate from mobile sources. The French manufacturers were threatened by German proposals in the 1980s to establish EC motor vehicle emission limits which would require the fitting of catalytic convertors to all cars.

The government had financed research on anti-pollution car technology between 1971 and 1978, but this had not resulted in usable industrial outcomes capable of satisfying future norms (Barrier-Lynn et al., 1988). The technology France had possessed in the 1970s was abandoned on the grounds that air quality standards had been achieved, according to one commentator (Roquelpo, 1988). Decisions made at the end of the 1970s over priorities in research and development had emphasised energy efficiency – the 'lean-burn engine' – at the expense of anti-pollution research – the catalytic convertor (Barrier-Lynn et al., 1988). The remnants of catalytic technology in France were sold off or abandoned. Just as vehicle emissions were seen as a problem, markets for the electronic engine controls which would enable catalytic convertor technology, were dominated by one German manufacturer, Bosch.

France tried to delay the imposition of catalytic convertors on small cars for as long as possible and withdrew from an agreement on small cars in 1989 because of pressure from Peugeot. Subsequently, however, the vehicle agreements have been extended to small cars and Peugeot and Renault have been able to comply. To understand what occurred, one must appreciate important links between the dominant Ministry of Industry and the partly state-owned car industry. This ministry would have seen its role in part as protecting the interests of manufacturers. As with the French Memorandum on sulphur, this illustrates France's ability to mobilise behind economic interests close to the state.

The Sofia Protocol, 1988   By the time negotiations for the Sofia Protocol were under way, France had gained a better understanding of NO_x pollution. The results of the DEFORPA study on forests were emerging, underlining the links between vehicle emission and forest damage, and the implications for the economy of NO_x regulation had been made clear. The 1985 EC Directive on air quality norms for NO_2 were transposed into French law in ministerial Circulaires.

Industrial leaders had no effect on the government's behaviour in the wider international arena, and France signed the Sofia protocol in November 1988 and approved it in July 1989. France's emissions of NO_x rose rapidly from under 800 kilotonnes in 1960 to peak in 1980 at 1,823 kilotonnes. There was a steady but slight decline to 1,615 kilotonnes in 1985 but emissions have changed little since then.

Under the terms of the Protocol, France undertook to stabilise NO_x emissions between 1987 and 1994, and then to reduce them by 30 per cent below 1985 levels by 1997. While the first commitment was seen as relatively easy for France to attain at the time, current preliminary figures for 1994 NO_x emissions suggest that France has just failed to achieve stabilisation. However, final figures have yet to be submitted to the UNECE. It appears very unlikely that the second, less formal commitment to a 30 per cent cut in NO_x emissions will be fulfilled.

In 1988, France also undertook to reduce NO_x emissions from major combustion sources by 20 per cent and 40 per cent in 1993 and 1998 relative to 1980 emissions. This was a requirement under the EC LCPD. These commitments will be met.

The VOC Story

France had already imposed regulations on classified installations prior to the 1991 UNECE VOC Protocol. These regulations were not designed specifically to deal with VOC emissions, but were conceived mainly to control sulphur and visible pollution. Article 27 of the new Arreté on Classified Installations published in 1993 gives detailed lists of VOCs and the hourly emission limits for Classified Installations.

In 1986, a covenant was signed between the Minister of the Environment and paint and varnish manufacturers to reduce the solvent content of their products. However, the proportion of emissions coming from industry was only 5 per cent by 1990, while an increasing proportion came from road transport. The VOC story is closely linked to that of NO_x.

Motor vehicle exhausts account for 59 per cent of French VOC emissions (ADEME, 1994). Catalytic convertors on cars reduce VOC emissions from 8.8 kg to 0.6 kg per standard driving cycle (ADEME, 1994). The negotiations which most concerned France regarding VOCs were those on motor vehicle emission limits conducted under EC auspices. Despite the imposition of catalytic convertors as described under the NO_x story, VOC emissions have risen steadily from 1,180 kiltonnes in 1980 to 1,260 kilotonnes tonnes in 1990. This growth was largely due to an increase in urban car use, but in addition, there has been a significant shift in goods transport. Heavy goods vehicle use rose by 70 per cent between 1970 and 1990. As catalytic convertors take up higher proportions of the vehicle stock, transport-related emissions will begin to fall.

France signed the VOC Protocol in November 1991, agreeing to reduce emissions by 30 per cent by 1999 relative to 1988 levels. However, it is thought that this cannot be achieved except under a scenario of limited economic growth. France approved the VOC Protocol only six years later in June 1997 at the same time as it approved the Second Sulphur Protocol. This suggests that there has been a reluctance to approve given doubts about compliance. Yet, there is a strong French wish to ratify agreements that have been signed.

VOC controls on stationary sources would affect many small-medium sized industrial plants. There is currently an internal debate between the Ministries of the Environment and Industry over whether priority should be given to economic growth or to achieving environmental goals. The Ministry of Industry argues that if France implements the measures necessary to achieve the reductions required under the VOC protocol, there will be adverse effects on the economy, whereas the Environment Ministry places greater value on human health.

Models of Domestic-International Linkages

The Unitary Rational Actor Model

The Unitary Rational Actor Model is based on the premise that the state maximises net national welfare. Under this scenario, the state pursues environmental protection up to the point where its marginal costs of abatement equals its own marginal damage costs. Compliance is seen as a function of the relationship between the actor's own abatement and damage costs and the costs incurred by defecting.

The rational actor model provides a simple but effective framework for explaining French policy on acid rain. However, the empirical basis for testing the model is somewhat weak because, for many of the international agreements studied, France experienced neither significant benefits nor significant costs. For most of the LRTAP measures, France participated but did not play a pro-active role, reflecting the weak benefits of acting as a good international citizen.

In the case of motor vehicle exhaust emissions, which provides stronger empirical evidence, the rational actor model remains reasonably effective. France did not perceive great benefit from controls but would suffer costs because its vehicle manufacturers had not been strong supporters of the winning technologies.

The rational actor model is also quite effective at explaining the French approach to the control of sulphur from the petroleum chain. France moved proactively in an attempt to secure benefits from new controls, but these benefits would be realised in the form of competitive advantage (or the alleviation of competitive disadvantage) rather than environmental advantage.

France is thought of as a highly centralised state, engaged in implementing national policies aimed at achieving a strong economy. It would seem, therefore, to be a good candidate for this model. However, France has so far signed and ratified all agreements, even where there was a clear understanding that these did incur costs. France has consistently followed the rhythm of European legislation. However, compliance has not been a compelling issue. In the case of the sulphur agreements, patterns of energy consumption have led, or leading to a considerable degree of 'over-compliance' with formal commitments. For NO_X and VOCs, compliance is far from assured. Some of the actors engaged in domestic policy-making procedures, such as the Ministry of Industry, clearly consider that the costs are too high.

Up to now, France has not been reluctant to sign international environmental agreements, even though it has sometime demanded delays. While at first this was because it did not feel very concerned about outcomes, later there was a real acceptance of the need for pollution reduction. As the measures increasingly bite into economic interests, divisions between two of the major ministries involved, Industry and Environment, become more apparent. There is an undeclared debate over what counts as net national welfare. An answer to this question points towards the 'domestic politics' model.

The Domestic Politics Model

The domestic politics model is based on the premise that values and beliefs regarding the costs and benefits of environmental poliy vary within societies, that actors have their own specific concerns and that the power to pursue individual and more narrow goals is unevenly distributed.

The domestic actors model provides a much richer framework for explaining French acid rain policy. However, again, the weak linkages between domestic policy and international agreements and the absence of significant benefits and costs in the case of the LRTAP agreements provides a weak empirical base. The domestic politics model, applied in the context of particular features of French political and economic culture, is particularly effective in explaining: a) the limited role of public concern and non-governmental organisations; b) the lack of perception of environmental benefits from action on acid rain due to the peripheral nature of the organisations involved with forest management; and c) the strength of the debate concerning motor vehicle emission controls. The vigour of that debate and the outcome is explained by the close relationships between state-owned vehicle manufacturers and the influential Ministry of Industry.

Table 9.5 shows the key actors who, in addition to governmental authorities, influenced, and were affected by, French policy with respect to the three pollutants controlled under the LRTAP Convention. Unusually for most European countries, the electricity industry has had little stake in the outcome of environmental policy-making. Sulphur policy for example, which was a key early arena for conflict in most countries, did not become an area of active international activity until the 1990s when the oil industry began to get involved. The motor vehicle industry became involved in the mid-1980s because of the prospect of EC Directives affecting NO_X and VOC emissions from motor exhausts. A brief discussion of the role of individual sets of actors illustrates the potential power of the 'domestic politics' model.

Table 9.5   Actors in French policy on transboundary air pollution

Actors	SO_2	NO_X	VOC
Electricity Industry			
Car Industry		+	+
Oil Industry	+		
Combustion Plant Operators	+	+	
Industrial VOC Emitters			+
Forestry Industry	+	+	
NGOs	+		

Note: A '+' sign indicates interest and influence

National government   Government policy in France may emerge from broad governmental goals, from public pressure or even from quite small interest groups. But whatever the source, key decisions are taken in 'multipartite' committees set up to include representatives from several ministries and bodies with relevant expertise. There is a tendency to treat problems as technical, and an idea that decision-making follows inexorably from consulting bodies of knowledge.

The 'multipartite' committees allow people in different ministries to get to know each other and become aware of thinking in other parts of government. In the environmental area, scientific evidence is constantly called on, with a tendency to emphasise the physical sciences. Undoubtedly the scientists selected are those whose behaviour and opinions conform best to the political biases of government, but there is great respect for evidence from scientific studies, from both France and abroad.

However, there is a deeper separation between the administrative and scientific cultures. These two groups of people are trained in separate institutions, each with their own ideas of prestige. It is not merely a case of the administrators keeping out the scientists, which probably does occur, but that the scientists themselves see such work as using not 'pure', but 'applied' science which for them has a lower status. These social divisions may impede learning on both sides. This can be a major handicap when highly technical issues are under discussion. These factors partly explain the slowness to appreciate the scientific problems underlying air pollution as shown by France's initial reaction to the first signs of long range pollution.

Although France is often considered to have a unitary administrative structure, the notion of 'the state' also needs examining. The France education system is designed to create a political elite. Selection of the top candidates into the Grandes Écoles and then on into the Corps d'État might be thought to create a unified ruling elite. However, the different Corps can work for the interests of their group rather than for the country as a whole. For example, members of the Corps des Ponts et Chaussées are often placed in the car and road-building industries and constitute a powerful road lobby. There are also close links between the Ministry of Industry and the car industry. Policies designed to restrict road transport will come up against this power bloc.

The existence of the Corps d'Etat and their links with ministries and major companies explains much of the behaviour observed in France's reaction to the acid rain story. The effects of this can be both positive and negative. On the one hand it can rigidify the country's response to new pressures. On the other hand, as Knoepffel and Larrue (1985) argue, it can operate at a regional level as a flexible system of regulation.

The rigid thinking of civil servants is sometimes held responsible for France's failure to benefit from new technologies. Members of the Corps des Mines are dominant in the field of air pollution administration and control. The Corps des Mines has adopted environmental issues as 'its territory', but their 'technocratic' orientation to problem solving may militate against practical solutions.

The Ministry of Industry traditionally plays a leading role in France's economic policy, through its relationships with key industrial sectors. The problem of air pollution and EC regulation has posed a challenge to this position and has provoked resistance. The coinciding emergence of the acid rain issue and the new socialist government in the early 1980s would have led to suspicion of government policies by key ministries. There was also a belief on the part of the major French car manufacturers that they could control the government, via the Ministry of Industry, and hence resist pressures from Europe. When they found that they could not, their reaction was immediate and strong, as shown by the reactions of Peugeot.

Ironically, in the field of air pollution, problems are emerging because of the very *absence* of a relevant elite structure. For example, an increasing source of air pollution comes from diesel. This fuel has traditionally been taxed less than petrol in France and is of major importance for both agriculture and commercial transport. The owners of lorry fleets form a powerful barrier to the intentions of the Ministry of the Environment which is to equalise the taxes paid on diesel and petrol in order to reduce overall diesel consumption. It is hard for the state to win over the lorry owners because they have no links with any Corps d'Etat.

As the pollution statistics show, road transport is now the main source of air pollution in France. The existence of powerful lobbies which fall outside the classic governmental structures therefore pose a real problem to the government in implementing measures which could have significant effects on pollution.

The energy sector The substantial reductions in French emissions of transboundary air pollutants during the 1980s can be attributed to structural change in the French energy sector. This springs, in turn, from French sensitivity to dependency on imported Middle East oil prior to the 1973 oil crisis. Until then, France had relied on its close links with the Arab world to underpin its security of energy supply. After 1973, there was a powerful drive to become more self-sufficient in energy. Self-sufficiency meant major investments in nuclear power stations which do not contribute to emissions of 'acid rain' pollutants.

Électricité de France (ÉdF), the state-owned, vertically integrated electricity company, has a virtual monopoly over French electricity supply. Administratively, ÉdF functions as a state within a state and its affairs have traditionally been immune from influences from the Ministries.

Throughout a period in which France learned to accommodate environmental goals into the over-riding aim of modernisation and rapid technological change, environmental policy-making did not involve the electricity industry. The industry and its strategies were accepted as a fact of life, rather than as a possible policy instrument.

France is also one of the EC countries with the largest net imports of crude oil. The petroleum industry has been tightly controlled by the state, mainly with a view to guaranteeing security of supplies. Energy policy in France in run by the Directorate General de l'Energie et des Matières Premières (DGEMP) whose task has been to ensure that France has sufficient supplies. However, it also promotes the interests of industry. The sector was governed by the Oil Law of 1928, by which government had to authorise all imports of crude oil and refined products. It also had extensive powers over all the industry's operations. These powers were eroded from the late 1970s onwards due to changes in the international oil market and EC legislation. In 1979, import quotas were ended and, by 1985, companies were freer to buy products from foreign refineries and to set up marketing outlets. Recent legislation has further loosened the government's hold over the industry, though it still retains powers to ensure supplies.

The industry has long been unable to meet demand for major products, especially diesel which is much favoured by differential taxation policies. Residual fuel oil is persistently in surplus because of low demand due to the existence of the nuclear programme. DGEMP and the Ministry of Industry acted decisively to promote the interests of the French petroleum industry with the 1990 'French Memorandum' to the EC on sulphur in liquid fuels.

The car industry The interests of the two major French car manufacturers, Peugeot and Renault have traditionally been promoted by government not merely for reasons of the trade balance, but because they are major employers. Renault is state-owned. These close ties to the state partly help to explain the initial reaction of the industry when 'acid rain' damage became associated with motor vehicle exhaust emissions at the beginning of the 1980s. Links between the dominant Ministry of Industry and the partly state-owned car industry were crucial in determining French policy. This ministry would have seen its role as protecting the interests of manufacturers.

NGOs The weakness of environmental NGOs in France is striking. Membership is low and environmentalism has a poor image. This is partly the result of a policy by which voluntary Associations can receive financial help from public authorities, making them less willing to challenge official views. Voluntary groups have increasingly been organised around leisure interests rather than to serve political goals. The exclusive nature of French policymaking, minimal indications of acid rain damage and the low cultural resonance of acid rain-linked environmental problems also serve to explain the lack of NGO influence.

As a result, the views of environmentalists were seen as far-fetched. The term 'ecologist' had negative connotations. Those opposing the French nuclear industry were seen as opposing the state itself, and as a result environmentalism became marginalised, and even portrayed as subversive. There has been relatively little coverage of environmental issues in the media, and what there is often focuses on the potential of French technology to overcome difficulties.

The French debate on acid rain was restricted to government, research bodies such as INRA, and 'insider' interest groups. The wider public was not admitted. Policy changes can be attributed to the state's acceptance that environmental goals were integral to France's modernisation programme. The motivation was certainly national self-interest, but the gains have to be measured in terms of prestige rather than simple cost-effectiveness.

The Social Learning Model

The social learning model builds on the domestic model but acknowledges the important fact that decision-makers do not have access to perfect knowledge about the problems with which they are engaged and that they search out and acquire information and ideas as policy formation takes place. This learning becomes incorporated into belief systems and operating procedures. Policy ideas can be diffused through networks of national experts.

The social learning model is less fully developed than either the rational actor or domestic politics models. But the French case study clearly marks out the social learning model as potentially having strong explanatory power. At the beginning of the LRTAP process, French scientific institutions and policymakers were sceptical about the consequences of acid rain. They participated in transboundary air pollution negotiations as good international citizens and because the structure of the French energy sector implied zero compliance costs. However, engagement in the negotiating networks forced French policymakers to take into consideration the costs and benefits of abating acid pollution.

This learning process became intense as policy attention moved on from SO_2 to NO_X and VOCs. Here, France did not enjoy any special advantages vis-à-vis other European countries and a tighter coupling of domestic and international policies was forced. French policymaking networks learned that traditional, health-drive air pollution systems would not by themselves meet international commitments motivated by ecological concerns. Policymakers also learned that French technical superiority was not automatic, and it was German vehicle manufacturers that set the pace with emission control. Even now, France faces compliance difficulties with the NO_X and VOC protocols. Good international environmental citizenship is not as easy as it appeared when sulphur was the pollutant under debate.

Finally, there is evidence that France learned to play EC-based regulatory games with some sophistication as it recovered from competitive disadvantage by setting in chain the sequence of events which led to the Liquid Fuels Sulphur Directive. France may still not be at the heart of the acid rain debate. But it is not the disinterested and self-consciously superior spectator that it was in the early 1980s.

Final Remarks

In one sense, the French case study is an example of the 'dog that did not bark'. It contrasts especially with the German and Scandinavian case studies where transboundary air pollution came to be perceived as a significant national threat. Nevertheless, the French study enriches our understanding of linkages between domestic and international regimes. It forces us to consider why a country with apparently little interest in the environmental issue in question became engaged and why, ultimately, it came to play a more active role.

Initially, France participated in the LRTAP Convention simply because it was part of the UNECE 'club'. This illustrates in itself that environmental issues cannot be divorced from the wider context of international relations. With no apparent interest in the acid rain issue, France still 'played the game'. It also illustrates a limitation on the use of the 'unitary rational actor' model. This statically conceived 'cost-benefit' model has some short-term explanatory power, but it fails to explain the more dynamic unfolding of events in France and the growing mobilisation of parties who were initially unconcerned. To explain this feature, we need to turn to the 'domestic politics' and 'social learning' models.

The final lesson is that economic interests cross boundaries as well as pollution. It was the engagement of French industrial concerns involved in the European trading system, more than any other factor, which turned France into an active player. This feature must also be integrated into any explanatory scheme.

Notes

1. Smith, Duval A. 'Quieter Flows le Traffic', *The Guardian*, 16.8.95.
2. M Ledenvic, Ministry of the Environment at a Conference: *Pollutions Atmosphériques a l'Échelle Locale et Régionale*, Cachan 7 - 9 Dec 1993.
3. The Ministry of Finance is a separate entity from the Ministry for the Budget. The Ministry of Finance deals with issues such as planning, distribution, competition and fraud, while the Ministry for the Budget ensures that the current state budget is adhered to.
4. Figures given by ADEME.
5. 'The French Memorandum: Tackling sulphur emissions throughout the petroleum chain', *ENDS Report*, 210, July 1992.
6. Quoted by P Roqueplo, a member of Mme Bouchardeau's cabinet.

References

ADEME, (1994), *Transport, Energie et Environnement*, Débat National Energie and Environnement.

Barrier-Lynn, C; Georgiades, Y. and Lambert, J. (1988), *Les Industries Automobiles Francaises et Allemand entre le Compromis de Luxembourg (juin 1985) et les Échanges d'Apllication de l'Accord (1988-1993)*, Bruxelles: Commission des Communités Éuropeennes.

Barthod, C.; Gauthier, O.; Landmann, G., and Ebner, P. (1993), Ecologie, débat public et politiques techniques: le déperissement des forêts dix ans aprés, Ministère de l'Environnement.

ERPURS (1994), *Impact de la pollution atmospherique urbaine sur la sante en Ile-de-France 1987-1992*, Observatoire régional de Santé d'Ile-de-France.

Knoeppfel, P. and Larrue, C. (1985), 'Distribution spatiale et mise en oeuvre d'une politique publique: le cas de la pollution atmosphérique', *Politiques et Management Public*, No. 2.
Roqueplo, P. (1988), *Pluies Acides: Ménaces Pour l'Europe*, Paris: Éditions Economica.

10 Italy: Learning from International Co-operation or Simply 'Following Suit'?

RODOLFO LEWANSKI[1]

Acid Depositions in Italy: A Problem?

For quite some time, 'acid rain' was not perceived as an issue in Italy neither by policy makers neither by the public opinion at large. Only in 1986 did this problem attract, for a brief moment, the attention of the ecological movement[2] and of the media, but soon fell out of sight again. In essence, there was no demand to the political-administrative system to intervene on the problem. To the extent that there was – and now is – a concern with the questions of air pollution, it was due to the perceived threats to human health caused by local industrial and vehicular emissions.

Even the scientific community ignored the topic for a long time. Research efforts began in the mid 1980s under the stimulation of the Ministry of Health in co-operation with national (Ministry of Agriculture and Forestry, Ministry of Defence and Air Force) and regional authorities (especially Veneto), public research institutions (*Istituto Superiore di Sanità* -ISS, Higher Health Institute, *Consiglio Nazionale delle Ricerche* - CNR, National Research Council-, *Ente per le Nuove Tecnologie. l'Energia e l'Ambiente*-ENEA – a research institution specialised in the fields of new technologies, energy and environment) and major sulphur emitters (such as the national electricity company ENEL). After its creation in 1986, the Ministry of Environment, partly in order to comply with EC regulation no. 3528/86 and others, set up the Italian Network for the Study of Atmospheric Deposition (RIDEP). The purpose was to co-ordinate the activity of the various organisations working on chemical deposition and to develop a common methodology for sampling and analysis so as to ensure comparability of data collected by the 124 'wet

255

only' stations and 26 'bulk' stations (more than half of which are located in Northern Italy) in existence at that time. At the same time, the new network expanded the number of the participants already involved in the monitoring work (Ministero dell'Ambiente, 1989, p.194, and 1992, p.241).

It is only thanks to research carried out over the last decade or so that the extent of the actual damage to the ecosystem in Italy has become visible. More than 40 per cent of trees display more or less serious damage (Elvingston, 1993, p.9), whereas in 1986 damaged trees were estimated to be only 1 per cent. Such data should, however, be treated with caution since it is not clear whether there has been an actual increase or simply the monitoring activities have become more careful. Twenty per cent of the trees showed a defoliation that exceeded 25 per cent according to a European survey of 1995 (Ågren and Elvingson, 1997, p.7). Damage is concentrated especially in two geographical areas: the North-Eastern Alps (Piedmont, Liguria, Aosta and Lombardy) and some of the central regions of the peninsula (Abruzzo and Molise) (Istat, 1993, p.117). Also, many lakes and groundwaters located along the Alps show rising levels of acidity (Mosello e Tartari, 1988, p.103; Conti e Melandri, 1994, p.290). Finally, ancient monuments, in which Italy is particularly rich, are also suffering from decay caused by acidification.

However serious, the acidification problem is certainly not as relevant and widespread as it is in some countries of Central and Northern Europe. This is due to two distinct factors: the prevailing alkaline character of soils and the geographic position of the Italian peninsula. As far as the second factor is concerned, the relative distance from other European countries and the protection offered by the Alps to the North, together with the dominant westerly winds, make Italy a net exporter of airborne pollutants (with a consistent share of its exports being deposited in the Mediterranean sea).[3]

It should be emphasised that while the level of knowledge about the nature of the problem is now satisfactory, in the 1970s and well into the 1980s when the subject of acid rain became a policy issue in Europe, Italian authorities involved in the international negotiations lacked any accurate scientific data. In this sense, the decision to join such agreements was taken in a situation of 'ignorance' as far as the implications of such agreements in terms of the actual abatement and damage costs involved.

Domestic Environmental Policy, with Specific Reference to Air Pollution

The development of environmental policy in Italy can be divided into four distinct phases on the basis of several major 'turning points' in its brief history. These different phases can be represented by the passing of relevant legislation and the entrance in the policy arena of new institutional actors.[4]

The birth of an environmental policy in Italy dates back to the mid-sixties. The first explicit piece of environmental legislation – Law 615 of 1966 – was designed to limit air pollution. The aim of this and several minor laws of the same period was to 'patch up' particular, geographically limited problem situations. This 'Anti-smog' Law, as it was commonly called, mainly dealt with intense air pollution phenomena affecting large urban areas in the Northern part of Italy where frequent thermal inversions occurring during the cold seasons, coupled with topographical conditions that hinder dispersion,[5] produced a serious threat to human health. A reduction of the sulphur contents in fuels used for domestic heating represented the core of the 'Antismog' Law's strategy.[6] Only to a limited extent did the Act enable local authorities – at least those willing and capable of doing so – to deal with emissions from industrial sources. Actual implementation and enforcement of air pollution control policies, however, only began in the early 1970s, essentially in a few areas of Northern Italy, when regulations were issued by central government (i.e. the Health Ministry), with considerable delay in respect to deadlines set by the Act itself.[7]

In this phase environmental issues had a very low political visibility, and the 'Antismog' Law remained for several years an isolated episode. In the absence of adequate legislation, attempts to respond to increasing pollution in the seventies came from local authorities and individual judges, both of whom tried to tackle specific problems by means of the available - antiquated - laws (such as the 1934 'Health Laws' or others actually meant for the protection of particular economic activities such as fishing or navigation (Lewanski, 1986).

Ecologists exerted very little influence; the only well-established organisation ('*Italia Nostra*') was mainly concerned with the protection of the national cultural heritage such as monuments, while other associations (for example, the Italian branch of the WWF, founded in 1966) were still in their infancy and had quite small memberships.

A second major turning point in the evolution of environmental policy occurred in 1976. It was marked by three significant events: 1) the passing of a second major piece of legislation, this time in the field of water pollution; 2) the upgrading of the role of a set of relevant institutional actors, i.e. the regions; and 3) the dioxin pollution accident in Seveso that deeply influenced the public's attitude towards the environmental issue.

From the mid-1970s to the mid-1980s environmental policy underwent a substantial acceleration as a number of previously missing legislative instruments were supplied to the 'tool box'. In 1976 Act 319, aimed at reducing water pollution, was finally issued after 10 years or so of debate. Several other relevant pieces of legislation, dealing with such topics as industrial and domestic waste disposal, sea protection and biodegradability of detergents, were passed as well in these years. The most relevant provision in the field of air pollution was Decree no. 30/83, transposing EC Directive no. 80/779, which introduced air quality standards for the first time. This provision, however, met with serious implementation difficulties. Indeed, even today many regions have still not drawn up the Clean Air Plans foreseen by the Decree. It should be noted that Italy introduced this legislation four years after signing the Geneva Convention and one year after its ratification (Act no. 289 of April 27, 1982).

Access to the policy arena in this period was still limited to a rather small number of actors, although new ones, both institutional and social, began to appear. As far as institutional actors are concerned, a first move in the direction of the establishing a central authority in charge of this specific policy area was the creation of an 'Interministerial Committee for Environmental Protection' which had the task of co-ordinating powers on environmental matters dispersed among 16 different ministries. Major institutional actors were created at the peripheral level, such as the local health units (about 630 decentralised units -USL- of the National Health Service instituted in 1978)[8] and the regions. The regions were originally instituted in 1970, but actually awarded substantial powers only in 1976; many of the regions' responsibilities are directly or indirectly related to environmental policy (nature and natural beauty protection, parks, hunting and fishing, pollution, health, land use, urbanistics, agriculture, mining).

With respect to social actors, a rather strong anti-nuclear movement contributed (together with opposition to the siting of nuclear plants from local authorities) in limiting nuclear power programmes. This movement, due to its ideologically leftist roots, also framed the environmental issue in anti-capitalistic terms. Beside the anti-nuclear movement, existing environmental organisations started growing (such as the Italian branches of the WWF, Greenpeace, the Friends of the Earth) and others were established such as the *LegAmbiente*. All together, ecological associations had some 1.5 million members in 1993, a figure that makes this the largest sector among all voluntary associations.

An indicator of the demand for environmental quality can be found in data provided by polls. These show that Italians are concerned with both local, national and international environmental problems at least as much as is the case of other European countries, if not more so (OECD 1993, 288-90). Such concern, however, does not necessarily translate into willingness to take action, as the rather limited electoral successes of the Greens in local and national elections seem to indicate. After having participated in local and regional elections since the early 1980s, the Green Party finally obtained seats in the national Parliament for the first time in 1987 with 2.5 per cent of vote. In the elections for the European Parliament of 1989 the Greens obtained 6.2 per cent, in the political elections of 1994 2.7 per cent and in the European ones of the same year 3.2 per cent. With the exception of 1989 (when two different lists including other smaller parties were competing), the Greens' share appears to have stabilised around the 2.5-3 per cent mark. There are significant regional variations to be noted: the Greens generally obtain better results in the Northern areas, and below average percentages in the South and on the major islands (Sicily and Sardinia).

The visibility and political relevance of environmental problems increased considerably in this period, as a result both of the popular reaction to the attempt of the central government to launch a large nuclear power plant construction program and of serious industrial accidents - especially in the chemical sector – that attracted media attention and deeply modified the public's previous attitudes on such topics. Particularly important in this respect was the accident which occurred in July 1976 at the Swiss-owned 'ICMESA' chemical plant in Seveso, a little town in the Lombardy Region.

A third turning point of environmental policy in Italy was the creation in 1986, after a number of unsuccessful attempts, of an authority specifically in charge of environmental policy at the central level, i.e. the Ministry for Environment. The new ministry was given responsibilities in the fields of air, water, soil and noise pollution, solid waste, mining, parks, sea and coast protection, though pre-existing ministries were successful in withholding other fields related to the management of natural resources (such as hunting, water management and land-use planning). Together with other elements such as the relatively limited staff, budget and technical expertise, this obviously has weakened the effectiveness of its action. Nevertheless, the establishment of this ministry 'officially' marked the upgrading of the status of environmental policy *vis-à-vis* other sectorial policies.

The new ministry's first priority was to foster previously missing or inadequate legislation concerning such topics as hazardous waste, landscape, soil protection, use of water resources and noise pollution. A number of provisions specifically dealt with air pollution. Decree 105 of March 10 1987, explicitly referred to the Geneva Convention (1979) and the Helsinki Protocol (that was still to be ratified by Italy at the time) as well as to the EC Directives of July 15, 1980 (SO_2) and of March 7 1985 (NO_X), setting limits to atmospheric emissions for fuel burning power stations. Decree no. 203 of May 24, 1988, transposing four EC Directives 80/779, 82/884, 84/360 and 85/203), regulated such aspects as emissions, properties of fuels, air quality standards, the distribution of powers between central government and regions, and the granting of permits for both new plants and existing ones. It also represented a major upgrading of air pollution policy because, in contrast to Act 615/66, it applied to the entire country. The Decree was approved only a few months before the signing of the Sofia Protocol (November 1, 1988) and the ratification of the Helsinki Protocol (October of 1988). Guidelines for the reduction of emissions and other aspects required for the actual implementation of Decree 203 were established only in 1990 (by Decree of July 12). Decree 124 of May 8, 1989, transposing Directive 88/609 and explicitly referring to the Helsinki and Sofia Protocols, established limits on the emission of pollutants from existing large combustion plants, and prescribed for reductions for SO_2 of 30 per cent by 1993 as compared to 1980 levels, of 39 per cent by 1998 and of 63 per cent by the year 2003. The Ministry also tackled the issue of air pollution in urban areas caused by vehicle traffic

(Decree of November 11, 1992; decrees of 1996 establishing maximum contents of benzene).

Public expenditure (of central, regional and local authorities) for environmental measures increased substantially in the late 1980s. In 1988 environmental expenditure reached 1 per cent share of GNP (Ministero dell'Ambiente, 1989, 332-337), internationally considered the minimum necessary threshold for a developed nation (though in the last years it has again fallen to 0.8 per cent, as compared to an EC countries average of 1.2 per cent).[9] It should be pointed out that the introduction of substantial distributive elements into the legislation represented an important source of consensus for the ministry and the policy itself. The actual capability of the ministry to spend allocated funds has, however, proved to be rather low (only approximately 1/3 of the allocated amount has been spent, the lowest rate among all central ministries). Although allocations for scientific research in the environmental field have certainly increased over the years (from 0.6 per cent of total expenditures for research in 1975 to 2.8 per cent in 1991), public expenditures in this area remain low as compared to many other industrialised nations.

During this period the visibility of environmental problems grew further, in part as a result of accidents like those of Bhopal and Chernobyl, which were widely echoed by the media. An increased demand for environmental quality by the public, partly related to the shock caused by these disasters, brought about the rejection of the Italian nuclear power program through a referendum held in 1987.

The Attitudes of the Business World towards the Environmental Issue

Although strongly opposing strict policies, many industries started to realise the inevitability, if not the necessity, of natural resource protection (Lewanski, 1992, pp.51-64). As a result of public policies imposing anti-pollution measures, some sectors of industry discovered that environmental protection could be a profitable opportunity. The emergence of the 'green business' can be explained in the light of three interwoven factors: 1) increasing regulation that forced polluting activities to modify technologies and install emission abatement equipment; 2) increasing amounts of public financial resources for environmental protection that created substantial business opportunities (such as construction of air and water depurators, facilities for waste management, etc.); and 3) growing environmental sensitivity among consumers in favour of 'greener' products. Furthermore, national producers have been stimulated to modify attitudes and strategies because they realised that manufacturers of other EC countries were enjoying a 'competitive advantage' due to the fact that they had started moving in this direction some years in advance. At present, the yearly income of 'green business' is in the range of approximately 6,500 billion lire, more than 4 billion US dollars (Malaman and Paba, 1993).

During the previous phase, environmental policy had without doubt asserted itself in the political arena, gradually gaining in status and influence. The break out of the 'Tangentopoli' (sidekick) scandal and the end of the Cold War, doing away with parties that had been in power for the last 50 years or so, ushered in a period in which the environment lost importance since attention was focused on the transformation of the political and institutional system and the country found itself in a difficult economic and employment situation. Presently, the representatives of the environmental movement occupy several important posts as members of the coalition supporting the Prodi Government (e.g. the position of Environment Minister was assigned to a 'Green'), but it is too early to judge what their actual capability to influence public policies might be. In this period the national environmental agency (*Agenzia Nazionale Protezione Ambientale*, ANPA)[10] was set up.

Summing up, environmental policy in Italy has developed rather late as compared to other industrialised countries. If one considers indicators of institutional capacity such as basic legislation and the creation of *ad hoc* administrations, the delay can be estimated to be approximately 10 years. One feature of Italian environmental policy style has been its reactive character, in the sense that policy measures are typically triggered by external inputs, be they emergency situations or international obligations. As far as the former are concerned, policy-making tends to react to issues as they break out, often dramatically, rather than anticipating them before they become policy problems. As far as the second aspect is concerned, very few legal provisions are entirely 'endogenous' in the sense that they are the result of a policy process occurring within the domestic polity. As pointed out above, in most cases such provisions are the result of requirements posed by extra-national commitments, i.e. international agreements represented, to a large extent, by EC/EU Directives. In other cases, Italy has 'imported' the legislation of other more advanced countries, especially Germany.

Nevertheless, although late and in response to external pressures, Italy has more or less closed the gap separating it from other industrialised countries, at least as far as the completeness of legislation is concerned. Financial resources allocated to the environment have gradually increased (despite the fact that the Ministry of the Environment is understaffed with respect to the quantity and complexity of its responsibilities) and public opinion has become more aware of environmental questions and more supportive of policies in this field. The major weakness of Italian policy in this field, as well as in others, is the very serious deficits that arise in the implementation phase due to the weakness of technical capabilities in the public sector and the prevailing formal-legalistic organisational culture according to which the concern for procedure prevails over considerations of effectiveness and results. Thus, the actual impact on environmental quality generally remains quite low (though there are notable differences between different administrative areas).

Italy's Role in the LRTAP and EC Negotiations

For the most part, Italy played a marginal role in the definition of LRTAP agreements. The interest of the Italian Governments in the preparatory work that would eventually lead to the Geneva Convention and the subsequent Protocols appears to have been very limited; that of environmentalists, political parties and public opinion appears to have been practically nil. The issue, was not perceived to be of interest by political actors (surprisingly, even when economic actors lobbied against specific commitments), and thus was left to technical actors to take care of. A minor, but significant indicator of such lack of interest is the fact that the Government did not even allocate financial resources to cover the costs of attending meetings. Thus, it is not surprising that Italy was represented only sporadically by competent representatives - namely, experts from the Central Commission against Air Pollution (CCIA)[11] at the Health Ministry, and in particular a member of the Commission belonging to the Meteorological Observatory of Brera, an institution also interested in long-range air pollution. In light of this interest, the Observatory occasionally covered the expenses incurred by its representative in taking part in the LRTAP meetings. The lack of financial resources continued to be a major obstacle to regular participation of Italian representatives at the negotiations of the EMEP and Helsinki Protocols until the mid 1980s. Also, from the reports of the sessions of the Executive Body for the LRTAP Convention and its working groups, the Italian delegation is notable for its persistence in not playing any active role even when it does take part in the meetings of the working groups: contributions by Italian representatives are never mentioned in the proceedings or in other official documents. Furthermore, Italian representatives are never assigned any relevant position (such as chairs of working groups or task forces), nor has Italy ever hosted a LRTAP meeting or conference. Frequently, meetings were attended by permanent officials of the Italian delegation to the UN in Geneva, who, however, were hardly in a position to appreciate the technical aspects and implications of the agreement being discussed. Somewhat paradoxically, however, in the end this favoured Italy joining all the agreements, although this did not occur for reasons connected to the protection of the environment, but rather on the basis of considerations connected to international relations policy. Especially in the initial phase, the Ministry of Foreign Affairs appears to have been the most active actor in determining Italy's position towards the LRTAP agreements and in attempting to respect the obligations deriving from them (for example by urging the Health Ministry to forward the required documentation to the Executive Body in Geneva).

Furthermore, Italy was in fact often represented by the officials of a powerful interest group potentially affected by the agreement, such as the State-owned electricity company ENEL, that had no difficulty in covering the expenses of its personnel. This situation also continued during the discussions of the Helsinki and Sofia Protocols. Although ENEL is a public body and was able to provide valuable technical expertise that the government otherwise would not have had, it is clear that the presence at the negotiation table of an actor whose interests were so directly at stake posed delicate questions of the adequate representation of the interests of the country as a whole.

Only after the creation of the Environment Ministry in 1986 did the Italian presence in the negotiation process became more continuous and substantive, especially in regard to the VOCs and SO_2 (Oslo) Protocols. The new Ministry saw in such negotiations an opportunity to assert its role on the domestic scene and to promote 'green' policies. Since it had a budget of its own, the Ministry was able to allocate the resources required for the participation of national delegations to the negotiations. This also reduced the role of interest groups, even though such actors were consulted in defining the Italian position at the negotiations. Moreover, representatives of affected interest groups continued to be included in the delegations to some extent, as was the case of the association of chemical industries *Federchimica* that took part in the VOCs discussions, or they were in fact represented by the Ministry of Industry.

It is interesting to note that consistent opposition to all the Protocols (with the exception of EMEP, which is not surprising if one considers that it does not introduce limits to emissions) came from potentially affected economic interest groups. In the case of the Helsinki Protocol ENEL, the oil companies and the association of industries *Confindustria* – backed by the Industry Ministry – strongly opposed the agreement. ENEL opposed the Sofia Protocol. In the case of the VOCs Protocol, strong opposition came from the oil, chemical, auto and power generation industries backed by the Industry and Transportation Ministries.

A notable aspect of the Italian case can be seen in the fact that the authorities had very few 'hard' facts or information, such as abatement and damage costs, at their disposal in order to determine what the interests of the country actually were, and thus to determine the Government's position at the negotiation table. A rough estimation of national emissions for the SO_2 agreement was elaborated only very late, in 1984, which was the year in which the Protocol was signed! A careful estimation of VOC emissions and of the reduction possibilities was carried out by ENEA (on the basis of data collected within the CORINAIR project), but it became available only in 1992, i.e. *after* the Protocol was signed by Italy (before that date only estimates of *industrial*, but not *vehicle*, emissions were available). A study of critical loads carried out by ENEA for the discussions concerning the new SO_2 Protocol became available only one year *after* the agreement was signed. Thus, even though the issue of feasibility of the agreements was raised, knowledge of the actual situation was insufficient due to the lack of trustworthy data. The obvious consequence is that there was no knowledge-based evaluation of the country's actual capability to honour the commitments it was undertaking. Only economic actors, such as ENEL and specific industrial sectors (chemical or vehicle manufacturers), seem to have had some idea of costs they would be called to bear, and on this basis tried to influence the government's position towards the agreements under discussion. However, such evaluations appear to have been of little relevance as far as Italy's positions in the negotiations and the final decision to take part in the international agreements were concerned.

Implementation and Compliance

Italy's substantive compliance record with the obligations of the LRTAP regime are, at the best, mixed. As shown in Table 10.1, the target of the Helsinki Protocol for sulphur oxides has been met, whereas it seems very unlikely that Italy will be able to achieve those for NO_X and VOCs.[12]

The reason for this outcome lies in the highly dichotomous character of the implementation process: whereas the administration has remained by and large inactive, several large emitters have achieved significant emission reductions.

Measures taken by administrative authorities with regard to the LRTAP regime have been almost non-existent. Even the ratification and transposition of these agreements into national legislation have occurred with considerable delays[13] (that, to be sure, is also true in transposing international and European provisions into domestic legislation in other policy fields as well). And once legislation is passed, the required administrative activities do not ensue. Analysis of the air pollution policies carried out in three different Regions (Tuscany, Lombardy and Veneto) has found no evidence that these have been influenced in any way by the international agreements under consideration.[14] The only activities directly generated by the LRTAP in the public sector have been the monitoring and research introduced by the EMEP Protocol. However, such 'technical' activities have encountered very substantial difficulties, not for reasons of political choice, but simply due to insufficient financial, technical and human resources. This is, once again, symptomatic of the lack of interest in the issue.

In such a context, it is not surprising that a reduction of NO_X and VOC emissions, generated by a large number of diversified sources and, therefore, requiring complex public policies, has not occurred. Vehicle traffic emissions can help illustrate this point: until now, authorities have been completely unable to produce strategies capable of reducing road transportation for both passengers and goods, whereas less polluting modes of transportation (urban and extra-urban) have continued to lose ground.[15] Notwithstanding concern for consequences of vehicle emissions on human health. Another example is energy supply, where a policy 'by default' has progressively developed: since authorities proved incapable of promoting a coherent strategy of supply diversification (be it based on coal, nuclear or renewable sources), the policy was *de facto* decided by SNAM (belonging to the State-owned ENI group) that promoted a widespread diffusion of methane gas throughout the country.[16] The use of methane certainly contributed to reducing sulphur oxide emissions, but at the same time caused an increase in those of nitrogen oxides. In this respect, energy policy acted as an intervening variable in the implementation of the LRTAP regime in Italy.

Table 10.1   Italy's compliance with international acid rain commitments

Pollutant	Reference level and date	Commitment	Situation in 1993 in Italy
SOx	3211 kt (1980)	30% reduction by 1993 (Helsinki Protocol)	Target achieved (1,988 kt by 1989)
NOx	1793 kt (1987)	Stabilisation by 1994 (Sofia Protocol)	Current levels probably higher than in 1987
	1704 kt (1986)	30% reduction by 1998 (Sofia Declaration)	Target difficult to achieve
VOCs	1879 kt (1988)	30% reduction by 1999 (Geneva Protocol)	Current levels probably higher than in 1988

Source: OECD 1994.

On the other hand, significant measures have been adopted by a number of large industrial firms aimed at obtaining substantial emission reductions: use of low sulphur content fuels, and desulphurisation and denoxification equipment installed at power plants (ENEL), use of methane gas instead of oil, VOC abatement thanks to the use of water coating rather than solvents in the painting of vehicles (FIAT), reduction of benzene and sulphur contents in petrol and adoption of new sealing and vapour recovery systems for storage tanks and petrol distribution (AGIP), new technologies for the production of fertilisers that reduced VOC emissions (Montedison).[17]

One interesting aspect in this regard is that many of these measures have often been adopted anticipating formal legislation or have gone beyond legal requirements. The reasons for such behaviour are to be found in the general strategies of such firms, that prefer to choose the timing of their investments themselves, or who are influenced by similar measures adopted by their competitors in international markets or by the necessity to create a positive 'image' vis-à-vis their customers (especially in the case of the chemical and automobile industry). Only in the case of ENEL, did the legal obligations derived from the LRTAP agreements directly influence the firm's abatement policies.

Finally, an evaluation of the effects of the acid rain agreements should take into consideration the fact that these, together with other European provisions, have played an important role, as previously pointed out, in upgrading national air pollution legislation. Even though the main aim of the latter is to control local air pollution and its negative effects on human health, these policies may well end up contributing indirectly to the solution of the 'acid rain' problem.

Explanatory Factors

The elements emerging from the analysis of the national policy process related to the LRTAP regime indicate quite clearly that Italy's positions were not determined by any aim of maximising net national welfare as far as air pollution is concerned. Such a conclusion is supported by the fact mentioned previously that, as the negotiation of the various LRTAP agreements proceeded, the involved Italian actors had almost no information at all about the damage costs implied by acid depositions. Only later would it become known that phenomena typically connected to acid depositions (water acidification, damage to plants and monuments) actually also existed in some areas of the peninsula, though to a lesser extent than in other parts of Europe and North America. For example, policy actors were not initially aware of the fact that Italian soils are in general less vulnerable to acid depositions. Moreover, the actors had no precise knowledge of the fact that Italy was a net exporter of pollutants, and that therefore from a 'rational' standpoint Italy had more to lose from agreements compelling it to reduce its emissions than it had to gain. Furthermore, in this connection it should be noted that a considerable amount of Italy's emissions, thanks to its geographic position, precipitated into the surrounding seas, therefore causing no 'offence' to neighbouring countries. Thus, even if such data would have been available, besides not having 'hard' reasons to join the LRTAP regime for its own welfare, Italy would have had little reason to tackle its emissions even for the sake of the welfare of *other* countries.

On the basis of such reasoning, one would well have expected Italy to be among the countries actively opposing the LRTAP agreements or at least trying to 'water' them down as much as possible during the negotiation process. Yet Italy *did* sign, and eventually ratify, all the LRTAP agreements, and actually went even further by joining the more exacting agreements, as in the case of the NO_X '30 per cent Club'.

An examination of the domestic politics process, and especially the role played by policy actors, i.e. those 'individual or collective entities that expresses unitary objectives identifiable inside the process',[18] helps shed some light on the reasons for this apparently irrational behaviour.

Empirical analysis indicates quite clearly that Italy's positions with regard to such agreements were elaborated by a rather small and closed policy network formed by: scientific experts, mainly belonging to public research bodies (CNR, ISS, ENEA); a political-bureaucratic actor, the Health (subsequently Environment) Ministry; and a very limited number of large firms with vested economic interests involved (ENEL, oil companies, FIAT, chemical companies) backed by the Industry Ministry.

Other actors were notable for their absence. No significant presence was found of political parties (including the Green Party) or of the ecological movement. A number of Ministries, such as those of Public Works, Cultural Heritage, Agriculture and Finance, with responsibilities potentially affected in some way by atmospheric pollution, also did not appear in the decision making processes. There was even a considerable indifference on the part of actors specifically affected by the problem of acid deposition, such as the National Parks administrations.[19] Finally, industrial firms specialised in the construction of abatement equipment failed to exploit the opportunity offered by the issue to promote their economic interests.

The international regime contributed to creating and stabilising such a network by offering an arena where actors could interact with each other.

However, within this network behaviour patterns and rules of the game were not yet consolidated and the actors involved did not share common views of the problems at issue. Moreover, its level of institutionalisation was low. Two opposing coalitions can be identified within the network: economic interests, on one hand, and environmental concerns, on the other. Thus it is not surprising that the process featured significant, though not extreme, levels of conflict.

Substantial opposition, as mentioned above, to the agreements emerged in relation to all the Protocols, with the exception of EMEP, from the powerful economic interests that would be called on to bear the costs of the required abatement measures. Lobbying was aimed at obtaining modifications to the measures under discussion (sometimes of those already in force), motivated by cost/benefit considerations rather than by forecasts of environmental impact. It is also interesting to note that there was some conflict among such economic actors themselves. When the Italian Government decided to sign the Helsinki Protocol, there was considerable disagreement as to which sectors should undertake the necessary measures and thus bear the costs. (The conflict was solved by asserting the principle that each sector – domestic heating, vehicle traffic, industries, power generation – should bear the costs of reduction of its quota of emissions.)[20]

On the other side, support for the agreements came from the competent Ministry (first Health, then Environment) and from the small specialised scientific community described above. As far as the former is concerned, international agreements were perceived as an opportunity to assert itself as well as to develop a still weak national environmental policy. To be sure, as has been pointed out previously, Italian policy in this field has developed to a large extent thanks to international inputs rather than from domestic factors. The Ministry played an active role keeping the process alive in the face of a lack of interest if not the outright opposition on the part of the other actors and directing it towards desired results. This was especially true after the Environment Ministry had been set up. This can be seen, for example, by the fact that when the Ministry started to deal with LRTAP policies, representatives of companies were no longer allowed to participate directly in international negotiations (though they were in fact represented through the Ministry of Industry).

Experts played a fundamental role by providing scientific support. Institutions connected to the Health Ministry, such as the above mentioned

Central Commission against Air Pollution (CCIA) or the ISS, but also other agencies like ENEA and CNR, were relied upon completely for assessing the proposals emerging from the international negotiations (though they were often unable to carry out such work in a timely fashion due to the lack of previous research in the field of emissions and acid depositions). However, a lack of co-operation between the various centres resulted in disorganised data and information that were hardly sufficient to serve as decision aids.

On some occasions the experts were able to act as a *liaison* between the environmental and the economic interests involved in the process. The CCIA (at least up until the Sofia Protocol) also played a co-ordinating role by bringing these actors together. The meetings of the Commission were in fact the time and place in the process where all the participants in the decision making process met together. National positions and interests were identified through this process.

However these two actors – the Health Ministry and the experts – were certainly not strong enough, as compared to their opponents, to carry the day. The scientific community was small, isolated and hardly influential. For the Health Ministry, although interested in asserting itself as the administration responsible for pollution policies, and especially for air pollution (it had promoted the 'Antismog' Act of 1966), the environment represented a very marginal concern in comparison with its more traditional medical tasks. The Environment Ministry created in 1986 was obviously much more focused on this as well as other environmental issues, but it was not a political 'heavy weight' within the Cabinet.

A decisive role in this situation was played by the Ministry of Foreign Affairs, whose position was determined by diplomatic and international relations considerations rather than by a substantive interest in the contents of the agreements and their consequences for the other national actors involved. It was mainly interested in seeing that Italy would fall in line with the positions of the other countries once *they* reached an agreement. This is shown by the repeated requests to the CCIA to appoint not just experts and minor officials to the Italian delegation but also higher ranking civil servants (senior executives from the Ministry of Health or government politicians) and also by the frequent written communications (memoranda, etc.) sent by the Minister of Foreign Affairs to the Secretary of the CCIA in order to keep him informed of the results of meetings. In order to achieve its aim, the Ministry occasionally played an active mediating role in resolving conflicts among the other actors, as in the case of the Helsinki Protocol. In the end all this proved successful in making the Italian Government sign. Incidentally, the influence of foreign policy can also be explained by a procedural peculiarity, i.e. the fact that it is the Foreign Affairs Select Committee, rather than the Environment Committee, in each branch of Parliament that has the responsibility of ratifying such international agreements.

Finally, it should be noted that the development of the LRTAP regime fostered a process of social learning by attracting the attention of national actors to the problem. In particular the regime was successful in fostering the creation of a group of scientific experts (within the CCIA), in involving sections of the national scientific community in the international debate on the topic and in generating at least some research and measurement activities in the field, albeit at a quite limited scale, that eventually produced evidence of the existence of the phenomenon of acid depositions and their damages also in Italy. In this respect, the international regime produced an increase over time of knowledge of the actors about the phenomenon, at least within the scientific community; the importance of epistemic communities, pointed out by P. Haas (1990), appears to be confirmed also in the case of acid depositions. The social learning process, albeit to a lesser extent, affected other policy actors, making them aware at least of the existence of the acid rain issue and, more importantly, of the concern of other nations about it. If we conceive social learning to involve 'a relatively enduring alteration in behaviour that results from experience' (Heclo, 1974, p. 306), it occurred to a more limited extent, as discussed above. Whatever policy measures were taken to curb emissions did not, however, originate within the domestic polity (since acid rain has never been perceived as an issue in Italy). They must instead be ascribed to the international regime, that induced national authorities to adopt some response to its obligations. Also, international commitments empowered domestic actors seeking opportunities to push environmental protection policies. Once the Environment Ministry was set up in 1986, it used international agreements as a persuasive argument to push its domestic policy agenda. Furthermore, in the case of a limited number of large economic actors (chemical and vehicle manufacturers) direct learning also occurred by imitation of emission limitation measures previously adopted by foreign competitors that were more advanced in this field because of earlier national policies. In this respect, one could say that the regime caused a process of ecological modernisation in specific sectors of the economy.

The Italian Policy Style: 'Following Suit'

The analysis carried out above indicates that a substantive internal policy process was in fact activated by the negotiations concerning the agreements of the 'Acid Regime'. This is surprising if one considers that:

- there was a general lack of concern for the acid rain issue;
- for a long time, there was no information indicating that the phenomenon caused serious damage to environmental or cultural goods within the national territory; and
- consistent opposition to such agreements from powerful economic actors emerged again and again.

Yet Italy, though hardly active in the negotiation phase in defending its interests, joined such agreements and went even further by undertaking voluntarily some of the most burdensome commitments foreseen by them. In the case of the Sofia Protocol, Italy joined the '30 per cent club'; in that of the VOC Protocol, the Government decided to take 1989 as a baseline, i.e. a stricter target for emission reductions.

Surprisingly enough, such decisions were taken neither for ecological nor for economic reasons, but were determined by the Ministry of Foreign Affairs on the basis of general considerations of foreign policy. The paramount factor that accounts for Italy's behaviour was the necessity of demonstrating 'cohesion and solidarity' with Western, and especially EC, countries, of maintaining good relationships with European nations in general and of safeguarding the country's international reputation and credibility. Italy could not 'lose face' and had to show that it was up to the standards of the most advanced European countries.[21] In essence, it was simply 'following suit'.

By way of contrast, the only case in which the Italian government took a strong position in an international negotiation process dealing with air pollution policy, was the EC policy making process concerning auto emission reductions. As is well known (see e.g. Arp, 1993) the topic was at the centre of a long and heated process among EC countries and vehicle manufacturers. In this case Italy, in contrast not only with the low profile kept in the LRTAP negotiations, but also with its weak capability to pursue national interests in many EC negotiations (as in the field of agriculture) took an outspoken stance that ensured that domestic interests were given adequate consideration. Such a difference in approach cannot be explained entirely on the basis of the interests at stake for one of the major national industrial sectors (vehicle manufacturers). One additional factor is that EC provisions were perceived as much more stringent than LRTAP agreements. Furthermore, and more interestingly for the purposes of this discussion, in this case Italy's opposition was strongly 'legitimised' by the similar positions taken by France and Britain. The Italian government could, therefore, at the urging of a powerful domestic interest group, adopt a strong stance without fear of being isolated or even criticised by the international community.

The specific mode in which the formulation of the decisions to join the agreements occurred has had, however, had serious consequences. The fact that these decisions were taken on the basis of considerations - international relations - totally exogenous to the specific policy sector affected by the agreements, caused the absence of adequate consensus and involvement on the part of the actors that would be called upon to implement them. The genesis of the Italian commitment accounts for the severe deficits registered in the implementation phase. Only a number of large firms adopted measures that allowed a reduction of emissions, but, as discussed above, this occurred for reasons that had little to do with the international agreements. For public authorities air pollution was a local issue, if it was an issue at all. The existence of international agreements as such was hardly felt as a significant obligation.

Notes

1. I would like to thank R. Florio and F. Paron for their assistance in carrying out field research and contributing to the drafting of the research report on which the present chapter is based.
2. Lega per l'Ambiente, *Aria pulita. Inquinamento atmosferico e deposizioni acide*, Atti del Convegno internazionale, Perugia, 8-9 July 1986, Rome, 1988.
3. In 1991 Italy exported approx. 216,000 tons of sulphur and imported 112,000; it also exported 97,600 tons of nitrogen whereas imports amounted to 47,900 (EMEP 1994).
4. For more on this see Lewanski, 1997.

5. The densely inhabited and highly industrialised Po Valley is in fact surrounded by mountains on 3 sides.
6. A number of subsequent provisions (no. 880/73, 393/75, 400/82, 792/82, 105/87, 240/88; decree of November 14, 1995) regulated maximum contents of sulphur allowed in different types of fuels used for household heating, industries, power plants and diesel vehicles.
7. For a more detailed analysis of the 'Antismog Law' and its implementation see Dente, et al., 1984.
8. USLs are responsible for management of hospitals and delivering health services in general, but also for guaranteeing the quality of the environment in which people 'work and live' (Act n.833/78). USLs include technical apparatus (*Servizi d'Igiene e Prevenzione* and *Presidi Multizonali di Prevenzione*) responsible for control and monitoring activities related to environmental conditions in general. Such responsibilites are now being passed over to regional environmental agencies (ARPA).
9. For purposes of comparison, in that same year public sector investments in pollution control and abatement (expressed as per cent of gross fixed capital formation) was 1.2 per cent in the USA, 3 per cent in Japan, 0.9 per cent in France, 2 per cent in Germany (OECD 1993, 295).
10. The Agency was formally set up only in 1994 (Act 61). Although it did not play a role in the policy under examination up to that year, it has become a relevant actor in more recent years (in part because it inherited personnel transferred from ENEA who had been previously involved in the acid rain issue).
11. Created by previously discussed Act no. 615/66.
12. Data produced by the Swedish EPA in 1993 showed that SO_2 emissions per capita and per sq km had also decreased from 89.4 in 1985 to 34.9 in 1989. Figures produced by the *Unione Petrolifera Italiana* (the association of Italian oil companies) estimated total SO_2 emissions to be 1,558 Kt for 1992 (against 3,716 Kt in 1980). NO_x per capita and per sq km emissions increased from 30.7 kg. in 1985 to 35.9 kg in 1989 and from 5.8 tons per km to 6.8 tons per km respectively.
13. The Geneva Convention of November 1979 required 30 months to be ratified (Act no. 289 of April 1982). The EMEP Protocol of 1984 required 4 years (Act no. 488 of October 1988); the Helsinki Protocol of July 1985, ratified by the same Act, required 39 months, the Sofia Protocol 38 months (Act no. 39 of January 1992), the VOC Protocol of June 1991 required 46 months (Act no. 146 of April 1995).
14. See Lewanski, Florio and Paron, F., Transboundary air Pollution and Acid Depositions, Italian Acid Rain Policies, research report to the EU, DG XII, Contract no. EV5V-CT92-0185, October 1995.
15. Density is 102 vehicles per Km. as compared to 74 in Great Britain, 67 in (West) Germany and 35 in France. Passengers transported by private vehicles in urban areas as compared to public means of transportation has constantly increased from 80 per cent to more than 95 per cent over the last decade (Conte and Melandri 1994, 317).
16. At the end of the 1980s methane accounted for approximately one fourth of total energy consumption.

17. Resulting in a reduction of VOC emissions by 27 per cent (from 3,027 tons to 2,211 tons per annum) between 1990 and 1992, by a further 5 per cent during 1993 and by 19 per cent in the following year.
18. Dente, B., *La valutazione dell'efficacia dell'intervento amministrativo nel Mezzogiorno: l'approccio dell'analisi delle politiche pubbliche*, IRS/Formez, August 1992, p. 5.
19. In only one case, that of the Stelvio National Park, was a station for monitoring acidification pollutants set up in 1984, but this occurred as a part of a research project financed and managed by the CNR.
20. The power (ENEL) and the oil (UPI) industries also disagreed on which strategy - installation of abatement equipment in the power plants or fuel desulphurisation - should be used in order to obtain abatement of sulphur emissions for electricity generation.
21. As indicated by an interview with an expert of the ISS involved in the negotiations, this attitude emerges also in the behaviour of Italian delegations who would refrain from reporting scientific data that did not conform with that presented by other delegations in order to avoid their disapproval.

References

Ågren, C. and Elvingson, P. (1997), *Still with us*, Special Report of the Swedish NGO Secretariat on Acid Rain.

Arp, H. (1993), 'Technical regulation and politics: the interplay between economic interests and environmental policy goals in EC car emission legislation', in D. Liefferink, P. Lowe, and J. Mol (eds.), *European Integration and Environmental Policy*, Belhaven, London, pp. 150-171.

Conte, G. and Melandri, G. (eds.) (1994), *Ambiente Italia '94*, Koiné Edizioni, Rome.

Elvingson, P. (1993), 'Younger strands now affected', *Acid Rain*, n. 5, pp. 8-9.

Haas, P. (1990), *Saving the Mediterranean. The Politics of International Environmental Cooperation*, Columbia University Press, New York.

Heclo, H. (1974), *Modern Social Politics in Britain and Sweden. From Relief to Income Maintenance,* Yale University Press, New Haven.

Istat (1993), 'Statistiche ambientali', Rome.

Lewanski, R. (1986), *Il controllo degli inquinamenti delle acque: l'attuazione di una politica pubblica*, Quaderni ISAP Saggi, Giuffré, Milan.

Lewanski, R. (1992), 'Il difficile avvio di una politica ambientale in Italia', in B. Dente and P. Ranci (eds.), *Quarto rapporto CER-IRS sull'industria e la politica industriale in Italia*, Il Mulino, Bologna, pp. 27-82.

Lewanski, R. (1997), *Governare l'ambiente. Attori e processi della politica ambientale*, Il Mulino, Bologna.

Lewanski, R. and Liberatore, A., (1990), 'Italian Environmental Policy: Approaching Maturity', *Environment*, June, pp. 10-15 and 35-40.

Malaman, R. and Paba, S. (eds.) (1993), *L'industria verde,* Il Mulino, Bologna.

Ministero dell'Ambiente, (1989 and 1992), *Relazione sullo stato dell'ambiente,* Rome.

Mosello, R. and Tartari, G. (1988), 'Deposizioni acide: il caso di alcune località dell'Italia nord occidentale', in Lega per l'Ambiente, *Aria pulita, Atti del convegno internazionale,* Perugia, 8-9 luglio 1986, pp. 103-111.

OECD (1993), *Environmental data. Compendium,* Paris.

OECD (1994), *OECD Environmental Performance Reviews, Italy,* Paris.

11 The British Case: Overcompliance by Luck or Policy?

SONJA BOEHMER-CHRISTIANSEN

Introduction

The aim of this chapter is to show the embeddedness of international negotiations in domestic politics. Demand and supply for environmental protection are not pre-determined by objective factors, though these clearly play a role. Technological and structural economic change induced by domestic politics were a major force for change in emission behaviour. The United Kingdom (UK) ratified the Geneva Convention (LRTAP) in 1983, but refused to join the '30 per cent Club' or sign the First Sulphur Protocol. The Sofia NO_X Protocol, on the other hand, was signed in November 1988 and ratified in 1991.[1] In June 1994 the volatile organic compounds (VOC) and the Second Sulphur Protocols were signed. Britain now feels well on course concerning overall compliance with LRTAP. The UK was subject not only to the UNECE regime, but also to European Community (EC) regulations as negotiated in Brussels during the same period. Except for NO_X, where the target is a 60 per cent reduction by 2010 from a 1990 baseline, this European supranational regime does not impose more stringent overall targets.[2] Rather it can be seen as the means by which LRTAP was implemented, as well as the major political driving force behind Britain's policy developments.

In 1994 it became clear that British compliance with the '30 per cent Club' for SO_2 had been achieved, the reduction in 1995 being given as 37 per cent. For NO_X, stabilisation of 1987 levels by 1994 was also achieved, though here baseline measurements are open to some uncertainty. Between 1980 and 1990, NO_X emissions were estimated to have increased from 2.3 to over 2.7 million tonnes and a 30 per cent reduction between 1988 and

1999 will be required under UNECE.[3] Nitrogen remains the main problem for both regimes. The number of NO_X sources is far greater than for sulphur (though less than for VOCs), with both power stations and transport as major contributors. NO_X emissions are expected to decline for a decade before rising again unless additional measures are taken. In current NO_X negotiations the UK desires 'a significant reduction'. For VOCs implementation regime is to remain much more 'voluntary' but is considered to be on course.[4]

In spite of earning the name 'dirty man of Europe' during the 1980s, by 1997 the UK had overcomplied for sulphur dioxide (SO_2) and could hardly be considered a dragger or laggard any longer. It achieved this with a minimal use of economic subsidies, which were only used for implementing the switch from leaded to unleaded petrol. Britain insisted that its bad image has been shed and sees itself as a world leader in combating air pollution.

The Problem Viewed from Within

The 'acid rain' problem was brought to the attention of Britain primarily by external actors at a time when British society believed to have solved its air quality problems. The problem of domestic acidification had first to be recognised and evaluated for its significance. 'Blame' for exporting damage (not just pollutants) had to be proven to the satisfaction of the British pollution control regime. This evaluation took time and led to the realisation of widespread, if minor damage to some natural but not agricultural eco-systems. British science never accepted any major causal links between acid deposition and 'Waldsterben'.

The British Isles are a densely populated, much deforested country with a long history of industrialisation. Many surface waters and soils are acidic, but these predominate in the least populated areas. This acidity, as clear lakes, was accepted as natural by many people until science persuaded them otherwise. The prevailing winds are clean, coming from the West and much deposition of pollutants takes place at sea. Britain is therefore a significant 'exporter' of air pollutants, receiving much less from abroad except during meteorological episodes. Lime rich soils predominate in the populated middle, east and south where few forests have been replanted and indeed reforestation had been resisted in the past.

The forest death hypothesis which so activated several other nations, had little salience at the time and was laid to rest to the satisfaction of scientists in the early 1990s. Acid peat moors enjoyed protection for grouse and deer hunting. Where damage to nature was observed and attributed to the acidification, such as for some surface waters and soils primarily in mountainous areas of igneous rocks, population densities were low, and in those areas where already in the mid-1980s, the ruling Conservative Party (1979-1996) had few supporters left, as they belonged mainly to the supporters of Celtic nationalism. Acid damage was only beginning to be measured and understood. It never attracted electoral support attention. The stage was therefore set for outsiders to intervene.

Britain has a long history of controlling air pollution from stationary sources dating back to the 14th century when the burning of 'sea' coal (as distinct from charcoal) was banned in London when Parliament was sitting. During the 19th century it became in 'vanguard' in cleaning the air around industrial premises, and in the post-war era legislation assisted 'fuel switching' by households (from coal to gas) and thus in banishing the infamous 'smog' from the streets of its cities.[5] For power stations it had in fact pioneered fluegas desulphurisation (FGD) in the 1930s. In 1979 the last of these in London (Bankside) was closed inspite of cutting SO_2 emission by 97 per cent. Costs and break-downs had convinced British engineers that FGD was not a technology of the future. The response to foreign accusations was scepticism and suspicion. Were not Norway trying to force its gas onto British markets and Germany to protect its competitiveness by spreading its environmental costs? Government called for 'sound' science and cost-effective solutions. Hence initially at least, there was no domestic demand for abatement. Few benefits could be recognised by the general public, though during the mid-1980s green pressure groups tried hard to weaken this complacency.

When widespread reluctance to accept domestic damage at a time when air quality was measurably improving weakened during the 1980s, this was largely due to their efforts supported by the British scientific community. Nevertheless, acid rain was never perceived by the British public or economic institutions as a serious problem, not even as one that will seriously damage old buildings - old buildings should show their age. Perceived environmental damage costs were low but would rise in time to include the political costs imposed by EC negotiations. The costs of preventing alleged and later recognised domestic damage by 'end-of-pipe' solutions were, however, considered quite unacceptable. Emissions were expected to decline over time, and time was what British negotiators needed most.⌐

The Domestic Negotiating Baseline and Subsequent Institutional Adjustments

By the 1970s, coal-burn in power stations was widely recognised as creating several environmental problems. An official body was set up in the mid-1970s to consider the problems and reported in 1978. This Commission on Energy and Environment revealed official perceptions that led to the acceptance of LRTAP in its 1979 form, and baseline policies which the Conservative Government inherited in 1979. It would take about a decade to turn these around. The report also reflects the plans and interests of the major political actors during the early stages of the acid rain negotiations in Britain: the Central Electricity Generating Board (CEGB) and British Coal (BC), both state owned monopolies, as well as the Alkali Inspectorate in charge of air pollution control which enjoyed non-confrontational relations with these industries. These bodies formed a powerful alliance against 'precaution' over acid rain until the mid-1980s.

The CEGB concluded that domestic coal had a bright future in Britain provided domestic environmental problems associated with spoil disposal, discharges to water, subsidence and open cast extraction had been solved. Coal and nuclear power were to be the fuels of the future with energy security, not cheapness, as main criteria. This agreed with the International Energy Agency's 'coconuc' policy adopted in response to the oil shocks of the 1970s.[6] CEGB had no reason or no internal incentive to consider acid emissions as damaging, or to seek for means to increase the price of coal as cross-subsidisation to nuclear was possible and remained hidden until privatisation. Prevailing levels of sulphur were considered satisfactory from the perspective of human health and agriculture. Nitrogen oxides from coal combustion were not identified as a pollution problem. The Commission endorsed the Geneva Convention, but concluded that :

> more research is required into the causes and mechanisms of the effects of acidification of lakes and rivers, into possible control technologies, into cost-effective technologies.... into the adverse environmental impacts of existing

emission control technologies and into techniques for ameliorating the effects of excess acidity in the problem areas.[7]

Air quality had been improving because of measures which included households and removed power stations emissions from areas where they might cause harm by emitting via tall chimneys. It was recognised that significant reductions over the next 20 years would be possible; these were not needed in the UK because they would make little difference to concentrations in the UK 'or on acid rain deposition elsewhere'. FGD was rejected because it produced too much waste and created other environmental problems, thus accepting the CEGB argument 'that environmental considerations' would militate strongly against its use in Britain.[8] The Commission was adamant that tall stacks did not enhance long distance transport, claiming that this was almost independent of chimney height, though it was admitted that on 1974 OECD data, Norway was receiving 25 per cent of its sulphur deposits from the UK, i.e. 2-3 per cent of total UK output. It identified oil as a major SO_2 source (50 per cent) and the displacement of fuel oil by coal was recommended to reduce local emissions. Referring to an OECD report on sulphur deposition and considering submissions from Sweden, it was noted that:

> most sulphur is deposited on or near the regions where it is emitted and that depositions in Scandinavia are many times smaller than those in the industrial regions of Europe.[9]

The CEGB was recommended for carrying out research on the movement and dispersal on major plumes over the North Sea. So far evidence was 'insufficient', 'beyond recommending the need to pursue the necessary research'. Since evidence had come mainly from within the CEGB research division, it would soon be challenged. In the meantime, however, government and public were reassured: sulphur emissions had declined and were estimated to fall. Air quality was improving, and total emissions, while temporarily up in 1980, were generally falling. This made 1980 a particularly unwelcome baseline for UNECE percentage reductions! The Commission expected that by the mid-1980s oil burning in power station would have ceased and transport have become the primary source of. No active control measures were called for, since the dispersal techniques required for sulphur dioxide were considered as effective for NO_x. Hence the baseline for British acid rain policy remained the 1974

Plan for Coal which boosted clean coal R&D and intended to develop British machinery manufacture and mining technology for coal.⌉

This energy strategy came unstuck when Margaret Thatcher, who was a firm supporter of nuclear power and private industry but an enemy of the coal industry, took over as Prime Minister in 1979. She saw the coal mining industry, especially the miners, as an enemy within for purely political reasons and did not recognise that 'clean coal' would have made nuclear power more viable. In her, the external political forces wanting British coal to be either cleaned or abandoned had a major ally. The 'national' cost-benefit balance was beginning to change for primarily political reasons. Fortunately for Mrs. Thatcher many new natural gas fields had been discovered in Europe and elsewhere during the 1970s. The stage was set for the taboo against the inefficient conversion of gas into electricity to be undermined.

So Britain could not have entered international negotiations to reduce acid emissions from a more negative baseline made up of energy policy considerations as well as regulatory weaknesses. Enacted legislation for stationary and mobile sources made no reference to foreign targets that might require protection from transboundary damage. Air pollution from large stationary sources was regulated by air quality standards related to 'best practicable means' as interpreted by a national inspectorate with considerable discretionary powers and a tendency for site specific decisions. This would make compliance with future aggregate mandatory limits, or 'national bubbles' difficult, unless, of course, these could be achieved without regulatory intervention. Mobile sources remained largely unregulated, apart from requirements under a permissive UNECE trade regime. The major British legislative instrument of the early 1970s, a major phase of environmental legislation in many countries, the Control of Pollution Act of 1974 (COPA) had remained weak with regard to air pollution.' New laws and institutions would therefore be required to cope with international and EC environmental demands before practical steps could be taken. These developments had to wait until the late 1980s because the British state had other priorities, including the privatisation of the fuel and power industries. ⌈A brief description will illustrate the profound changes which have taken place: a) a new legislative framework, and b) a new organisational structure. ⌉

a) The complex regulatory apparatus of the Environmental Protection Act (EPA, 1990) gives increased powers on emission controls to central government. Under the Part I industrial processes could be required to upgrade their equipment, thus allowing government to ensure future emission cuts. However, cost effectiveness would remain the primary criteria. However, compliance was to be brought about by the EA only for large emitters (part a processes), with local authorities responsible for part b processes. The industries thus regulated would include fuel and power, the underlying principle being best available technology not entailing excessive cost (BATNEEC). The applicable regulations under EPA are primarily European:

- Air Qualities Standards Regulation to implement 80/779/EEC on air quality limits and guide values for SO_2 and suspended particles, and 85/203/EEC for nitrogen dioxide, making the Secretary of State responsible for ensuring compliance;
- The Control of Industrial Air Pollution Regulation (Registration of Works), which implements directive 84/360/EEC on combating air pollution from industrial plant and without which the LCPD could not be incorporated into UK law;
- The Health and Safety of Emission into the Atmosphere Amendment, which amends regulations defining types of processes and emissions that are scheduled, i.e. are regulated by the Inspectorate and takes in 84/360/EEC on combating air pollution from industrial plants.

Subsequent environmental regulations and legislative adjustments are dealt with in section 4.

b) Powerful new 'regulators' were set up for the privatised industries who, though ultimately responsible to Parliament, are without ministerial supervisions and possess an inordinate amount of discretion. However, they are responsible only for economic, not environmental matters, though this may be revised in future.[10] The environmental regulators did not escape reorganisation and partial privatisation. The pollution inspectorates were first merged into Her Majesty's Pollution Inspectorate (HMPI) and then absorbed into a new, much larger body, the EA under the label of 'integration'. Talked about for many years, the EA started operating in 1996. As a quasi-governmental body with a statutory footing, it must raise as many of its own resources as possible. The 1995 Act which set up the EA began its passage through Parliament in November 1993 with the added purpose to consolidate clean air legislation. It deals with criminal proceedings, disclosure of environmental information and increases the enforcement powers of inspectors and the Environment minister. Courts, which traditionally have played but a minor role in the enforcement of air pollution regulations had potentially enlarged their role via prosecutions brought by the EA.

With new powers derived from EC legislation, air pollution control has therefore been significantly centralised by taking powers from inspectorates and local government. The state, or rather the Department of the Environment (DoE, since 1997 DoE, Transport and Regions) could now draw up annual emissions plans which require EA authorisation for each industrial sector. The powers of Local Authorities for small emitters were also increased, subject to 'guidance notes' from the DoE/HMIP/EA which indicate air pollution control standards. The potential for tension between the DoE, the EC and EA is large and would be tested by political pressure for more stringent regulation. By the mid-1990s, Britain possessed a national emission plan and a separate emission programme for the EC Commission. The programme would ensure compliance with EC acid rain regulations for the Energy Supply industry; the national plan is incorporated into the national programme, but does not say how emission reductions are to be achieved and does not include any FGD commitments. Rather, these reductions are to be achieved by replacing current coal based technology with gas turbines and imported low sulphur coal. Plan and programme were therefore not identical, but both were threatened by proposals for the SSP (see Section 4). For NO_x, the regulation of mobile sources has also become easier but is now almost entirely a matter for the EC. In the light of the profound changes to the domestic legislative and environmental regulatory system, one could expect Britain to negotiate, during the period of change and adjustment to national and European priorities, not only for lower targets, but primarily for time to do so.

Britain's Role in Intergovernmental Negotiations

The Government had to play a four level game: inside 'the corridors of power', at home, inside the EC and with UNECE. In the EC Britain had far less freedom, especially when majority voting became possible in the environmental arena, than in UNECE. Bargaining at UNECE became increasingly influenced by EC legislation which tended to produce concessions through linkage politics, including commercial considerations and the political ambitions of the EC Commission. The baseline of these four negotiations as outlined in Section 2 accounts for the initial 'dragging' role of the UK, that of sceptic but willing researcher, of procrastinator but skilful negotiator.[11] It does not explain British eventual compliance and even overcompliance. Since the same public servants and scientists were involved in both UNECE and EC acid rain negotiations it became impossible in practice, to distinguish between the two regimes. The former can be seen as defining the broad objectives and targets, the latter the measures used to implement the former. A major factor explaining the British negotiating role is therefore the importance attached to maintaining environmental sovereignty. Viewed as a legal condition essential for the implementation of environmental and economic rationality, as well as for the sake of technological flexibility, this led to a British preference for LRTAP as negotiating forum over the EC. While cautious and negotiating for time to allow low cost or cost free adjustments at home, Britain nevertheless remained a willing and co-operative partner in Geneva, fighting its major battles over acid rain in Europe.[12] It made major contributions to strengthening the scientific basis of the UNECE regime. An irony of this approach is that by relying on scientific evidence to reduce its culpability through the critical loads approach, this in fact revealed greater than suspected damage at home and thus also assisted in creating the domestic conditions for policy change.

The role of Britain as laggard is explained in more detail below, but one of the factors deserves mentioning here. It is the effect on intergovernmental negotiations of the domestic constitutional tradition of collective responsibility. This means that no minister may 'speak' out publicly for the environment above what has been agreed in Cabinet, which in turn reflects the importance of consultation with interested parties in domestic policy formulation and implementation considered to be a single, continuing process.[13] Demand and supply of environmental protection are brought into balance by administrative effort between the signing and ratification of treaties. Given this nature of the domestic policy process and the profound organisational changes that affected not only it, but also the industrial structure, meant that Britain needed time to bring external demands into harmony with domestic perceptions and capacities. Lengthy domestic bouts of bargaining preceded international ones. For example, co-ordination with the VOC emitting industry alone involved consultation with 16 trade associations, with the DTI being the other 'interested Government department'.[14]

Domestic policy systems are difficult to turn around, as are energy technologies and hence policies that are meant to be effective. The negotiations for the SSP illustrate these broader principles clearly. The tradition not to sign a treaty unless implementation is felt to be feasible requires a lengthy domestic consultation process with domestic polluters, experts and regulatory agencies. This continues in the period after signing an IEA when technical experts and trade associations in particular are asked to devise more detailed implementation strategies, in our case, for example, the two emission reduction plans, one for the EC and the other for UNECE. This stage may require revisions in domestic legislation and involves much 'scenario' building, i.e. taking into account trends in the economy and technology. The UK therefore makes a difficult negotiating partner who tends to know what it wants and how to get it by tough negotiations. In brief, the UK strategy in response to foreign pressures was to negotiate lenient requirements and gain time. Adding economic to scientific modelling became attractive to the UK when it was realised that critical loads too would lead to unacceptably high emission reduction demands.[15]

The UK strategy paid off handsomely from the perspective of limiting public expenditure:

- investments in FGD have largely been avoided
- the introduction of cleaner vehicles was progressing steadily without subsidies
- the costs of regulating new pollutants was devolved to industry
- implementation for small and mobile sources has been devolved
- the scientific community was enabled to make major contributions
- economic instruments were developed.

Implementation and Compliance Strategies

These are outlined in three stages: from the mid-1970 until the mid-1980s when 'dragging' predominated; the latter half of the 1980 when the foundations for compliance were laid; and the period when compliance was achieved, the SSP was negotiated and urban air management becomes the new strategy.[16]

Mid-1970s to Mid-1980s: Resistance and Research

In 1976 the coal miners brought down the Conservative Government of Edward Heath, but an embattled Labour Government lasted only until 1979. In early 1982 the hugely popular but expensive Falklands war was won. The major political event in the acid rain story was the very bitter coal miners' strike of 1984-1985. This largely coincided with the peak in the attention paid to the issue and helped to strengthen the interpretation that acid rain campaigns were directed against the coal industry. Acid rain abatement did not feature in any political party programme during this period, nor did it attracted any leading political figure or institution. For government scientists and the DoE, however, LRTAP began a lasting involvement in intergovernmental negotiations. As there were no immediate policy implications until the EC demanded action, signing LTRAP caused no difficulties. The Departments of Energy (DEn) and Trade and Industry (DTI) came to oppose the convention only later. But tension was already 'preprogrammed'. In a reply to a parliamentary report the DoE promised that it would aim to reduce both SO_2 and NO_x emissions by 30 per cent of their 1980 level by 1990 and that it would pursue these targets 'with urgency and vigour'. It was not allowed to do so. However, policy relevant research in the natural sciences was removed from the CEGB/industry and became a task, in terms of management and funding, of the DoE - a major step in the 'greening' of a ministry largely concerned with local government.

The DoE was supported by young researchers, soon to emerge as the leaders of green pressure groups, who had become suspicious about the hundreds of 'dead' lakes in Scotland that were believed to have been fished out. Suspecting acid rain damage and importing claims from Germany, acid rain became a campaigning issue and enjoyed a brief period of notoriety in 1984/5. The media was attracted, but the practicalities of air pollution control remained in the hands of a demoralised the Clean Air Inspectorate accused of a 'cosy relationship' with industry. Its inspectors prescribed neither FGD nor any NO_x abatement technology the 'best practical means'. Official responses were required, but these came still largely in response to external political pressures. Industries undergoing privatisation were demanding protection against loss in the value of their assets. A fierce battle began against the EC's Large Combustion Plant Directive (LCPD), especially to its application to existing plants.[17] The threat to the British car industry arising from German proposals to tighten vehicle emission standards so that they would require catalytic converters for all cars, was not yet recognised.

The national economy remained in serious difficulties with rising unemployment, a major balance of payment problem and restless trade unions. There were few public resources available for taking air pollution control towards practical implementation. Environmental 'slippage' began relative to North European countries. In Brussels London was fighting for its interests in the coming Single Market along a very broad front ranging from agriculture to environmental regulation, much of the latter perceived as tricks by the Germans to weaken the British economy. An ideology devoted to 'rolling back the state' prevailed and distaste for consensus became the context in which regulations would be decided. Environmental protection was practised opportunistically with strong presumptions against regulation and the idea of green Kensyanism. Competition, value for money and efficiency, not ecological modernisation, were the 'buzz' words.

Mid-1980s to 1990: UK Agrees to Emission Cuts and 'Greens'

Acid rain remained, to a lesser extent, an external issue to be negotiated at UNECE and in Brussels, but environmentalism within government and its agencies began to grow fast. Privatisation, the cutting of public expenditure and controlling inflation, however, remained the priorities of politics and high economic growth rates were achieved through private spending booms engineered by tax cuts. Public opinion also greened as the 'quality of life' improved for many. Official concern for the environment found expression in a rhetoric that greened as the Earth Summit approached and political leaders realised that 'the environment' needed more official support if only for both electoral reasons. In 1987, after much heart searching, Britain signed the Single European Act and legislative and rhetorical greening became official after a significant vote by the Green Party in the European elections. Adjustments to the regulatory demands of the EC had also become unavoidable and official rhetoric had to change to make new regulations palatable. However, deeply shaken by the coal-miners strike and its political consequences, British Coal and its supporters continued to view acid and greenhouse gas emission controls as attacks to be fought with better science rather than technology, and understandably so. The clean coal R&D programme started in 1970s which, if adopted, would have reduced acid emissions significantly without FGD, lost government funding. The potential of gas as a 'clean fuel' was beginning to be realised in research circles and communicated to government, but held no promise for miners and their families. Concessions on acid rain would, and could be made, but they would not be popular.

The political aim of the Government's greening is revealed in Prime Minister's wish to defeat socialism. Expressing her belief in global warming, international environmental diplomacy and nuclear power (acid rain received a passing mention), she chided the environmental lobby for not supporting nuclear power and accused it of using environmental problems:

> ... to attack capitalism, growth and industry. I sought to employ the authority which I had gained in the whole environmental debate, mainly as a result of my Royal Society Speech, to ensure a sense of proportion.[18]

The civil service seriously searched and found positive aspects of environmental regulation, having recognised that fundamental changes in the UK environmental regime had to be made. Relationship with environmentalists warmed as this would assist not only with gaining needed popularity for the minister, but also in 'turf battles' with arch enemies, the DTI and DEn/CEGB, both about to disappear. In 1988 when the LCPD was finally agreed with considerable concessions made to Britain, i.e. half the emission reductions initially proposed in 1983. In the same year the privatisation of the CEGB was announced, initially including its nuclear component (though this proved impossible until the mid-1990s). Remnants of the DEn would be absorbed by the DTI. Car pollution began to attract attention as Germany urged the 'cat car' upon the EC. The public reacted positively to the introduction of unleaded petrol after its price had been cut by reducing excise duty. The privatisation of the British car industry had also virtually been completed. The high stakes of government in the regulation of both the car industry and electricity generation were disappearing fast. Passing environmental costs onto the consumer would become easier, to the relief of government. Official promises to reform the quasi-governmental environmental regulatory system were made.

However, implementation of emission abatement targets immediately ran into difficulties as the investment climate worsened, recession began to bite deeper and professional demoralisation spread. Two industrial branches with impacts on 'acid rain' remained unsold and would continue to restrain policy. BC, in spite of spectacular productivity gains, remained too unprofitable for a quick sale, and the Forestry Commission lands remained under the control of the Scottish Office, their sale likely to be less attractive if considered damaged by acidification. Some incentives not to recognise major acid rain damage therefore remained, but in November 1988 Britain signed the NO_x protocol without any public debate. At a conference hosted by the Goethe Institute and Friends of the Earth and aimed at persuading government to do more, a junior Minister for Environment, Countryside and Water argued that:

> ... it is wrong to disparage the UK's intention virtually to halve SO_2 emissions simply by comparing... the effort being asked for throughout Europe is by no means equal... our policy aim is to achieve a 30 per cent reduction on 1980 levels by the end of the 1990s - representing a 46 per cent reduction on the peak year of 1970. The Government does not have a target for 1997.[19]

A senior DoE official in charge of acid rain policy in Europe dates the change in British policy on acidification inside government to 1985, but sees 1989 as the year when Britain 'switched on' and stopped seeing Brussels and Germany primarily as interfering in British domestic affairs. In January 1990, the DEn began negotiations with PowerGen and National Power, the two major generating companies resulting from the break up of CEGB, to find less expensive alternatives to FGD.[20] Discussion papers were published which suggested that compliance with LRTAP/EC could not be expected until the late 1990s. New legal powers would be required and economic instruments more closely investigated. Government rejected grants and taxes to promote cleaner diesel engines at the advice of the Treasury, though showing more interest in raising revenue through energy taxes. Late in 1990 British opposition to the directive on vehicle emissions proved in vain, Europe insisted on the most stringent NO_x standards for all cars and thus helped to destroy British investments in leanburn engine technology. The fate of a British owned car industry was sealed. The PM herself was dethroned in December by Tory rebels fearing to lose their seats in 1992.

Early to Mid-1990s: SSP, VOCs and More Regulation

In 1991 it was noted that 'the present government has left the next one facing a Europe increasingly concerned about the dangers of air pollution and blaming Britain for a substantial slice of it'.[21] The Conservative Party led by John Major was returned to power in summer 1992 but had to govern with a very narrow majority. In spite of much rhetorical enthusiasm for Rio, this remained confined largely to within the DoE, the battles with the DTI and Treasury about emission policy became less fierce. The House of Commons Select Committee on Energy was also abolished and the environmental protection administration had come to rely on more non-governmental organisations as 'customers'. Environment was an established policy, though acid rain hardly featured outside the administration.

Fortunately for the government, a steep decline in acid emissions had already began and by late 1993 a 30 per cent reduction of sulphur emission was announced, well above the 20 per cent required under the LCPD. Forty per cent had been achieved by the mid-1990s and a 47 per cent drop from LCPs alone was expected by 1998. This was achieved through using a very limited FGD programme, increased nuclear output and, primarily, switching to natural gas for electricity generation. However, the government could not be sure of the ability of private industry to deliver in future. Emission planning and new powers were therefore needed to implement any further reductions. New negotiations also started, for the Second Sulphur Protocol (SSP) and volatile organic compounds (VOCs). The existing policy network handled these issues without major public involvement, though the approach of Maastricht produced and increasingly indecisive administration on European matters. At home, however, planned institutional and legislative changes were becoming realities. When in late 1992 Britain left the EMS and the value of the pound fell sharply, widespread tax increases were becoming a necessity and deregulation emerged as a new policy goal overriding.

New debates started in Brussels about the sulphur content of gas oil, including diesel fuel for which sulphur content was to be cut.[22] With a major interest in the oil industry the UK again protected industry, but in July 1992 it was announced that the oil industry now accounted for half of the EC's sulphur dioxide emissions, making this industry rather than the declining coal industry, a major target for future regulation. If EC proposals were to be implemented, massive investments by the oil sector (or its departure from Europe) were unavoidable, given the choice between fitting FGD to all combustion plants using oil, or desulphurising fuels, including diesel and heavy fuel oil widely burnt at sea.

A national emission programme was announced which included a commitment to a 8,000 MW FGD retrofit programme to protect British coal, something the private industry never wanted and which it has successfully resisted since.[23] Yet the building of a second FGD plant at Ferry Bridge (200MW) had been announced in 1992 as part of Britain's EC sulphur programme which announced a rise significant rise in FGD capacity by the year 2000.[24] Sulphur emission trading also began to be investigated by the DoE/HMIP and as part of the Government's deregulation initiative, though this was later abandoned. In November 1993 Britain claimed that EC targets under the LCPD had been achieved. However, new stringent regulations were being called for under SSP. Industry protested about these unexpected demands, this time with less success.[25] SSP was signed in 1994 and industry warned that additional obligations were likely to arise after 2005.[26] Ratification took place in 1996.

UNECE pressures further to reduce sulphur dioxide emissions grew from the 'critical loads' approach which the UK had championed as a delaying tactic, but which had demonstrated domestic widespread damage to surface waters and some soils, that is change from pristine conditions. Contemplating the costs of 90 per cent reductions by 2005, government argued for less demanding 'acceptable loads' and an extended timetable. These negotiations reveal much about British attitudes and procedures because emissions plans painfully negotiated with privatised industry for EC requirements were now thrown into disarray. Model calculations had suggested that the UK would have to spend DM 2894 million a year on FGD to meet this target. UK stakes were therefore, it was claimed, very large: a new regulatory system would be tested as well as the rate of closure of coal-fired power stations. The allocation of emission quotas between the various sectors of industry would have to be renegotiated. A report by the House of Commons Trade and Industry Select Committee argued for resistance with reference to a widely perceived 'coal crisis', thus revealing UK negotiating strategy. These arguments did not impress government, but were used in negotiations. The precise reasons for lenient regulations were:

- the continuing higher price of British coal compared to imported coal
- the share of other fuels, including gas, nuclear, orimulsion and imported French
- nuclear electricity had or were about to take a significant market share
- the new electricity market had resulted in relatively high prices for coal-fired
- power, and the expectation that the regional electricity companies would
- not be willing to contract coal-based power
- the environmental regulation hampering the development of markets for coal
- that the build-up of coal stock was aggravating the problem.

The UK delegation was therefore instructed to delay and reduce commitments, though both UNECE and national experts were calling for cuts over 70 per cent.[27] The initial UK starting position was ratification by 1998, with an offer to reduce emission by 50 per cent by 2000. Britain therefore stalled negotiations in September 1993. Cabinet then decided that it would promise only 60, then 70 per cent. In December an 80 per cent target by 2010 was accepted. The negotiations were considered more successful than previous ones, but proved difficult because Britain again needed time for negotiations at home, but with industry 'at arm's length'. So SSP was signed after lengthy consultations with HMIP and industry and officials mindful that during the negotiation of the First Protocol, 'a good deal of international opprobrium' had been attracted.[28] Nevertheless, UNECE had to back off the full critical load requirements as deposition targets, and instead negotiated 'gap closures'. In absolute terms, even by 2010, British emissions will remain the highest in the EU, after Spain, SSP will only begin to 'bite' after 2000 when the British fuel mix might consist of nuclear power, gas and renewables and more combined heat and power (CHP) rather than coal plus FGD. The National Emission Plan for the LCPD had now become redundant.[29]

In preparing a new emission plan, the following regulations had to be taken into account and made mutually consistent, not an easy task:

- the Large Combustion Plant Directive (LCPD:EC -1988) which is to be revised further. This required the splitting of the UNECE 'bubble' between the electricity sector, other large plants (including oil refineries), smaller boilers and firms, and mobile sources. This cannot be done without accurate data on sources and emissions;

- the EC Integrated Pollution Prevention and Control Directive (IPPC); which seeks to implement best available technology, while British Integrated Pollution Control 1 (IPC) seeks BATNEEC,[30] best available techniques not entailing excessive costs;
- a directive on the sulphur content of fuel oil;
- a domestic proposal for quota switching (emission trading for S and N quotas, later abandoned as there was nothing left to trade);
- a faltering energy efficiency programme, which includes a tax increase (15 per cent of VAT was intended but only 8 per cent accepted by Parliament in 1994);
- changes in ambient standards for benzene published in February 1994;
- VOC guidelines, which also affect the oil industry (though here the battle between who is to pay: the fuel supplier or the car industry, continues).[31]

When authorisations under IPC were eventually granted in April 1993, hopes of transparent decision-making had to be abandoned as the generating companies refused to be seen as 'public utilities' and hence publish their emission data. However, the utilities had already negotiated a set of ceilings for acid emissions that will not constrained their operations at least until 2003. Rising costs of compliance, including internal transaction costs of regulation, worried government which agreed with OECD that excessively fast compliance schedules discourage learning by doing and lock firms into obsolete technology. Inadequate consultations between regulators and firms might lead to uncertainties and unrealistic demands. The DoE saw a silver lining to the acceptance of the SSP; it promised to make LCPD redundant. Achieving this became a UK diplomatic objective and this prospect is likely to have made Cabinet more flexible about UNECE.

Based on more voluntary principles, the *negotiation of the VOC protocol* created much work for civil servants preparing for its implementation prior to signing. The target industries are offshore oil, solvents coatings and paints, glues etc., and once again motor vehicles. DoE guidance notes were issued for consultation which aimed to control

VOCs by largely voluntary means with the help of local authorities and their environmental health officers, and trade associations. The British VOC strategy was then developed in consultation with DoE scientists, the NSCA and trade organisation after the protocol had been signed, the time of ratification being determined by the government being reasonably sure that it could achieve compliance. Negotiations for a second round of VOC reductions started soon after entry into force of the first protocol, which is by far the most complex of the Geneva Convention to implement, but not necessarily the hardest to negotiate because of the absence of strong export/import imbalances or powerful commercial pressures, even though suggestions made in 1991 VOC guidance notes on the manufacture of paints and other coatings had proved controversial and provoked intense lobbying by the industry to get proposed EC standards weakened. Reports of the official inventory of VOC emissions available since 1992 indicated a substantial upward revision of earlier figures, i.e. an annual estimate of 3.3 million tonnes compared to 2.39 mt, with estimates for oil, chemical and food industries still uncertain, again suggesting that bubbles are fixed so that implementation becomes feasible without major reductions. Implementation is currently under way under EPA Part 1, but considered very slow for small and medium sized emitters.[32] It was recognised that existing controls might only just ensure LRTAP compliance, promising a 32 per cent drop of 1988 emissions by 1999. Given slow progress at home, by early 1993 the UK was also opposing Stage I and II proposals on the 1993 EC directive for petrol refuelling and sought a derogation for small rural petrol stations. The oil industry lobby (UK Petroleum Industry Association) proved persuasive and experienced. The oil industry wanted to pass the costs to the car industry (as large carbon canisters), the cheaper option which, it was claimed, would also reduce VOCs.

In October 1994 new standards for petrol, diesel and heating oil came into force which will reduce SO_2, as well as VOCs and particulates and thereby implement the 1993 EC directive. This was achieved without much public debate, but remained unpopular in the oil industry, which refused to accept VOC reductions on leaded petrol as this is being phased out. The costs for the UK were low however, as only little fuel above 0.2 per cent of sulphur in weight was sold. In early 1995 the UK was facing prosecution by the EC for failure to monitor air pollution from NO_X. The DoE admitted the ground level ozone levels are bound to continue rising until catalytic converters begin to have an effect, though it was also claimed that any faith in the test cycle associated with these catalysts was misplaced. The European Federation for Transport and Environment announced that new cars, while complying with the test cycle, breach controls on the road.[33] NO_x will clearly test LRTAP as well as the EU.

As matters stood in the mid-1990s, the national plan for complying with UNECE and LCPD was so permissive, (because based on the assumption of continuing coal burn) that compliance became easy and there has in fact been over compliance rather than free-riding. The political interest in finding some escape to the regulatory 'overdetermination' through economic instruments and the idea of quota switching for sulphur emissions was therefore still born. What economic theory had tended to forget was that all interventions in the market involve government and hence involve political costs as well.[34] Public support for more measures would be needed, however, for future compliance. Government began acting early.

In early 1995 the public was informed that national emissions of NO_x, VOCs and carbon monoxide (CO) and particulates had stabilised or were already falling.[35] Even tighter emission standards would be needed after 2000, and were achievable in the energy sector by combined cycle gas turbines (CCGT) and combined heat and power (CHP) and, perhaps in future, clean coal technology (using pulverised coal and producing two streams of electricity), as well as selective catalytic conversion (ammonia reacting with NO_x over a catalyst). All these promise to reduce acid emissions as well as carbon dioxide - acid rain abatement had merged with the mitigation of global warming. However, small and medium sized firms and the transport sector remained major weaknesses in the abatement of emissions generally, but especially in urban areas. So urban air management emerged as strategy central government devised for local governments, potentially with positive implications for LRTAP. Complaints about urban air quality had been rising, with traffic getting most blame as generating ground level ozone, benzene and particulates. However, as the producers of lead additives pointed out, unleaded petrol used in cars with catalysts, was a major source of benzene and other aromatics, both now seen as serious threats to human health.

The UK was by now credited with more car miles per head of population per year than any other European country, with traffic expected to double by 2025 according to official projections. In February 1995 the Transport minister quietly discarded the earlier idea of road pricing so strongly promoted a year earlier by the environmentalists and the RCEPT, but an initiative on 'managing air quality' began in 1994 and was still being debated in 1997 when New Labour promised an 'integrated transport revolution'. Though dating back to a DoE consultation document on 'Improving Air Quality' from November 1993, the air quality strategy is targeted at 'hot spots' and proposes improvements by making local government responsible for managing traffic subsequent to setting up monitoring networks for air pollutants. Pressure for this had been coming from domestic sources as well as the EC's proposed framework directive on air quality. The Environment Act, in force since 1995, provides the framework legislation for this initiative.

Cleaner air for Britain's cities is now publicised as part of the government's Sustainable Development strategy. The Department of Transport even, merged with the DoE in 1997, appears concerned about the physical and political 'sustainability' of road traffic growth. People would have to be persuaded or taxed to use the car less. Having fought the catalyst car for many years, this 'device' was now praised as the future salvation to several compliance problems. In practice, there have been major cuts in public expenditure on new roads. Responses from industry were not enthusiastic; it fears the economic impacts of increased congestion on the roads, while environmentalists complain that the proposed air pollution control measures do not possess the force of law and expressed doubts about local government' enforcement powers. Inconsistencies in public policies clearly remain. Looking at this scene in general, a veteran environmental journalist concluded, that 'a thick fog of complacency, tinged with secrecy and contempt for the public, still swirls through Britain's air pollution policy'.[36] Probably an unfair assessment giving the conflicting pressures only single issue lobbies can avoid.

Explanatory Perspectives

This section must account in some depth both for the initial 'dragging' and subsequent achievements, including a more positive role in international negotiations. Four sets of domestic factors, described in some detail above, will be used to explain UK behaviour in relation to the LRTAP/EC acid rain regime:

- initial perception of very high compliance costs in both economic and political terms;
- initial very limited, if any, net domestic damages arising from acid emissions(excluding VOCs and increasingly NO_x);
- limited ambitions, in government, to become a leader in emission control technology making abatement for its own sake unattractive and an effects-based environmental policy.
- Weak domestic demand for abatement was a major secondary factor; for strong pressure here could have weakened resistance by government justified on economic grounds.

During the late 1980s perceptions changed, largely because of:

- the impact of privatisation on abatement and other costs to government;
- the impact of privatisation on costs to industry, and hence the incentive for fuel switching.
- the inclusion of pollutants with weaker transboundary effects and more balanced domestic costs and damages.

In essence this British 'conversion' was due to the disappearance of industrial resistance to emission regulation enabled, even forced, by government policy. This policy was not in motivation environmental. To explain why this should lead to emission reductions from stationary and to a lesser extent mobile sources, one has to look closely at structural changes in the British economy as well as changes in its regulatory systems, at energy markets and changes in the ownership of emission sources.

Emission reduction, for the purposes of implementation must clearly be allocated to specific sources which tend to be subject to different regulatory regimes and ownership. In the early 1980s, all major emission sources in Britain were in public ownership, by the mid-1990s this picture had completely changed. With it came the desired emission reductions, something requiring detailed explanation if only to raise doubts about the simple deduction that privatisation automatically leads to emission abatement. Rather, once emission sources were no longer in the public sector, the interests and commitments of government changed. Regulation imposing higher costs, ultimately on consumers, had become politically more feasible. There was, however, a large social price to be paid.

Prior to privatisation, the Secretary for State for Environment could expect major opposition from several ministries largely representing industrial interests relying to some extent on public money. In these circumstances, major decisions were made by Cabinet and the PM herself.[37] Cabinet approval for major emission reductions would be needed in cases of interministerial disputes, disputes which the DoE lost until the SSP when the scientific work done in support of the critical loads approach appeared to bear fruit and Cabinet was in the end persuaded to accept the new proposals for the SSP. Support for the leanburn engine also reached Cabinet level, though here the more detailed story is not known, though support for this technology developed by British Leyland/Rover and Ford, came from the DTI. For stationary emission sources, a closer look a changes in the energy sector as well as government are needed, for vehicles at the car industry.

For government, some change in the balance of 'environmental' power probably began with the Rio Conference and in response to the sustainability debate, but acid rain is mentioned only briefly in government reports of that period. British environmental policy remained devoted to 'sound science' and cost effectiveness, and continued to avoid the deliberate use of environmental regulation to force technology. Also, the government tends not to adopt specific targets until practical procedures for their implementation are in place and the capacity for enforcement been developed.[38] Domestically the DoE was not the primary regulatory authority for air pollution and had to reach agreement not only with other ministries, but also with the pollution inspectorate for minor changes. The environmental regulators were part of domestic negotiations, but given the various constraints outlined above, could not play a major role. Divided among themselves, threatened by loss of competences and subject to continuous reorganisation, they did not make for a powerful influence in favour of more stringency. New regulatory powers obtained during the early 1990s helped to turn British emission behaviour, but were not a primary cause.

During the 1980s the DoE remained isolated and poorly prepared for the clean air agitation which swept across Europe. Though the leading official actor in acid rain negotiations, it did not possess direct influence over energy or transport policy. So officially domestic acid damage was not admitted until 1988 when the LCPD was finally signed and planning for some FGD began. The UK had lost several battles in Brussels and some adjustment in 'perception' was needed. The DoE in response to the Environment Committee's first report on air pollution stated that

> further research had shown that 'acid rain' was not solely a long-range transboundary phenomenon but that acid depositions were in some measure affecting the local environment in the UK.[39]

Domestic and foreign pressures combined to change the perception of 'national interest' only when an increasingly clear and confident picture of British emission reductions from stationary sources emerged. Such change was forecast by energy modellers, which meant that political decisions could now be made on the basis of emissions reduction scenarios. This prospect was beginning to emerge from within the DoE in the early 1990s, for in its second report of 1992 to the 1990 White Paper on environmental policy, it stated that the national emission reduction plans made in 1990 would be achieved by 'switching to new low-polluting plant, notably those able to use natural gas, as well as to low-sulphur coal in existing coal fired power stations'.[40] Since by 1990 UK SO_2 emissions had declined by only 10 per cent for the power sector, the need to target this sector had become unavoidable.[41]

Table 11.1 illustrates the drastic reductions required after privatisation for large combustion plants and the need to control 'private' small plant in future. Table 11.2 shows a scenario for sulphur emissions which demonstrates that while major reductions are 'planned', this national plan itself was not ambitious. In future sources outside the power sector will need to be controlled.

Table 11.1   Sources of UK acid emissions in 1993 (per cent of total) (NM means non-methane, i.e. not from evaporation of petrol)

Emission Source	SO_2	NO_X	NMVOC
power stations	71	26	< 1
other industry	19	11	56
road transport	2	51	36
Domestic	4	3	2
Contribution to UK total	96	91	94

Source: National Atmospheric Emissions Inventory.

Table 11.2   Scenario of future SO$_2$ emissions (k/tonnes/yr)

	DoE estimates		National Plan + extrapolation		Projected	
	1980	1985	1992	2000	2003	2005
Total LCP quota	3895	3043	3246	1858	1452	?
of which power industry	3006	2627	2852	1563	1202	?
Small plant	902	683	587	477	436	409
Total	4899	3726	3833	2335	1888	?
UNECE target				2449.5	1861.6	1469.7

Note: Small plant figures based on linear extrapolation of trends between 1985 and 1992
UNECE target for 2003 based on interpolation between 2000 and 2005.

Source: SPRU, University of Sussex.

Energy and transport policy objectives, while represented in Cabinet by ministries, reflect 'exogenous' developments and demands, which governments need to consider but over which they have but limited control. These considerations include changes in price of fossil fuels, vehicles and abatement technologies. These prices are affected, at times significantly, by emission regulation and therefore influence competitiveness in the market place, especially for traded products and for basic inputs such as energy. Price changes also challenge internal plans and commitments of major monopolies. A main actor in the early story was therefore CEGB, a 'state' within the British state. Required to protect its environment since the late 1950s, this applied only to the local area. By the early 1980s, it was not a popular institution because of its pursuit of nuclear power. Its pronouncements on air pollution were believed less than those by foreign pressure groups. Not allowed to borrow privately, the capital expenditure for FGD would have meant a major burden for the Treasury. (Gas was not yet available as a fuel.) Only after privatisation and regulated for competition, could the emission behaviour of the new electricity generators change significantly.[42]

The British fuel mix had moved out of coal far more rapidly than anybody could have imagined even in 1982, the decline beginning in 1990. With it came a significant reductions of acid (as well as greenhouse gas) emissions per unit energy produced; generation suggest that all UK coal-fired power stations remain in a vulnerable position. By late 1992 the 'dash for gas' had already pre-empted nearly half the market for British coal, whose future would depend on the relative prices of imported and British coal and natural gas, which is by no means assured to remain low (Parker and Surrey, 1993). The incentive to use natural gas may decline as its price rises or that of other fuels falls, hence the desire by industry for long-term contracts with gas suppliers to remove this uncertainty. Some gas is already imported (between 20-30 per cent since mid-1980s), the UK having ceased to be a net exporter of fossil fuels and is now more interested in promoting energy efficiency, another reason for changes in 'national interest' calculations. This rapid decline of coal industry has altered the landscape and politics of Britain. The closure of 31 coal pits was announced in October 1992 with the loss of 30,000 jobs, many more were expected in 1997 as existing contracts were running out. Opposition from 'coal' as an institution was therefore over by mid-1990 and the hope that the New Labour government would act in its advantage has not materialised. The dash towards gas seems unstoppable for the time being and while less gas capacity is likely to be built than originally anticipated, the danger of creating significant overcapacity is real. Norway is considered to be the only secure gas supplier in 10-50 years, the range of years over which domestic supplies will run out. Recession and industrial 'restructuring' were contributory causes as they caused a decline in energy demand, though this was rising again in 1997.

Recognising the potential for emission reduction of the dash for gas, the electricity industry successfully avoided existing FGD commitments; first by simply not submitting plans and later abandoning them thanks to the lenient targets negotiated by government! But why this 'dash to gas'? In the mid-1990s fossil fuel prices were less than half their real 1975 value, the main fall having taken place in 1986; the 1980s were therefore a period of falling prices in strong contrast to the 1970s. This favoured the greater use of fossil fuels, including cheaper imported coal and gas. Hence also the calls for higher fossil fuel prices for competitive as distinct from environmental reasons. When the EC directive against burning gas in power stations was repealed on 'environmental grounds' in March 1991 the way was free as well as rational to switch to gas.[43] Natural gas generates power more efficiently and with fewer emissions, and the building of power stations using gas and turbines is fast and relatively low in capital costs. So the coal industry in the UK virtually disappeared after the privatisation of the CEGB by the early 1990 and that of BC in 1995. Nuclear power, which remained in the public sector until the mid-1990s was also enabled by government to increase its contribution to electricity generation and hence benefit to acid emission abatement. Provided with a subsidy, the 'non-fossil fuel levy', actually a tax of 15 percent on all electricity bills, debts and future costs could be managed. The coal industry argued that these subsidies should have been used on FGD and maintaining a larger role for coal.[44]

Competitive pricing structure further weakened the ability of government to persuade the electricity generators to install FGD to coal-fired plant. In general, generating costs are lowest for new combined cycle gas turbines, up to one third higher for new coal-fired plant with FGD, and highest for a new nuclear plant. With individual power stations now bidding into the national grid on a daily basis, new plant gas win at current prices and emission reduction is a side-effect of commercial decisions.[45] Nevertheless, future reductions from recent baselines will require the intensifying regulation of sources other than LCPs, making transaction costs to government larger. Additional reduction measures will be required by 2003 even for sulphur and NO_X negotiations are likely to support future measures in the transport sector where 'the state' is in difficulties, though bound by EC regulations.

The car industry, especially Ford and British Leyland, now owned by BMW, also became 'actors' against the UNECE/EU regime. The British car industry by not selling to the USA had not acquired catalytic emission technology and had largely escaped any emission regulation at home. During the 1980s it had argued against catalytic converters because, in response to the 'oil crisis', British car manufacturers had developed more efficient engines burning 'leaner' which now needed protection against higher NO_X emission standards. The emission behaviour of this 'leanburn' engine would become a crucial issue in the acid rain story as the DTI tried to protect this technology against German proposed legislation. This battle was lost by Britain once fully privatised official protection was no longer available to the car industry.

Conclusions

By 1996 the primary reason for the decline in British emissions revealed itself as market-driven technological change initiated by government policy. Legally enforced competition then required British utilities to escape from the FGD commitments accepted in 1988. A long and politically difficult process of structural change was accompanied by much institutional development which led both to fuel switching to a cheaper option which happened to be less polluting, as well as to the intensified use of nuclear generation, and to a new regulatory regime. So compliance was associated with at least some degree of environmental problem shifting – to solid waste and nuclear waste disposal/emission issues. Combined with recession and deindustrialisation, a degree of emissions reductions was therefore brought about which few had foreseen in the early 1980s. Regulations were then adjusted to emission scenarios, rather than vice versa; acid emission reduction in the UK so far aimed primarily at large stationary sources over which strong regulatory powers already existed; transport emissions are proving much more difficult, which may explain a trend towards more 'voluntarism', demand management and local regulations. This suggests that the setting of official targets and timetables is not consistent with using economic instruments and demand management. Cheaper and more effective reductions are promised, but cannot be guaranteed by law. Fixed targets may create allocation problems which may be politically expensive for government, especially when many small emitters and a large variety of pollutants have to be controlled.

The conversion and achievement of the UK should not be exaggerated, however. The outcome is in part pure luck, and underlying characteristics making for a cautious and sceptical approach to IEA remain. A high value put on low energy prices and concern about costs to industry make it likely that UK policy will continue to stress uncertainties in the scientific knowledge base and seek both to delay and weaken international agreements that are costly to the national economy.

With respect to the theoretical objectives of the project, the British case allows the rationalisations suggested under model I only with difficulty because it cannot explain the time dimension, nor how and why utility maximisation is achieved. The pollution exchange balance was clearly important, but had to be 'discovered' first, which gave rise to a lengthy period during which 'dragging' was both rational and required by domestic policy-making tradition. There is no evidence for defection because strong implementation institutions have an incentive to implement irrespective of overall costs. Indeed, in the end there is not such thing 'as the state', only the ease with which the state can pass costs on to industry and consumers. Utility is a broad category which can only be given full meaning by 'on the ground' research. The case therefore supports the hypothesis that the domestic politics model possesses the greater explanatory powers, but in this case appears to have required a strong external 'push'. However, it is just as easy to argue that British emission reductions would have taken place anyway, and that a great deal of time and money was wasted in intergovernmental negotiations when the 'freed' market might have done better earlier. But this would surely have been a chance outcome. The domestic actor model is therefore likely to be more predictive provided the many factors and processes included in the analysis can be estimated with regard to their future behaviour. The case itself suggests that this is not likely.

For linkage politics, one may conclude that identical (emission) baselines may require very different compliance costs in different countries. Under pressures for 'fairness', or 'burden-sharing', this tends to produce 'equal cost' rather than equal abatement regimes. In the UK acid emission developments also had side-effects on the carbon dioxide front, which now allows government to present itself as a leader in game of climate politics.

Abbreviations

BATNEEC	Best available technology not entailing excessive costs
BC	British Coal
BPEO	Best practicable environmental option
bpm	Best practical means
CCGT	Combined cycle gas turbines
CEGB	Central Electricity Generating Board
CHP	Combined heat and power
CLRTAP	Convention on the Long-range Transport of Air Pollutants

CO	Carbon Monoxide
COPA	Control of Pollution Act
DEn	Department of Energy
DoE	Department of the Environment
DTI	Department of Trade and Industry
EA	Environment Agency
EC	European Community
EMEP	European Monitoring and Evaluation Programme
EMS	European Monetary System
EPA	Environment Protection Act
FGD	Fluegas desulphurisation
FSP	First Sulphur Protocol
HMIP	Her Majesty's Pollution Inspectorate
IEA	International Energy Agency
IIASA	International Institute for Systems Analysis
IPC	Integrated Pollution Control
IPPCD	Integrated Pollution Prevention and Control Directive
LA	Local Authorities
LCP	Large Combustion Plant
LCPD	Large Combustion Plant Directive
mt	Metric tons
MW	Mega Watts
NERC	Natural Environment Research Council
NGOs	Non Governmental Organisations
NO_x	Nitrogen Oxide
NSCA	National Society for Clean Air
OECD	Organisation for Economic Co-operative Development
ppb	Parts per billion
RCEPT	Royal Commission on Environmental Pollution and Transport
SO_2	Sulphur Dioxide
SSP	Second Sulphur Protocol
UK	United Kingdom
UNECE	United Nation Economic Commission of Europe
VOC	Volatile organic compounds

Notes

1. In 1992 the official figure was still 3.5 mt with 2.7 mt of SO_2 from large combustion plants (LCPs) with almost 70 per cent of these emissions still derived from electricity generation. This sector did not deliver significant reductions in emissions until after

1992. Total NO_X emission estimates showed a steady increase until 1970, remained roughly level until 1985, then increased to 2.86 mt in 1990, but fell to 2.75 mt in 1992. The main sources here are power stations and vehicles, with agriculture being implicated in some areas.

2. The EC aspiration is a 65 per cent SO_2 reductions by 2010 from a 1990 baseline.

3. These emissions are being cut by the introduction of three way catalytic converters since 1991 and improved combustion technology in power stations. Britain had no difficulty with promising to stabilise NO_X emissions after it had significantly raised its own baseline estimate. The accurate measurement of baseline emissions is therefore critical to compliance. In the UK it became clear that there were no such measures good enough for regulatory purposes even for sulphur until 1993.

4. VOCs emissions rose slightly between 1980-1990, though an upward revision was expected as much data remained uncertain. The most significant sources are road traffic, solvent use (paint), the oil industry and arable farming. Estimated emissions were 2.6 mt in 1990 and 2.56 mt in 1992. The official estimate is 2002 kt/a for stationary and 972 kt/a for mobile sources, with road traffic seen as the source of 41 per cent of VOCs. On its own estimates, the DoE expects to be able to reduce UK VOC emissions by 32 per cent by 1999.

5. Britain was first country to adopt national air pollution legislation with the 1865 Alkali Act introduced on the best practicable means principle.

6. Many economic and political developments of this period relate world oil price increases by 13 times.

7. Commission on Energy and the Environment, *Coal and the Environment*, HMSO, London, July 1981, p.165.

8. FGD means adding a low tech chemical plant which requires a vast amount of limestone which needs quarrying, transporting and generates much waste, or by-products such as gypsum or sulphur.

9. Commission on Energy and the Environment, op.cit. p. 159.

10. The gas regulator, for example, is under pressure to 'protect the environment' through price changes, but has refused to raise taxation without parliamentary approval.

11. Britain managed to gain advantages in negotiations (with reference to protecting its coal industry) when this industry had already been doomed by a new economic regulatory regime.

12. These tended to arise because of the direct effects that EC directives (on combustion plants and vehicle emissions) would have on technology choice, commercial competition and legal competence.

13. Interview with Chris Wood, DoE, 12 January 1994. The deregulation initiative of DTI would make the drafting of new environmental legislation more difficult and privatisation created many complexities.

14. DoE, *Reducing Emissions of VOCs and levels of ground level ozone: a UK strategy*, October 1993, p.77. A large number of guidance notes to LAs and prepared by HMIP were issued in 1991.

15. Reports of 11th and 12th meeting of the Task Force of Economic Aspects of Abatement Strategies, 1994. The idea of defining Somas (sulphur management areas) was discussed. Multi-impact assessment is the new goal of modellers, requiring five

large sets of data (energy and scenarios, emissions, impacts/critical loads, cost data for different technologies, side effects). Impacts should combine the joint effects of all air pollutants.

16. For more details see Boehmer-Christiansen and Skea, (1993), and Boehmer-Christiansen and Weidner (1995).

17. Large combustion plants are a cluster of boilers with an aggregate input capacity exceeding 50 MW of electricity. The LCPD was agreed after 4 years of negotiations in 1988.

18. Thatcher (1993, p.640). In September 1988 she admitted to scientists that she believed in global warming and initiated a generous funding programme of basic research through the DoE.

19. House of Commons, Environment Committee Session 87-88, First Report, Air Pollution, vol. II, HMSO, London, 1988: Quoted from letter to Chairman of Committee, p.458.

20. *The Guardian*, 9 February 1990.

21. *The Guardian*, March 4, 1991 p.2 Paul Brown, 'Patten tries to link green aims with power policy'. He argued that there was still no true competition in electricity generation as prices remain propped up by fixed contracts, holding back what the market would really want, namely CHP and minipower stations.

22. From 1994 (to 0.02 per cent) and again in 1996 to 0.05 per cent of weight.

23. Skea, (1993, pp.92-95). This book makes no reference to the Geneva Convention.

24. *Environment Business*, December 1993.

25. Of the 11 million tonnes of heavy fuel oil estimated to be burnt in the UK annually, about half is used for electricity generation, the rest in small industrial boilers which are regulated locally.

26. Cambridge Econometrics saw only a 52 per cent cut by 2005, with 80 per cent needed by 2010 on the forecast of growing demand for coal after 2005 due to rising demand for electricity. Other experts were more optimistic.

27. J F Skea, in Minutes of Evidence, *Select Committee on Sustainable Development*, 14 June 1994, Session 1993-94, pp.93 -96.

28. DoE, *Fourth Annual Report*, HMSO; London, March 1994, p.20. The monitoring station in Central Hull is as far from emission sources as possible.

29. DoE, *The UK's programme and national plan for reducing emission of S02 and NO$_X$ from existing large combustion plants*, London, December 1990.

30. IPPC is likely to supersede the EU Air Framework directive from late 1995 which introduces BAT as long as this is 'industrially viable from a technological and economic point of view', thus offering the possibility of getting rid of LCPD, with existing plants due to comply with BAT before 2005.

31. Interviews with Jim Skea, DoE officials and Steve Sorrell, between 2nd Nov. 1994 and February 1995 and documents from SPRU, DoE and library of the NSCA.

32. For details see ENDS 212 and 214, September and November 1992. Emitters in order of importance were solvent use, oil industry, combustion, food, waste disposal and many others.

33. *The Times*, 4th March 1995, p. 2, Car 95 Report. Car pollution and deteriorating indoor air quality both offered credible causation.

34. Additional VAT on domestic fuel was not accepted, and regulators refused to implement energy efficiency taxes because these had not been agreed to by Parliament. VAT was abolished by Labour in 1997.
35. DoE/DT, January 1995; *Air Quality: meeting the challenge*, HMSO, London , p.23.
36. Geoffrey Lean, 'Where did all the fresh air go?', *Independent on Sunday*, Special report compiled in association with Friends of the Earth, 5 March 1995, pp. 4-5. The accusations made were widely discussed in the press and included blaming 'the government, for thousands of deaths, including 160 on a single 'smog' day in December 1991.
37. These include the refusal to sign the FSP and stay out of the '30 per cent club'. In May 1984, shortly after the House of Commons Environment Committee had warned government that that financial decisions had to be taken to apply technology to the reduction of acid emissions, she called her advisors together in preparation for the important intergovernmental meeting in Munich in June. The meeting took place outside the Whitehall committee system and, according to Lord Marshall, revealed the 'mistaken' DoE strategy of using the forest death argument to obtain her support. When asked about the scientific evidence for this German claim, she was told that there was very little and retorted 'then why are we bloody well here? Let's have lunch'.
38. British negotiators complain that other countries promise before they know that they can comply.
39. The DoE based its view on reports from its Terrestrial Effects Review Group, contract research sponsored by it and NERC. From Governments Reply, December 1988, to House of Commons, Environment Committee, First Report, 1987-1888, *Air Pollution*, May 1988, HMSO, London.
40. DoE, *This Common Inheritance: Second Year Report*, HMSO, London October 1992 para 17. Since the mid-1970s, world and national gas resources had expanded rapidly and the privatised electricity utilities, under competitive and environmental pressures, were indeed able to fulfil Conservative hopes. The UK had began pushing for withdrawal of 1975 EC directive banning gas use in power stations, and achieved this in 1992. The European gas delivery/distribution, more difficult than for oil, was improving rapidly, while Chernobyl (1986) and rising costs had further damaged the prospects for nuclear power.
41. This compares with recommendations made by the DoE in 1989 of 66 per cent for refineries, 50 per cent for manufacturing and 10 per cent for the power sector. LCP industries outside the power sector had reacted angrily then, pointing out that while they had already substantially reduced their emissions, CEGB had only reduced by 6 per cent between 1980-1987, confirming the governments policy of protecting its 'own' polluters compared to private ones.
42. The FGD burden arising from the international obligations was split between PowerGen and National Power, with the former applying to retrofit 2GW and National Power deciding to add FGD to Europe's largest power station (4GW) Drax. FGD for Drax cost £680 million with quarries in Derbyshire supplying 600,000 tonnes of limestone annually, which is most unpopular with local people. A lot of gypsum is produced as 'waste', more than the UK building industry requires and gypsum mining

elsewhere has largely ceased.

43. This happened after the oil companies had exerted considerable pressure.
44. Since 1991 Nuclear Electrics received over £3bn in subsidies, over £2 bn for current operations.
45. To measure and identify the costs of electricity generation and the role played by different fuels, in a competitive market, is however extremely difficult, as it depends on many factors and remains highly controversial.

References

Boehmer-Christiansen, S.A. and Skea, J. (1993), *Acid Politics*, London: Belhaven.

Boehmer-Christiansen, S.A. and Weidner, H. (1995), *The Politics of Vehicle Emission Reduction in Britain and Germany*, London: Cassell.

Parker, M. and Surrey, J. (1993), 'The Genesis of the UK Coal Crisis', *Nature*, vol. 362, 25 February 1993, pp.677-680.

Skea, J. (1993), 'The Environmental Dimension', in S. Thomas et al., *UK Energy Policy: An Agenda for the 1990s*, Brighton: SPRU.

Thatcher, M. (1993), *The Downing Street Years*, London: Harper Collins.

12 The Implementation of IEAs in Spain: From National Energy Considerations to EU Environmental Constraints

JOAQUÍN LÓPEZ-NOVO AND FRANCESC MORATA*

Introduction

Spain's air protection policy can be seen as a political game that takes place within – and is constrained by – two inter-linked games, one domestic and one international. The domestic game is the one of energy policy, which is a sort of higher order political game that constrains both the problem definitions and the options that shape the air pollution policy. On the other hand, the international game includes two rings: the International Environmental Agreements (IEAs), and, after 1986, the environmental policy of the European Community (CEP). Spain's air pollution policy is a product of the interactions between these different games. While it carries the environmental and economic constraints imposed by Spain's energy choices into the arenas of IEAs and the CEP, it also brings policy commitments and targets back from the international and European levels into the national arena of energy policy.

The usefulness of the idea of nested games as an heuristic device lies in the fact that it allows us to analyse air quality protection policy in the context of its complex and evolving relationships both with the energy production policy and consumption – and the IEAs. The idea of games nested within games allows for the conceptualisation of similarities and differences across countries. Though energy policy and air pollution policy are always somehow nested, the precise nature of their relationship can vary widely according to contextual contingencies. As the Spanish case shows, local contingencies, such as the structure of domestic energy resources and the quality of those resources, play a large role in shaping the nesting of the two policy games. For example, two countries using the same combustion technologies and burning similar quantities of coal, may differ widely in their total emissions, depending on the quality of their coal.

Furthermore, the metaphor of nested games should not be understood in static terms as just a one way street from energy to environmental choices. As a matter of fact, as environmental legislation grows, imposing new environmental commitments, environmental policy becomes a constraining game for energy policy-making. Spain provides an illustration of this evolution. Prior to 1991, National Energy Plans (PENs) did not take environment considerations explicitly into account. However, the last PEN 1991-2000 clearly asserts that the energy choices are compatible with the environmental commitments that Spain has assumed as an EC member state. It is a sign of the new times that energy policy is now packaged in an environmentalist rhetoric.

To understand the nesting of energy and air pollution policies in Spain, it is necessary to highlight some attributes of energy policy-making (Jiménez-Beltran, 1993). The first one is that, as energy policy copes with a vital problem for a country's survival and independence, energy choices are framed in the strong rhetoric of national interest, and such a rhetoric gives them both prominence and priority in the face of conflicting policy claims. Armed with the argument of national interests, energy policy makers will tend to perceive IEAs that appear to be in conflict with – or to impose constraints on – national energy choices as threats to the national interests. From this perspective, until 1994, IEAs were viewed with suspicion as the instrument of foreign interests.

Second, once made, strategic energy choices that entail the construction of an expensive power infrastructure cannot be easily reversed, or their reversal entails prohibitive costs – as Spain's nuclear moratorium shows – and the waste of valued resources.

Third, in countries that still lag behind in economic development, energy planning is oriented not only on maintaining or renewing the country's energy supply capacity, but also on expanding the capacity necessary to sustain the ongoing process of economic development. In such countries, the issue of energy supply is closely linked with the issue of development and economic growth potential. Spain is, at least in the

Western European context, a country in a transitional stage with an economy that still is supposed to grow considerably in order to achieve levels of employment and welfare similar to those of their more developed neighbours. Therefore, energy choices in Spain are made with an eye to ensuring that the country can realise its growth potential, or at least that this potential is not endangered.

We can expect that the different actors that intervene in the policy arena of air pollution control will frame environmental problems in different ways, and that these variations will - to some extent, at least - reflect the particular position they occupy in the policy arena. Thus, for example, while the environmentalist actors will tend to frame the issues related to air pollution control primarily in terms of damage costs, business actors will tend to frame those same issues in terms of abatement costs. Government agencies operating in the policy arena will tend to align their frames of reference with those of their clienteles. But, at the same time, they will also share a concern with questions of national interests and they will be very sensitive to the impact of policy choices on the country's potential for economic development.

The Problem: Spain's Emissions Profile

SO_2 Emissions

According to data provided by the CORINE-AIR program, that registers standardised data on SO_2 emissions to the atmosphere in the EU member states, Spain is the third biggest emitter, behind the United Kingdom and Germany. Spanish emissions represented 12.3 per cent of the total in 1980 and 13.6 per cent in 1990. Following official estimations, the tendency will be increasing until 2000. Moreover, emissions *per capita* rose from 116 per cent of the Community average in 1980 up to 128 per cent in 1990. They could reach 208 per cent in 2000. Around 70 per cent of Spain's emissions were produced by large non-industrial combustion plants (power plants), while industrial combustion accounted for 20 per cent and transportation represented 6 per cent (MINER, 1995).

Two factors account for Spain's relatively high level of SO_2 emissions: the reorientation of Spain's energy policy towards the consumption of domestic coal after the energy crisis of the 1970s and the low quality of most domestic coal used. The domestic coal, particularly brown and black lignite are bad and dirty. They have a very low thermal potential and they are rich in sulphur. While clean coals have less than 1 per cent of sulphur, some domestic coal may have as much as 6 or 7 per cent. The use of domestic coal to produce power unavoidably yields relatively high levels SO_2 emissions.

Table 12.1   Total SO_2 emissions per capita in the EU (in kg/SO_2)

Country	Population	1980	1990	2000
Germany	80,058	93.6	72.5	16.2
Austria	7,884	50.4	11.4	9.9
Belgium	10,025	82.6	44.4	24.7
Denmark	5,170	87.2	34.4	17.4
Spain	39,085	84.9	59.3	54.8
Finland	5,042	115.8	51.6	23.0
France	57,372	58.4	21.0	15.1
Greece	10,300	38.8	49.5	57.8
Ireland	3,547	62.6	47.4	43.7
Italy	56,859	66.8	29.6	23.4
Luxembourg	390	61.5	43.6	25.6
Holland	15,178	30.7	13.6	7.0
Portugal	9,858	27.0	28.8	30.8
United Kingdom	57,848	84.7	65.3	42.3
Sweden	8,678	58.4	15.0	11.5
Total EU	367,294	73.5	46.5	26.9

Source: *Estrategia Energética y Medioambiental*, MINER, 28 March 1995.

Thus Spain scores as one of the biggest SO_2 polluters. However, several caveats are in order here. The first one is that, if emissions are expressed in terms of a ratio to country surface, Spain, with 4.3 tons per square kilometre and year, ranks as one of the lowest emitters, far behind the UK (15.4 tons/km2), Belgium (12.9), Germany (9.3), Denmark (7.7), and Italy (6.9). Secondly, the bulk of Spain's SO_2 emissions are produced by power plants using domestic coal of low thermal potential and high sulphur content. Thirdly, more than half of Spain's SO_2 emissions are produced by two coal combustion power plants which are among the biggest polluters in Europe (See Table 12.2). It is worth mentioning here that those two power plants are owned by the state company ENDESA, which depends on the Ministry of Industry, which is one of the main actors in the policy arena of air protection.

A fourth consideration is that, in spite of the high level of total emissions, the dispersion of power plants across Spain's territory prevents the concentration of high levels emissions in the ambient air. Consequently, the Spanish map of critical masses, which assesses SO_2 emission levels on the basis of the carrying capacity of the territory, has only identified one quadrant with a concentration of SO_2 above its critical mass threshold. This quadrant corresponds roughly with the province of Coruña, located in the North Western corner of the Spanish peninsula, where the As Pontes power plant is located, together with a smaller power plant and a large refinery. According to public officials interviewed, around 20 per cent of the emissions produced in this area are deposited within it, while the remainder is carried into the Atlantic ocean.

Finally, the dry climate that prevails in most of Spain, which yields relatively low precipitation, favours dry depositions of sulphur and VOCs. However, it is only in wet depositions that, by chemical reaction, SO_2 is transformed into the sulphur acid carried by acid rain. This means that acid rain is not a major environmental problem in Spain, although it is not completely absent from the agenda. Still, the problem has not achieved the critical mass necessary to make the issue of acid rain a major environmental concern. Until now, acid rain has remained a local issue, without gaining much attention in the national media and public opinion.

Besides the fact that climate does not favour the formation of acid rains, the alkaline nature of Spanish soils makes acid depositions less damaging than in other countries. Alkaline soils are less vulnerable to acid corrosion than non-alkaline soils. The main threat faced by Spanish soils is erosion, not corrosion. According to official data, in 1990, medium and high erosion affected 44 per cent of the Spanish soil (MOPTMA, 1993). Since the country loses approximately 1,000 million tonnes of fertile soil per year, Spain is the only European country included in the United Nations Program for the Environment Risk Map (UNPERM). Moreover, this problem is intensified by the frequent forest fires, which affected almost 400,000 hectares in 1989 and 1994 and nearly 10 per cent of the total forest area from 1988 to 1994.

NO_X Emissions

Spain's NO_X emissions reached 839.000 tons in 1985, which made it the fifth largest EC emitter, after Germany, the UK, France and Italy (Table 12.2). The ratio of emissions to surface was only 1.7 tons per square kilometre, compared to an EU average of 4.6 tons. Transportation generated 54 per cent of total emissions, non-industrial combustion generated another 31 per cent, and industrial processes around 8 per cent. This situation did not change substantially during the period 1985-1994, in

Table 12.2   **Total emissions of NO_X per capita in the EU (in Kg/NO_X)**

Country	1987	1990	1994	2000
Germany	36.6	32.5	23.1	18.6
Denmark	59.0	54.7	43.3	34.2
Finland	53.6	57.5	51.6	45.6
Greece	72.4	82.5	69.9	66.0
Ireland	32.4	38.1	36.7	36.4
Luxembourg	48.7	64.1	46.2	43.6
United Kingdom	44.0	47.2	43.4	35.0
Sweden	49.7	46.6	44.9	40.3
Holland	36.8	36.4	27.8	17.7
Austria	29.7	28.2	21.7	19.7
Spain	23.4	31.9	23.3	22.8
France	28.4	30.5	26.8	23.9
Belgium	29.6	33.3	24.9	23.9
Italy	29.9	31.3	29.4	28.0
Portugal	11.8	13.2	19.0	19.6
Total EU	34.9	36.3	30.6	26.3

Source: see Table 12.1

which Spanish NO_x emissions increased in relation to other EC/EU members. In 1994 they represented 76 per cent of the EU *per capita* average and 8.1 per cent of the total.

VOC Emissions

Spain's emissions of VOCs reached 2.141.000 tons in 1995, ranking it as the fifth largest emitter of the EU. Emissions can be classified according to whether their sources are anthropogenic – 60 per cent of total emissions – or biogenic. About 15 per cent of anthropogenic emissions come from transportation; coal mining, gas distribution and other activities yield 15 per cent of emissions; and the evaporation of solvents, more than 60 per cent.

CO_2 Emissions

A look at CO_2 emissions, which are the most active contributor to the *green house* effect, provides some interesting information about patterns of energy consumption. Since CO_2 emissions are produced by the consumption of all fossil energies – coal, petrol and natural gas – they are a good indicator of energy consumption. Table 12.3 shows the per capita emissions of CO_2 within the EC. A closer look to the table shows the existence of wide disparities in emission rates across countries, and that such disparities accord well with the level of economic development of the different countries (France would be the only exception, due to the weight of nuclear energy in the balance sheet of energy consumption in this country). The wide gap in per capita CO_2 emissions between Southern and Northern European countries indicates that in the former countries per capita energy consumption has still to grow and get closer to the EC average. This disparity helps explain the reluctance of Southern European governments towards IEAs seeking reductions in emissions levels.

Table 12.3   **EC: CO_2 emissions per capita in 1990**
(in Tons per capita - EC=100)

Country	Index
Belgium	133
Denmark	122
France	79
Germany	137
Greece	84
Ireland	103
Italy	81
Luxembourg	400
Holland	118
Portugal	45
SPAIN	64
United Kingdom	122

Source: MINER, 1991.

Summing up, Spain's emissions profile of major air polluting substances corresponds with its level of economic development. The only meaningful deviation concerns SO_2 emissions, and concerns the low quality of the domestic coal used to fuel local power plants. Spain's relatively remoteness from the densely industrialised areas of Central and Northern Europe, the presence of large sparsely populated areas and the dispersion of the major sources of emission across its territory, produces a pattern of air pollution that differs markedly from that of Central and Northern Europe. We may express that difference metaphorically by saying that while the Spanish pattern resembles a white cow with a few dark spots of different intensities, the Central and Northern European pattern resembles a bear with black and/or dark brown fur. That means that in Spain air pollution tends to be seen as a local problem which is particularly intense in a few local-industrial areas (Avilès, Basauri, Tarragona, Cartagena and Huelva) and which, depending on weather conditions, may also become temporally intense in the big metropolitan areas (Barcelona, Bilbao, Madrid). Due to the significant gap that separates Spain from the more developed Western European countries in terms of per capita energy consumption and per capita emissions, public authorities have been, until very recently, reluctant to accept across the board reductions in emissions, particularly when those reductions were seen as imposing a loss in terms of potential for growth.

The IEAs' Domestic Policy Arena

Until 1996, the main actors of the policy arena were two governmental agencies: the State Secretariat for the Environment (SEMA), part of the former Ministry of Public Works, Transport and the Environment (MOPTMA), and the State Secretariat for Energy (SEE) of the Ministry of Industry (MINER). The national parliament, regional governments and social actors, such as political parties, environmental groups and mass media, played only a secondary role – apart from the electricity industry – at the national level in this issue.

Governmental Actors

The Secretary of State for the Environment Until 1996, under the socialist government, the Secretary of State for the Environment (SEMA) was the main environmental agency of the Spain's central government and its representative at the EU level. The SEMA was set up in 1991 in response to EC pressures to raise the rank of the Spanish environmental administration. At that time Spain had a very poor record regarding the implementation of the European environmental directives and the low administrative rank of the environmental administration was considered as one the causes of this problem.

Spain's entry into the EC played a paramount role in the consolidation of the position of environmental policy on the government's agenda and in the creation of an environmental administrative infrastructure. Though a law on air pollution had already been enacted on 1972 – and is still in force – Spain lacked a coherent environmental administration until its integration in the EC. Administrative competencies for the environment were widely dispersed among eight ministries and co-ordination among those agencies was left to a process of mutual adjustment by persuasion.

As a sub-unit in a multifunctional ministry, and lacking agencies and a large staff of its own, the SEMA did not appear to be a 'strong' bureaucracy. However, neither was it was a marginal one. Its close working relation with the European Commission (DG XI) and its responsibility for the implementation of EC policy endowed the SEMA with a very important administrative resource: the capacity to assess the environmental impact of major industrial and public works projects and to veto them on the basis of their alleged environmental inadequacy. This veto power gave the SEMA some political clout, particularly in its relations with the Ministry of Industry, the other major governmental actor in this policy arena.

The SEMA managed the network of the EMEP stations monitoring transborder air pollution and it was Spain's representative in the EMEP governing body, which has already set up ten stations in Spain. In addition, the SEMA supervised the national network of air pollution stations (789 manual and 303 automatic) that measure pollution levels throughout the country.

Together with the SEE, the SEMA represented the Spanish government in the working groups established by the different Protocols within the framework of the Geneva Convention. The two agencies worked out the position of the Spanish government in the negotiations leading to the signing of the Protocols. On the whole, in case of disagreements, the SEE was able to make its position prevail. Thus, as we have seen, during the negotiations of the first SO_2 Protocol, the SEMA was ready to assume the commitment of a 30 per cent reduction on emissions levels while the SEE considered such a commitment too ambitious and opposed signing the Protocol. In the end, Spain did not join the Helsinki Protocol on the SO_2 emissions.

The Secretary of State of Energy　The Ministry of Industry and Energy (MINER), through its Secretary of State of Energy (SEE), controls the public electricity utility ENDESA – in process of privatisation at the time being – which in turn is the strongest member of the electricity industry association UNESA. The latter clearly influences the SEE's position on air pollution policies. The greater resources of UNESA allow it to act as the intelligence unit of the SEE in assessing abatement costs and in choosing abatement technologies, as well as in designing policies addressed to curtail emission level in large combustion plants. Thus the SEE is a major player in the policy arena of air pollution, allied with powerful sectoral interests. First of all, it has a very closed link with the electricity industry which was placed under the control of the MINER when the first socialist government nationalised the power distribution network, forced a nuclear moratorium and established a complex system of payments to remunerate power producers. Moreover, the production strategies of the electricity industry are determined by energy plans drawn by the SEE. In this sense, the electricity industry and the SEE may be seen as the two sides of the same coin. Secondly, it is worth mentioning here that the MINER is one of the largest emitters of air pollution, for Spain's main producer of thermal power and the owner of the most polluting power plants is the public enterprise ENDESA, a member of the State holding National Institute of Industry (INI) which is under the control of the MINER. In addition, the SEE runs the CIEMAT, a public research centre that carries out research on the environment and power technologies, which is partly financed by UNESA.

The SEE controls the network of stations that measure the emission levels at emission points. These stations are installed in the major combustion plants, both industrial and non-industrial, and provide crucial information on emission levels of air polluting substances. The SEE used to handle this information with extreme caution, even treating it as privileged information . The information provided on emission levels at the point of emission was aggregated at the national level, thus making it impossible to pinpoint specific sources of a particular pollutant. The reluctance to release this information on time and in a disaggregated form gave Spain a bad reputation as an unreliable informant in international environmental forums.

The very different institutional environments of the SEMA and the SEE affected the way they defined the issues of air pollution control. Obviously, the SEE has been very sensitive to abatement costs, not only real costs, but also anticipated costs. As mentioned before, in Spain the issue of energy is closely linked with the issue of growth capability, and the MINER has tended to see the energy constraints imposed by the environmental policies as losses in terms of a lower capacity for economic growth. In addition, the SEE has been also highly sensitive to the issue of technological dependence. Therefore, it has fiercely defended environmental policies that guarantee the widest choice of abatement technologies, and opposed policies that go beyond fixing global reduction targets. From this perspective, IEAs should not attempt to determine the policy instruments to be used, the choice of which remains the preserve of national sovereignty.

As expected, the SEMA has been more sensitive to considerations of air pollution damage costs, and was more ready to commit itself to environmental targets. Its policy style has also been less secretive and defensive since the EU's environmental directives provide for the general public to get access to the available environmental information. However, the SEMA was not blind to the issue of national interest and it also subscribed to the view that there was a certain amount of hypocrisy on the part of the Northern European countries when they pushed for across-the-board reductions in emissions levels, irrespective of a country's level of economic development.

The Electricity Industry

UNESA, the trade association of the electricity industry, is a very active player in the air pollution policy arena. The tight regulation of the electricity industry requires close co-ordination between UNESA and the SEE, and sets the stage for a joint partnership in environmental matters. Thus, as noted, UNESA and the SEE jointly define the ministry's stance on air quality policies, particularly those that most impinge on SO_2 produced by electricity industry.

UNESA was the only private actor that paid heed to and followed closely the negotiations leading to the Geneva Convention and the subsequent IEAs signed within the Convention. In the mid-1970s, UNESA was already involved in an international commission on 'Electricity and the Environment' set up by the United Nations Economic Commission for Europe (UNECE).

As already mentioned, ENDESA, the owner of the most polluting power plants in Spain, is a public enterprise, integrated into UNESA. ENDESA manages four large coal combustion power plants located in As Pontes (A Coruña), Andorra (Teruel), Compostilla II (Ponferrada), Carboneras (Almería), as well as a coal mine in Puertollano (Ciudad Real). Since 1987, the company was involved in a judicial process for an alleged ecological crime caused by SO_2 emissions from its Andorra power plant. As a consequence of legal and environmental pressures, in December 1993, ENDESA presented a 1994-1998 'Environmental Plan' in which it committed itself to 'go further than the strict compliance with existing laws, to intensify its efforts and to take specific measures to guarantee the rational use of resources and the minimisation of waste, while contributing to the sustainable development demanded by society'.

The Environmentalist Movement

The environmentalist movement was born in Spain in the mid-seventies, propelled by anti-nuclear feelings. Up to today it has remained local and issue oriented, not able to forge a successful political organisation, either at national or at the local level. As a social movement, the Greens wax and wane, following an opportunity structure of local conflicts whose texture varies from place to place. As a political movement, it is thoroughly irrelevant, unable to establish itself in national or regional politics. Only in the recent local and regional elections of 1995 were some activists elected on the lists of the IU-the Green coalition.

According to official data (MOPTMA, 1995), there are almost 700 environmental organisations in Spain. Traditionally, the strongest environmental groups formed around territorial issues and especially nature area protection. Until recently, other issues such as water and air pollution as well as toxic wastes have been relatively neglected. Due to their ideological and political fragmentation, the Greens have repeatedly failed in setting-up a single organisation at the national level. The problem seems to have been overcome after the creation in 1991 of the confederal CODA (Coordinadora de Organizaciones de Defensa Ambiental) which co-ordinates many local and regional organisations and is probably the most representative organisation at the national level. In addition, most international environmental organisations have set up branches in Spain: Greenpeace, ADENAT WF, Friends of the Earth. However, it is impossible to get reliable information on the membership and financing of environmental organisations. Although Greenpeace officially claims to have 70,000 members, it is not a mass movement, but rather a highly active lobby enjoying some respectability among the mass media. Besides undertaking spectacular protest actions, its main resource remains information provided by national experts and other national offices abroad.

The Spanish environmental movement as a whole has ignored acid rain problems due to the low salience of the issue at the national level. The only exception was the attempt by Greenpeace to put pressure on the

Spanish government during the negotiations of the EC Directive on Large Combustion Plants. Greenpeace explains its neglect of the Geneva Convention by pointing to the voluntary nature of the agreements and the lack of an authority structure able to enforce the policy commitments. From this perspective, IEAs are irrelevant. The only policy arena worth the effort of political mobilisation is the EU, for it has the formal authority necessary to guarantee the enforcement of its environmental directives.

At the local level Greenpeace has been very active, together with other local environmental groups, leading campaigns against Andorra's and As Pontes' SO_2 emissions. In 1988, Greenpeace issued a dossier entitled *Acid Rain and* joined environmental local groups to support the cause initiated by the public prosecutor. In 1992, it denounced ENDESA before the Spanish Ombudsman for misleading publicity in its campaign *Living Energy*. It also released press communiqués denouncing the degradation of the environment provoked by ENDESA and its reluctance to carry out improving measures. Finally, ENDESA agreed to reduce Andorra power plant SO_2 emissions up to 90 per cent by 1998. Greenpeace considered this reduction target as a great success and withdrew its suit against the enterprise, while stressing the necessity to reduce other power plants emissions.

Public Opinion

Compared with the Northern European countries, in Spain there is a lack of environmental awareness. The fact that Spanish public opinion and pressure groups supporting the environmental protection have been traditionally weak explains why this policy has not been relevant, either in terms of financial or organisational resources. In 1989, only 3.9 per cent of population gave first priority to environmental protection (Eurobarómetro, 1989); among the EC member states, only Greece and Portugal were less sensitive in this respect. Furthermore, a survey conducted in 1990 pointed out that 62 per cent of Spaniards still were not sufficiently informed by the media about environmental problems (EP, 20/12/90). However, in the last few years opinion polls have detected an increase in the ecological sensitivity of the public (Eurobarómetro, 6/96). However, whether this will be incorporated into everyday behaviour is another matter since the predominance of consumerist values in the Spanish society operates as a powerful obstacle to environment values (Pridham, 1993). In any case, most of Spanish public opinion still ignores the main causes and effects of acid rain pollution.

Negotiating and Implementing IEAs

The LRTP Convention

Spain was one of the signatory countries of the Geneva Convention in 1979, and ratified it three years later. At that time, air pollution was not perceived as a serious environmental problem – let alone transboundary pollution – and the environment itself did not feature in the politics of the time. Within this context, the rationale for joining the Convention was twofold. On the one hand, there was the positive motive of partaking in the politics of *détente* between East and West. The Geneva Convention provided the new Spanish democracy with a valuable opportunity to show its new political credentials in the arena of international politics, and to regain the international recognition that had been denied to the Francoist dictatorship. On the other hand, there were no clear reasons not to join the Convention, for it did not entail any immediate policy measures. The Convention called for international co-operation and research on the issue, and left for future negotiations the determination of specific policy targets for reductions of emission levels.

Those reasons were still valid in September 1984, when the Protocol for financing the EMEP network was signed in Geneva. Since the establishment of that network was one of the stated goals of the LRTP Convention, not participating in the design of the institutional arrangement to implement that goal could have been interpreted as sign of inconsistency and unreliability, which would have seriously damaged the country's reputation as a faithful partner. IEAs such as the Geneva Convention may be viewed as a closed institutional network that makes possible monitoring reputations and to detect both reliable partners and unreliable opportunists.

In any case, Spain did sign the EMEP Protocol without other pressures than the mere pressure for consistency and good faith, which came out of its previous commitment in the Geneva Convention. It seems that, at least so far, Spain has contributed its share to support the development and the maintenance of the network. At present, as already noted, EMEP has ten stations installed in Spain. In addition, the Spanish network includes regional and local stations. A station of the BAPMON network (Background Air Pollution Monitoring Network) created by the UN is also located in Spain. These stations are under the direct control of the central government, while some regional governments have set up their own monitoring networks in co-ordination with the national one.

The Helsinki Protocol on SO_2

As we have already seen, Spain did not sign the Helsinki Protocol on the reduction of SO_2 emissions, though, afterwards, it reached the Protocol's target of a 30 per cent reduction in the level of SO_2 emissions. By the end of the negotiations, the Spanish delegation was split, with the environmental bureaucracy favouring the agreement and the MINER opposing it. This opposition reflected the shift in Spain's energy policy to domestic coal in the early eighties, and the constraints that such a shift imposed in terms of emissions levels. However, since then the trend has been a slow, but steady decline in the weight of domestic coal in the national energy balance and its substitution with natural gas. This trend will continue in the future as Spain's domestic coal is replaced by cheaper, and cleaner, imported coal. Today, the main obstacle to further reductions in domestic coal consumption is the high dependence of some regions on coal mining.

The Impact of the EC Directive on Large Combustion Plants

The battle around SO_2 emissions was continued in a different institutional arena a few years later, in the negotiations leading to the 1988 European Community's Directive on Large Combustion Plants (LCP). This time, however, the EC institutional setting left no way for Spain to opt out and the Spanish government had to fight in earnest to protect its interests. The main strategy was to get the Commission to grant Spain an exception that would leave room to accommodate both its energy needs as a laggard country and its peculiar endowment of energy resources. In the end, Spain succeeded in reaching the preferential treatment it claimed. Thus, for all combustion plants, the Directive fixed a global reduction target which started with the freeze of the 1980 emissions level by 1993, to be reduced subsequently by 24 per cent by 1998, with a further reduction to achieve a global 37 per cent reduction with respect of the level of the baseline year by the year 2003. For new combustion plants, Spain obtained an important concession: a derogation lasting until the end of 1999 which allows the Spanish government to approve the construction of new combustion plants with a output equal or superior to 500 MW – up to a maximum of 2000 MW, if they use domestic coal – which have to be operative before the end of 2005. New combustion plants using domestic coal are allowed to operate with desulphurization index of the 60 per cent, instead of the 90 per cent applied to the rest of the countries.

With respect to NO_X emissions, the Directive fixes a reduction of 20 per cent by 1998 – taking again 1980 as the base year – for existing plants; and 650 mg/Nm for new combustion plants. The Directive also struck a compromise between the German position, that pushed for the universal adoption of the best available technology, and the position of Spain and the UK that defended the freedom for each country to choose the technology to be applied.

The EC Directive was transformed into national legislation by a Royal Decree enacted in April 1991 (Real Decreto 646/1991) and, later on, its targets were incorporated by the National Energy Plan (PEN) for the period 1991-2000. Since the production of electricity is a major source of emissions of polluting substances to the atmosphere, it is convenient to examine the PEN forecasts on the increase of power supply for the period mentioned in order to appreciate the impact this will have on emission levels. The PEN foresees an increase of electricity supply of 8377 MW for the entire decade, a prediction based on the unrealistic estimate that the electricity demand would grow at a yearly rate of 3.44 per cent. However, the power plants burning domestic coal represented no more than 15.97 per cent of the total new power generation capacity (1,338 MW), significantly below the threshold of 2,000 MW for domestic coal established by the EC Directive 88/606.

For what concerns the reduction of SO_2 emissions, the PEN sets three goals. First, a reduction of 42 per cent of SO_2 emissions of existing LCPs by the year 2000, taking the 1980 emissions as the baseline. This goal exceeds amply the target of a 24 per cent reduction by 1998 set in the Directive. Second, a reduction in the emissions of new power plants below the limits imposed by the Directive. Third, the realisation by the year 2000 of a 30 per cent reduction of the overall SO_2 emissions, including all kinds of emissions - fixed and mobile sources.

Since in Spain at least two-thirds of all SO_2 emissions are released by LCPs, the targets for the reduction of emissions have mostly to do with the electricity sector. It is expected that the 42 per cent reduction of sulphur emissions will be achieved by applying two types of measures:

- Reducing the sulphur load of the coals used in LCPs. This may be achieved by increasing the share of foreign coal and natural gas in some electric groups. Some of the existing power plants are to be retrofitted to use gas, and this fuel represents 25 per cent of the new electric equipment (2.135 MW).

- The introduction of installations to wash coal before combustion, desulphurisation equipment, and clean combustion technologies. In this respect, it should be noted that the new combustion plants will use cleaner technologies: two of them pulverised coal and two other atmospheric fluid beds. In addition, the power plant at Puertollano has developed an innovative combustion technology based on the gasification of coal and coal washing has been introduced in the power plant of Andorra. In the near future, as the new power plants come into operation, the production mix will change in favour of the less polluting plants.

Further reductions in SO_2 emissions are expected to be obtained by applying the programs of the Plan for Energy Saving, with the options of minimising the number of new power plants, extending the life of already existing power plants and building new equipment into old power plants. However, it is still difficult to assess the extent to which those different measures have been or are being implemented.

The Sofia Protocol on NO_X

Spain subscribed to the Protocol on the reduction of NO_X emissions, signed in Sofia in October 1988 and officially ratified it in March 1991. The Sofia Protocol provided for a freeze of NO_X emissions by 1995, allowing the signatory countries to take any base year they wished between 1980 and 1986. NO_X emissions do not pose for Spain the same problems confronted with SO_2 emissions. LCPs release less than one third of the total NO_X emissions while transportation accounts for almost all of the remaining two thirds.

The Spanish government has not devised specific measures to implement the Sofia Protocol as it is considered that the implementation of the European Directives on LCPs, on emissions of mobile sources, and on the quality of petrol derivatives are enough to reach and even surpass the target of a freeze of NO_X emissions levels. With regard to LCPs, the PEN 1991-2000 sets out to achieve a level of emissions of 263 Kto by the year 2000 - a 29 per cent reduction with respect to 1980. This target is in line with the Community Directive 88/609, which assigns Spain the target of 277 Kto of NO_X emissions by 1998 - a 25 per cent reduction with respect to 1980. The measures to achieve the reduction of NOx emissions consist of the introduction of special burners, combustion by phases, and the re-powering of some plants from fuel to natural gas.

Spain also signed the 1991 Protocol to reduce VOCs emissions, joining the group of countries that made a pledge of a 30 per cent reduction. However, this Protocol has not yet been ratified by the Spanish Parliament, and no specific measures to implement it have been devised.

The Negotiation of the New Protocol on SO₂ Emissions

Finally, it is worth mentioning here, the position of the Spanish government on the negotiations of the second Protocol for further reductions of SO_2 emissions. In these negotiations there was, for the first time, the opportunity to base the new reduction targets on the information provided by maps on 'critical masses' elaborated for that purpose. This methodology allowed for a more flexible approach to the determination of targets for the reduction of emissions, differentiating such targets on the basis of the environment's carrying capacity. Such an approach was particularly favourable to a country like Spain since it allowed for discrimination between highly polluted and less severely polluted areas. Nevertheless, the Spanish authorities were very suspicious of the actual maps drawn because - they argued - they were not consulted during the process of map-making, and the information used to determine the critical masses of Spanish soils and ecosystems was unclear. An official interviewed in the course of the field research complained that 'the province of Alicante had been treated with the same parameters they applied in the Netherlands'.

In spite of those suspicions, the Spanish authorities provisionally accepted the RAINS map of critical masses. In fact, the map only records one quadrant in Spain as having emissions above its critical mass threshold. As mentioned before, this quadrant corresponds to the Atlantic province of Coruña, located in Spain's North Western corner, which contains the power plant of As Pontes, by that time, the largest single SO_2 polluter of the European continent.

In the first round of negotiations, the Nordic front – the Scandinavian countries, Germany, the Netherlands, etc. – put forward a proposal for a 55 per cent reduction for the countries with emissions levels above their critical masses. The Spanish delegation considered this proposal unacceptable arguing that, if it were applied in Spain, it would entail extremely high costs and would not solve the problem in the single quadrant above the critical mass. The counterproposal of the Spanish delegation was to accept the target of 55 per cent in the quadrant of Coruña (As Pontes), and to continue with the PEN targets in the rest of the country. By the end of the first round of negotiations, it seemed that Spain had convinced the other countries to accept its proposal.

In the second round of the negotiations, which took place in August 1993, the Nordic front re-ignited the controversy by tabling the issue of LCPs. This time, the Nordics wanted to impose emission limits for both old and new combustion plants, a proposal which would mean for Spain that it would lose the concessions it had negotiated in the EC Directive on LCPs. Again the Spanish delegation strongly opposed this proposal arguing that, for new combustion plants, the target should be in line with those contained in the Directive, while for the old plants the targets should be only indicative, and not binding.

However, in the third round of negotiations, that took place in Oslo, Spain succeeded in reaching an agreement with the other countries which took its special characteristics into account. Given its situation, Spain accepted a higher reduction of SO_2 emissions (55 per cent) in its North Western cell, where the thermal plant of As Pontes is located, and a lower one (35 per cent) for the rest of the country by the year 2000. In this way, the Spanish argument against a general reduction was accepted, and each cell is considered by itself. This includes, however, taking interactions and interregional flows of air pollution into consideration. By 1995, Spanish maps of critical loads by cells were elaborated by experts from the SEMA and the SEE, using data provided by the CIEMAT and UNESA. A cell by cell evaluation required specific data for each one of the several plants, and a ministerial order was addressed to the electricity sector in order to collect such information. In addition, the MINER negotiated a new royal decree with the electricity sector and the MOPTMA in order to fix the same limits for each thermal plant and for the petrochemical sector as a whole.

Thus, unlike what had been the rule in the past years, Spanish authorities followed a new strategy of providing specific information for each plant in each cell. According to public officials, such an attitude is much more effective during the negotiation of IEAs since in this way the Spanish arguments could be clearly understood by the other countries, i.e. the adoption of measures fitted to the specific conditions of a particular case instead of general criteria.

Policy Evolution in the 1990s

The Impact of the Economic Recession

The Energy National Plan (PEN 1991-2000) has been unable to foresee the new economic, energy and environmental context of the 1990s. As a consequence, the initial environmental targets have been recently reviewed.

From 1990 to 1994, the Spanish economy did not grow at the rate of development on which the PEN was based (the real yearly growth rate was 1 per cent, instead of the anticipated rate of 3.51 per cent). During the same period, the energy demand rose by 2.3 per cent per year, a rate far greater than the growth of GNP, which seems to indicate low energy efficiency during the same years. It is worth noting the important increase in the demand for petrol and natural gas. Moreover, primary energy consumption has risen of 1.9 per cent per year, while the PENs forecast was of 2.1 per cent (Table 12.4).

Table 12.4   **Energy demand and GNP increasing** (1991-1994)

	Annual rate				Annual Average	
	1991	1992	1993	1994	1990-94	Forecast
PEN FINAL ENERGY	3.4	0.6	0.3	5.0	2.3	2.4
- coal	-3.2	-13.1	-11.4	-3.5	-7.9	-1.5
- oil	3.3	1.6	1.3	5.0	2.8	2.1
- gas	10.3	3.1	-0.6	12.0	6.1	4.8
- electricity	3.6	1.0	0.6	3.6	2.2	3.6
PRIMARY ENERGY	3.0	2.8	-1.2	3.0	1.9	2.1
GNP	2.2	0.7	-1.1	2.0	1.0	3.5
Electric energy	2.4	1.5	0.6	4.6	2.3	--

Source: *Estrategia Energética y Medioambiental* (ESEMA), Dirección General de Planificación Energética, MINER, March 1995.

The New Policy Approach

The Plan on Energy Saving and Efficiency (PAEE) 1991-2000 is being implemented through four specific programmes on Saving, Substitution, Co-generation and Renewable Energies. The Energy Saving programme is intended to reduce oil and electric energy consumption in the industry, transport sector and services. The Substitution programme seeks the shift from oil and coal consumption to natural gas. The Co-generation programme provides incentives for projects on the joint-production of heat and electricity in order to improve energy consumption in industrial processes. Finally, the Renewable Energies programme aims to introduce alternative energies as a way of counteracting the greenhouse effect and to foster, in the long term, energy security and diversification.

Total private investments linked to the implementation of the PAEE represented 220,000 Mio pesetas during the period 1990-1994 while public investments (PAEE, ERDF and others) amounted to 44,000 Mio. pesetas. Emissions avoided through the several projects were calculated at 4,500 Kto of CO_2, 120 Kto of SO_2 and 16 Kto of NOx. In addition, in order to improve energy efficiency, a royal decree was enacted to enforce the EU

Directive on reduction of CO_2 emissions. The decree introduced new concepts such as: energy certification applied to the new buildings, periodical energy inspections, and audits in high energy consumption industries.

Regarding the transport sector, we have already noted that the EU Directives on motor vehicle emissions have had positive effects, particularly on SO_2, CO_2 and VOCs emissions. Nevertheless, due to the continuing growth of the transport sector in Spain, the main problem remains NO_x emissions. There is also a clear lack of public control on catalytic converters (cars).

As already said, special burners are being introduced in the coal power plants in order to reduce both for SO_2 and NO_x emissions. A UNESA-MINER monitoring committee was set up to achieve the targets of reduction fixed for public plants. In addition, the new National Electricity System Organisation Act of January 1995 gives first priority to the environmental protection, which includes the development of co-generation processes and renewable energies.

The petrochemical sector requires greater investments due to the multiplicity of sources. Instead of introducing more strict legislation which, as the Spanish experience shows, is not the best way for reducing air pollution, the MINER has been promoting voluntary agreements with the big refineries, such as REPSOL, CEPSA and BP.

Once again, the EU legislation determines the Spanish adaptation and provides financial resources which enable public and private sector to fulfil international requirements. Regarding the enforcement of the new Community Directives (Combustion Plants Framework, etc.), national plans were drawn up for each sector in order to apply for financing from the EU Cohesion Fund (200,000 to 250,000 Mio pts.). The advantage is, of course, that domestic consumers will not have to pay the whole costs.

Last, but not least, the new Spanish policy strategy has also had an impact at the administrative level where a new Sub-general Direction for Energy and Environment Planning, located in the MINER, was created in 1994 with the explicit task of defining a new strategy linking energy with the environment.

Conclusions

According to the theoretical framework used by the research project, the Rational Unitary Actor Model conceives compliance as a function of the structure of incentives. Until 1994, Spain could be viewed as a laggard country in the field of acidification policy, mainly because of the lack of external incentives to modify the initial perception of Spanish 'national interest'. The dry climate prevailing in most of Spain, which yields relatively low levels of precipitation and favours dry depositions of sulphur and other suspended particles, together with the alkaline nature of Spanish soils meant that acid depositions were less damaging in Spain than in other countries. As a result, Spain has been less vulnerable to problems associated with acidification. In addition, due to the significant gap that still separates Spain from the more developed Western European countries, in terms of per capita energy consumption and per capita CO_2 emissions, the Spanish government has been reluctant to pay abatement costs that would have resulted from emission reductions. Such reductions were seen as both imposing a loss of potential economic growth to the country and creating social problems in the coal-mining sector, which supplied the energy sector.

Spain has only been ready to assume abatement costs linked to the implementation of EC environmental directives on air pollution control where its interests have been taken into account and European funds have been available to co-finance the energy sector's reconversion in order to make it more efficient.

The Domestic Politics Model 'pictures governments as complex organisations, where sub-actors pursue multiple and to some extent conflicting objectives' which can cause *implementation failure,* despite the official willingness to comply with IEAs. The Spanish case only partially fits with the assumptions of this model. As a result of the opposition of the State Secretariat of Energy (SEE), which was a major player in the arena of air pollution policy, Spain did not sign the Helsinki protocol on SO_2 emissions even though the State Secretary for the Environment (SEMA) was ready to accept the agreement. The Ministry of Industry and Energy was of the opinion that the energy constraints resulting from the environmental policies imposed unacceptable limits on Spain's growth capability. In addition the SEE was also very sensitive to the issue of technological dependence. Therefore, it pushed for environmental policies that guarantee the widest possible choice of abatement technologies. As the IEAs were closely linked to energy policy, which was a matter of high national interest, there was an implicit consensus on this issue between the government and the main opposition parties.

In addition, the Spanish society has given little priority to environmental quality irrespective of economic prosperity or recession. This attitude can be explained by three related factors: the high priority given by public institutions to economic growth; the weakness of environmental parties and groups; and the lack of environmental awareness on the part of the general public. With a few exceptions, the Spanish media and the main environmental organisations have ignored acid rain problems due to the low salience of this problem at the national level.

Apart from UNESA, the trade association of the electricity industry, NGOs have not been directly involved in the formulation of the clean air policy. As a constitutional principle, the central government is exclusively responsible for formulating national positions and negotiating international agreements, even though it is supposed to inform regional authorities when regional competencies are concerned. Even though this is the case with air pollution control, regional representatives did not show any particular interest in participating in the formulation of Spanish positions. Despite the fact that sub-national governments are responsible for the implementation of national clean air legislation, their possibilities for action were limited by the reluctance of central government enforce the provisions of the IEAs. Thus on the whole, a small policy community, made up of central governmental actors (SEE and SEMA), UNESA, and especially the public electricity board (ENDESA), has dominated the policy making process on this issue. Neither the political parties nor the environmentalist movement played a significant role at this stage. The main environmental organisation (i.e. Greenpeace) paid little attention to the Geneva Convention and the several protocols because of the voluntary nature of the agreements and the lack of any structure capable of enforcing the policy commitments. In contrast, Greenpeace was very active during the negotiations of the EC large combustion plant directive (LCPD).

Finally, the Social Learning Model is 'concerned primarily with processes of *learning* and transnational diffusion of ideas'. As decision-makers do not have a perfect knowledge about the problems they are dealing with, the policy process becomes a learning process. Scientific knowledge, information and persuasion can play an important role in bringing reluctant governments to redefine their initial positions. This approach seems to fit the Spanish situation during the negotiations leading up to the signing of the second SO_2 protocol in 1994. Although the Spanish negotiators initially disagreed with the making of the maps which were produced in order to determine the critical loads, a consensual on the relevant scientific knowledge, based on shared values and norms, did gradually emerge. On the basis of a common methodology, which accommodated the peculiarities of the Spanish situation, Spain's negotiators were ready to provide - for the first time - detailed information on its main emission sources of pollutants in order to keep negotiations going and to reach satisfactory agreement. Spain was willing to go along with the agreement because its scientific arguments had been accepted. In the meanwhile, Spanish energy policy had been reformulated and European funds made available to co-finance the energy sector's reconversion in order to make it more efficient. Since 1994, the central government has taken the initiative to promote gentlemen's agreements with the more relevant economic polluters (energy and transport) in order to reach the EU environmental standards.

However, in the last analysis, the effective implementation of the obligations of the acidification regime in Spain seems to have been more the result of domestic considerations and EU constraints than of the closeness of fit between national positions and the targets of the IEAs. As a result of the fact that domestic energy choices have traditionally prevailed over concern for environmental damage, the implementation of IEAs was ineffective until 1991, when the central government transposed the LCP directive, with its special concessions to Spain on SO_2 emissions. The ambitious Environmental Plan subsequently adopted by ENDESA in order to reduce SO_2 emissions by 90 per cent during the period 1994-1998 was also related more to the need to implement the LCP directive than to the Geneva Convention. In the new context of budget restrictions to reduce the public deficit, the Ministry of Energy launched a 'Saving and Energy Efficiency Plan' in 1991 and, in 1995 a new 'energy and environmental strategy'. Combined with a new formula for sharing of the abatement costs with the EC and private target groups, it is felt that this policy should enable Spain to comply with most of its international environmental commitments by 2007.

References

ENDESA (1993), *Plan de Medio Ambiente de ENDESA*, Sintesis del Plan, Madrid.

Greenpeace: 'Dossier Lluvias Acidas', July 1988.

Jimenez Beltrán (1993), 'Implicaciones Ambientales de la Política Energética Española', *Energía*, vol. XIX, pp. 41-49 (1993).

MICYT (1993), *Informe al Congreso de los Diputados sobre las Actuaciones Energéticas en 1990-91*, Ministerio de Industria, Comercio y Turismo, Madrid.

MINER (1991), *Plan Energético Nacional 1991-2000*, Madrid.

MINER (1995), *Estrategia Energética y Medioambiental, Documento de Síntesis*, Ministerio de Industria y Energía, 28 March 1995.

MOPTMA, *Medio Ambiente en España*, Ministerio de Obras Públicas, Transportes y Medio Ambiente (different years).

MOPTMA (1993), *Calidad del Aire en España*, 1990, Madrid.

MOPTMA (1994), *Estaciones y redes de vigilancia de la contaminación Atmosférica en España*, Madrid.

OCDE (1989), *Instrumentos económicos para la protección del Medio Ambiente*, Paris.

Pridham, G. (1993), 'National Environmental Policy-Making in the European Framework', Paper for the Workshop *European Policy and Peripheral Regions in the EC*, ECPR Joint Session, University of Leiden, 2-8 April.

13 Comparative Analysis: Accounting for Variance in Actor Behaviour

ARILD UNDERDAL

Introduction

The time has come to try to integrate the main observations and findings reported in our nine case studies into a comparative analysis that examines patterns of variance in actor behaviour across countries and protocols.[1] This is the task to which we now turn.

To recapitulate, our *dependent* variables are (1) the strength of regulations advocated by the various governments in negotiations about LRTAP protocols, and (2) the national records of implementation and compliance with signed protocols. In chapter 3 we outlined three different *models* designed to predict and explain negotiating positions and implementation records. These models point to different causal mechanisms and different (sets of) *independent* variables. We will now use these models to organise our comparative analysis.

There are basically two kinds of empirical tests that we can undertake to evaluate conceptual models. One is to judge them on the basis of their 'fruits', that is, compare specific propositions derived from each of them with observed actor behaviour or outcomes. The better the propositions derived from a model fit the empirical observations, the more confident we can be that the model captures the essence of the system or process it is designed to represent. Another approach focuses on the *heuristic* value of models. From this perspective the essential question becomes whether the model correctly identifies and describes the important causal mechanisms through which actor behaviour and outcomes of political processes are formed.

These two approaches do not necessarily lead to the same conclusions. It is conceivable that we could find, for example, that model I predicts actor positions and levels of compliance better than any of the other models, and yet would have to conclude that the intellectual or political processes that lead to this behaviour could be better *described* or *understood* in terms of some other model. This suggests that the two kinds of 'tests' should be seen as complementary in the sense that they relate to different analytic *functions* that models can perform. This conclusion is supported by the fact that the models we use in this study have in fact been constructed for somewhat different analytic purposes. In general, rational choice theory can be seen as conditionally *prescriptive* in the sense that it aims at identifying what constitute(s) the best option(s) available to an actor in a given situation. The domestic politics and the social learning models are primarily tools for *de*scription and explanation. More precisely, their principal objectives are to help us identify the major social mechanisms through which the behaviour of complex actors is formed and understand how these mechanisms work. Now, rational choice theorists would argue that their models have descriptive value, too; since an actor generally can be assumed to choose the best option available, the choice *prescribed* by the model will also be the choice that he will in fact make. Conversely, advocates of the other two models would argue that since collective decisions are made through particular kinds of institutions and political processes, a good description and understanding of these institutions and processes is necessary to predict and even more so to explain outputs. Moreover, they would argue that even though their models have not been designed for prescriptive purposes, they nevertheless have interesting 'implications' for policy. The important lesson here is that each of the two approaches to evaluation can contribute something that the other cannot. In this chapter we will therefore try to supplement attempts to test specific propositions or hypotheses with an attempt to evaluate the three models also as heuristic devices, that is, on the basis of their usefulness in guiding us towards the important mechanisms through which negotiating positions and implementation records are determined. In doing so we rely heavily upon observations and interpretations offered by our case-study authors, including a set of more comprehensive reports providing more detailed information than the chapters included in this book.

The first sulphur protocol and the NO_x protocol are the only emission control agreements where sufficient time has passed to evaluate compliance and implementation records. We will therefore focus the analysis in this chapter on these two protocols.

Model I (Unitary Rational Actor)

To recapitulate, model I assumes that the positions governments take in international negotiations will be those that (they expect will) maximise net national utility or welfare. The basic assumption is that a country will *accept* only regulations from which it expects to reap a net benefit (or at least not lose), and *comply* with an agreement it has signed only as long as compliance costs do not exceed the costs it would incur by defecting.[2]

Applied to a case of transfrontier pollution, the model depicts state positions as a function of three major determinants: damage costs, abatement costs, and what we have called 'the pollution exchange balance', meaning the ratio of pollutant export to import. In its most simple form, the model predicts that the positions governments take towards international emission control regulations will correspond to the pattern summarised in table 13.1.[3]

Table 13.1   Actor positions as predicted by model I

Abatement costs	Damage costs	Export/import	Negotiating position
Low	High	Net importer	Pusher
High	Low	Net exporter	Dragger
High (Medium)	High (Medium)	Balanced	Intermediate
Low	Low	Balanced	Bystander

With regard to the implementation phase, the model predicts that the parties most likely to defect would be (a) countries with high abatement costs, low damage costs, and a substantial net export of pollution (i.e. 'draggers'), and – perhaps – (b) countries with high abatement costs, significant damage costs, and net import of pollutants. The former have at best weak incentives to contribute. If they sign a pollution control agreement they would probably do so reluctantly – after having made concessions bringing them close to their resistance points (perhaps even after having been misled or coerced to make concessions that turn out not to be in their best real interests). Defection could in such cases come simply as a belated correction of previous over-commitments. Countries in category (b) may well support the introduction of relatively strict control measures (particularly if their damage costs are high), but since the damage is inflicted upon them from the outside, cutbacks by 'exporters' would be more important than reductions in domestic emissions. These are the countries that could have most to gain from deliberate 'cheating', that is, from making insincere commitments in order to induce others to contribute (more). Whether they will in fact succeed as 'cheaters' depends upon the ability of important partners to detect defection and their will and ability to retaliate. The less sensitive important partners are to defections, the more likely that 'cheating' will succeed.

Before embarking upon the analysis we should point out that at the time when the first LRTAP protocols were negotiated accurate and reliable figures for national abatement or damage costs were not available. In the case of the first sulphur protocol it seems that few governments actually even tried to have elaborate cost/benefit calculations undertaken as a basis for decision. This is, of course, not to say that they were not concerned with costs and benefits; in fact, most of our case studies strongly indicate that they were. The point is simply that in most cases policies seem – particularly in the early stages – to have been developed on the basis of rather crude and tentative perceptions of damage and abatement costs (and in some cases primarily on the basis of coincidence with other policy concerns). In particular, estimates of *damage* costs were uncertain. Even though abatement costs were easier to estimate, conclusions depended heavily on the assumptions made about important exogenous developments, such as future advances in relevant technologies and fuel shifts or other changes in energy production and consumption patterns. The knowledge base has been strengthened considerably over time, and we see governments making increasingly more systematic use of cost/benefit estimates. However, even today quite large margins of error persist.

The important implication for our analysis is that in choosing our methodological approach we should give priority to *robustness* over numerical precision. To enhance robustness we will examine to what extent our conclusions hold up if we use somewhat *different* cost estimates, and adjust our analytic ambitions to the quality of our data. We will rely mainly on ordinal rather than cardinal level measurements, and look only at aggregate profiles rather than try to determine the relative impact of each component of the cost/benefit calculus.[4] In practical terms, we will proceed in two steps. As a first cut, we simply examine how well national positions and compliance records for SO_2 and NO_X conform to the picture hypothesised in table 13.1. For this purpose we use for SO_2 the abatement cost estimates of Amann and Kornai (1987), and for NO_X those reported by Amann (1989). We use two different measures of damage costs – one measuring the extent to which actual depositions exceed critical loads (total acidity, fifth percentile), the other measuring damage to forests and waterways (in per cent of GDP).[5] Country profiles are produced simply by aggregating ordinal level rankings for the three independent variables listed in table 13.1.[6] This approach should enable us to determine the relative strength of incentives to establish emission control regulations. It does not, however, enable us to determine precisely who will win and who will lose (and, by implication, who will sign a particular agreement and who will refuse to do so).[7] For the latter purpose we need to know the 'resistance points', and these points can be determined only on the basis of cardinal level estimates of *net* benefits.[8] As a second cut we will therefore use an estimate of net economic benefits (per capita) from a 30 per cent general reduction in SO_2 emissions (based on Mäler 1989, p.27). No similar comparative estimate has been found for NO_X. In pursuing both these approaches we look first at national positions, then at compliance records.

At least one note of caution is in order before we look at results. National environmental policies are formed through a *series* of decisions, and the cost/benefit estimates available at the beginning of this sequence (time t_1) were in important respects different from those presented later (at decision points $t_2..t_n$). This is most obvious when we look at the policy process as a whole (from agenda-setting to implementation), but – as the case of FRG clearly illustrates – it applies also to single phases such as the formulation of national negotiating positions. If we were to do a more sophisticated analysis we would try to replicate these series of decisions. Here we simplify matters by focusing mainly on the negotiating positions taken towards the *end* of the negotiations and on aggregate compliance records. In doing so there is, however, a price to be paid in terms of validity and accuracy. Moreover, in the tests provided we implicitly assume that each problem and policy option is considered on its own merits only; in other words, that there are no issue-linkages affecting behaviour. The real world is certainly more complex. LRTAP itself is a sequence of related policy games addressing overlapping environmental problems. In such a sequence an actor concerned with long-term or aggregate benefits may, for example, accept a loss on one occasion in exchange for a (larger) benefit on some other occasion. Furthermore, since environmental problems to a large extent occur as a by-product of perfectly legitimate human activities such as industrial production, transportation etc., environmental policies will – in order to be effective – have to penetrate other policy areas as well. Environmental policies can therefore rarely be considered *only* in terms of their environmental impact. These are some of the reasons why we should not expect a perfect 'fit' between model predictions and actual behaviour.

Negotiating Positions

In the case of the first sulphur protocol, the country best fitting the profile of a 'pusher' would be Sweden. Other candidates would be Switzerland, Norway, and Finland. Spain came closest to the profile of a 'dragger', followed by the UK and Italy. Germany has fairly high damage costs, relatively high abatement costs and – compared to other countries – a fair balance between export and import of pollution. Thus, FRG best matches the profile we associate with an 'intermediate' position. No actor fits well the description of a 'bystander'; the closest we could get would be Italy. In the NO_x case, the prime pushers should be Switzerland and Sweden, while Spain again best fits the description of a dragger (so do, to a lesser extent, also Italy and the UK). None of the countries included in our sample correspond well to the profile of an intermediate (a possible candidate would be France) or to that of a bystander.

Using Mäler's (1989) calculations of net benefits from a general 30 per cent reduction of SO_2 emissions, we would predict that seven of the ten countries included in our study would in the end sign the agreement, while three (UK, Italy and Spain) would not. Using estimated net benefit per capita as our indicator, we would predict that the prime pushers would be the Netherlands, Switzerland, and Norway (other candidates would be Sweden and Germany (FRG)).

In tables 13.2 and 13.3 we have compared our predictions with actual national positions. In both tables countries have been categorised on the basis of model predictions. Different typographical means are then used to indicate correspondence or deviance. Note that some governments – that of Germany (FRG) being the prime example – changed their positions significantly during the negotiations, thus ending up in a different category than the one they started out in.

Table 13.2 Predicted and actual national positions, first sulphur protocol

Country	Predicted position (I)	Predicted position (II)	Actual position	Deviance Pred. I-actual	Deviance Pred. II-actual
Finland	Pusher	Intermediate	Pusher	0	+
France	Intermediate	Intermediate	Intermediate	0	0
Germany (FRG)	Intermediate	Intermediate	Dragger→ Pusher	–↑+	–↑+
Italy	Dragger (bystander)	Dragger	Dragger (→ intermediate)	0 (+)	0 (+)
Netherlands	Intermediate	Intermediate	Intermediate +	0 (+)	0 (+)
Norway	Pusher	Pusher	Pusher	0	0
Spain	Dragger	Dragger	Dragger (bystander)	0	0
Sweden	Pusher	Intermediate	Pusher	0	+
Switzerland	Pusher	Pusher	Pusher –	0 (–)	0 (–)
UK	Dragger	Dragger	Dragger	0	0

Legend:
Predicted position (I) = predicted on the basis of country 'profiles'.
Predicted position (II) = predicted on the basis of Mäler's abatement and damage cost estimates.
+ = more 'progressive' than predicted.
– = less 'progressive' than predicted.

Table 13.3   Predicted and actual positions, NO$_X$

Country	Predicted position (I)	Actual position	Deviance predicted-actual
Finland	Intermediate	Intermediate	0
France	Intermediate	Intermediate	–
Germany (FRG)	Pusher	Pusher	0
Italy	Dragger	Dragger ➞ Intermediate	0 ➞ +
Netherlands	Intermediate +	Pusher	(+)
Norway	Intermediate	(Pusher ➞) Intermediate	(+) ➞ 0
Spain	Dragger	Dragger	0
Sweden	Pusher	Pusher	0
Switzerland	Pusher	Pusher	0
UK	Dragger	Dragger	0

Legend: see table 13.2

The overall conclusion, then, is that at least our national profile approach predicts negotiating positions quite well. Both approaches correctly separate pushers from draggers and separate – although somewhat less accurately – both these groups from the middle categories of intermediates and bystanders. Using Mäler's net benefit estimate we would predict end positions in the SO$_2$ case correctly for nine out of our ten countries – the exception being Italy, who 'should not' have signed. To the extent that we can talk about deviant cases, we would point to Italy and – though less pronounced – also FRG, both being somewhat more progressive in their final negotiating positions than we would predict. In both these cases positions changed towards the end of the negotiations.

Compliance Records

The predictions derived from model I are by and large confirmed also when we turn to compliance records. The Spearman rank correlations between predicted interest in international regulation and actual reductions of SO$_2$ emissions range between .63 (for Mäler's estimate of net benefits, transformed into per capita figures) and .87 (when damage costs are measured in terms of damage to forests and waterways). With damage costs measured as depositions in excess of critical loads the correlation is .72. In terms of rank order comparisons, the only really deviant case is

Switzerland, whose *relative* performance comes out as somewhat weaker than predicted. The only country which consistently does somewhat 'better' than our indicators would predict is Germany. Looking at absolute figures, the most striking observation is that the overall level of emission reductions is higher than we would have predicted on the basis of Mäler's estimates. Also the three predicted 'losers' have in fact achieved the 30 per cent target, to which two of them refused to commit themselves. This clearly reflects one of the conclusions that stand out from our case studies: that a substantial fraction of the emission reductions observed has been achieved as a side-effect of other, exogenous developments. For example, one Swedish estimate attributes about ¾ of the reduction in SO_2 emissions from Swedish sources to various exogenous factors. Most of the cutbacks in SO_2 emissions from British sources seem to have come as a by-product of privatisation and the introduction of a more market-oriented policy in the energy sector. And in Eastern Europe a substantial fraction of emission reductions have come as a consequence of economic recession and restructuring of industry rather than deliberate abatement efforts.

In the case of NO_X, the rank order coefficients between 'national profiles' and actual emission reductions are .82 (when damage cost is measured as damage to forests and waterways) and .90 (when the critical loads indicator is used). Here we cannot really talk about deviant cases, but again the only country that consistently comes out slightly 'better' than predicted is Germany. In absolute terms the overall level of emission reductions is, as we would expect, less impressive than for SO_2, but the difference is larger than we would have predicted. For Spain, Italy and UK, who all signed the NO_X protocol, emissions seem in fact to have increased – in the case of Spain the increase is substantial. Of the countries that signed the more ambitious declaration proclaiming 30 per cent reductions as their agreed target, only Switzerland is, as of the time of writing, close to 'delivering', while Germany is roughly on schedule to meet its target.[9]

Since both protocols deal with politically malign problems, model I suggests that at least two categories of parties will have incentives to defect. One of these categories includes parties with high abatement costs, low or moderate damage costs, and a high rate of export of own emissions. Of those who signed the first sulphur protocol, a prime suspect would be the Netherlands (although, according to the critical loads indicator, its damage costs appear to be too high); in the NO_X case it would be the UK.[10] The other category consists of countries with high abatement costs, significant damage costs, and net import of pollutants. In this case the strength of incentives to defect critically depends on the link between own compliance and the compliance of important partners. The prime suspect in this second category for both SO_2 and NO_X would be Norway.

In order to get a crude measure of the extent of deliberate defection we have constructed new country profiles and matched these with actual emission reductions. Only in the SO_2 case is the ranking on the per-cent-of-pollutants-exported variable sufficiently different from the *net* import/export variable used above to permit a real strategic test. The results are clear and consistent: the predictions derived on the basis of the new country profiles were significantly *less* accurate than the ones reported above.[11] This applies not only when we look at correlation coefficients but also when we examine each of the main 'suspects' individually. We can, in other words, find no evidence of (this particular kind of) 'cheating' at the governmental level.[12] Instances of non-compliance seem generally to be more in line with the over-commitment hypothesis, supporting the proposition that non-compliance often comes a result of involuntary implementation failure rather than deliberate defection.

Conclusions, Model I

Given the quality of data available to us, we doubt that much could be gained by pursuing the evaluation of model I much further, through more sophisticated tests. Let us therefore stop at this point and try to summarise what we have learned from the crude cuts reported above.

Evaluating the model in terms of the tenability of the hypotheses derived from it, we can conclude that we were able to predict the general patterns of behaviour reasonably well. The negotiating positions observed corresponded roughly to the pattern hypothesised in table 13.1, and so did records of compliance. National positions seem at least to some extent to have been modified in response to new estimates of damage costs. Although the dramatic turn-around of the FRG position in the early 1980s seems to have a complex explanation, media discovery of scientific evidence indicating significant ecological damage was certainly one important factor. Even though correlation coefficients varied with different estimates of damage costs, the overall picture seems fairly clear. At the same time it is obvious that we are not talking about a perfect fit. For example, the difference between the two protocols in terms of emission reductions achieved seems greater than the scores on our independent variables would suggest. Comparing the various countries, we would say that Italy seems to have adopted an optimistic approach in signing on to emission control regulations.[13] The contrasting case appears to be the UK, who was concerned not to accept obligations that could prove hard to fulfil later. After its turn-around in 1982–83, the FRG consistently stands out as slightly more 'progressive' than our national profile data would lead us to expect.

Evaluating model I as a heuristic tool, we can first of all conclude that costs and benefits have indeed been key concepts in national policy-making processes. The burden of proof clearly rests with anyone who would argue that actors do not consider policy options in those terms. That being said, however, it is equally clear that with one partial exception (the second sulphur protocol) none of the emission control targets set in the protocols were the products of sophisticated model calculations identifying what constituted, in some precise sense, a collective optimum. Nor were the negotiating positions and implementation policies of governments derived from elaborate calculations of national costs and benefits, although as the knowledge base improved and institutional capacity increased more sophisticated estimates seem to have been made.[14] The concept of critical loads, which was introduced as a principal criterion in the second sulphur protocol, is itself a precise and sophisticated (although to some extent disputed) evaluation standard. Moreover, the case studies provide ample evidence that societal actors and to some extent also government agencies pursue interests that can hardly be considered truly *national* concerns. It is therefore important to remember that as used here model I grossly simplifies a complex world. Its advocates could rightly point out that the model itself can be refined far beyond what we have done in this study – even to include domestic politics as well as issue linkages.[15] Any specific application is likely to have limitations that are not inherent in the basic assumptions themselves. Even so, Schelling's (1967, p.238) remark that what is rudimentary and conceptual about the unitary rational actor model (and more particularly game theory) probably remains its most useful elements is probably valid in our case as well.

As we now move on to examine the other two models, we do so not primarily from the perspective of having to account for some 'residual' that remains to be explained (cf. Moravcsik, 1993). Rather, we shall proceed on the assumption that the patterns that emerge from our case studies may be consistent with more than one model or set of propositions. There is a real possibility that even observations that are fully consistent with model I may be equally well or better accounted for in terms of some other model or theory.

Model II (Domestic Politics)

Societal Demand and Support

To recapitulate, model II leads us to see negotiating positions as well as implementation and compliance records as functions of societal demand/support and governmental supply of environmental policy. In re-examining this model in the light of our case studies we now proceed in two steps. First, we examine – as far as our data permit – the main general hypotheses formulated in chapter 3. Second, we try to aggregate scores on the various dimensions into a more general assessment, asking to what extent societal demand and governmental supply can help us account for variance in negotiating positions and implementation records.

In chapter 3 we assumed societal demand to be heavily influenced by 'objective' national damage and abatement costs. In contrast to model I, however, we see the amount of political energy generated by a particular problem as a function of other characteristics as well. More specifically, we pointed to, inter alia, the following aspects as important dimensions:

- the domestic distribution of abatement and damage costs;
- the kind(s) of social values affected by environmental degradation;
- the basic values, beliefs and attitudes of the (attentive) public; and
- the presence and strength of intermediate agents articulating and to some extent aggregating the concerns and interests of various segments of society (non-governmental organisations, the media etc.).

In table 3.1 we also pointed to a set of cognitive aspects, the argument being that damage that is seen as certain, close in time, highly visible, and comes as a 'shock' tends to generate more concern than damage with the opposite characteristics. However, the acid rain problem provides too little variance among countries on these dimensions to enable us to explore those propositions systematically here.

The domestic distribution of abatement and damage costs In chapter 3 we formulated two hypotheses about the impact of the domestic distribution of abatement and damage costs on societal demand and support of environmental policies:

> H_{II2}: Damage or abatement costs that hit the social 'centre' of society will be better articulated and carry more weight than those that hit the social 'periphery' only, and the bias will be stronger the greater the distance in social and political resources between centre and periphery.

> H_{II3}: Under 'business-as-usual' circumstances the amount of 'political energy' infused into a particular concern tends to be higher when costs and/or benefits are concentrated in a specific and important sector of the economy than when they are widely dispersed throughout society or indeterminate. For issues generating substantial political mobilisation, the opposite 'bias' can be expected.

We have no solid basis for testing the former hypothesis. Even though damage costs and abatement costs to some extent hit different segments of society, there seems to be no case in which those who pay the lion's share of abatement costs systematically belong to different social strata than those who suffer from acidification. Nor can we point to clear-cut differences among our ten countries in this regard.

We are in a somewhat more 'fortunate' position with regard to H_{II3}. The concentration of costs or benefits in specific sectors of the economy shows a sufficient range of variance – among pollutants (for example, abatement costs for SO_2 are by and large more concentrated than for NO_X and VOCs) as well as among countries. However, one striking observation is that the actual distribution of abatement costs has been substantially affected by government policies. Thus, in Norway about 70 per cent of total abatement costs in the industrial sector have been paid by the state rather than by the polluters themselves – through various investment support schemes, tax reductions etc. A similar picture appears in the case of Sweden. The German approach to emission control, with emphasis on the development and introduction of new technologies, helped create a constituency of R&D institutions and segments of industry with positive interests in abatement that the British approach did not, or did to a much smaller extent (Boehmer-Christiansen and Skea, 1991; see also Sprinz,

1992). The approach taken in international pollution control efforts may itself have important implications for the distribution of costs and benefits. One general impression is that regulations prescribing the use of particular instruments or means (e.g. the introduction of catalytic converters) tend to generate a more determinate distribution of costs and benefits and therefore also more conflict than agreements that set only a general goal (reduce by x per cent by year y) and leave each government free to determine how that goal is to be achieved.

Consistent with our hypothesis, we find strong indications that spreading abatement costs through government intervention helps to soften the resistance of polluters, without generating equally strong resistance from the public at large. There is also some evidence to indicate that concentration of tangible benefits in a specific and important sector of the economy tends to strengthen demand for and support of environmental policies. The case of FRG/Germany probably offers the best illustration; the fact that international pollution control agreements in some cases could help strengthen the competitive edge of major producers of 'advanced' technology seems to have been one source of domestic support for Germany's active role in international environmental diplomacy.

However, our case studies also suggest that under certain circumstances a wide-spread distribution of abatement costs may well spell political trouble. This is so particularly when regulations have to deal with multiple and diverse sources of emissions. Such complexity tends to make the design of emission control policies as well as monitoring and enforcement schemes more difficult, and reduce the availability of solutions through 'technological fixes'. This is clearly illustrated by comparing the cases of NO_x and VOCs, where the sources of emissions are multiple and diverse, to that of SO_2, where the major emitters are fewer and less diverse. This observation should not lead us to reject H_{II3} as falsified; rather, what it suggests is the need for a supplementary proposition focusing on the *nature of targets*. To be more precise, we would – other things being equal – expect environmental policies to be more 'progressive' and also easier to implement when targets are few, homogenous, wealthy, and have access to advanced (production) technology than when they are many, heterogeneous, poor and technologically less advanced.

Our case studies indicate that the domestic distribution of abatement and damage costs makes a difference in other respects as well. Suffice it here to point to two examples. Particularly the Swiss, German and French case studies show substantial differences among different cantons, *länder* or regions in terms of their demand for and support of emission control policies. The pattern seems to correspond reasonably well to what we would expect on the basis of a map showing the geographical distribution of abatement and damage costs. Thus, in Germany southern *länder* like Bavaria, (relatively rich in forests and also home of major producers of 'cleaner' cars) were internal pushers while the heavy-industry states favoured a less demanding approach. In Switzerland, major urban agglomerations served as pace-setters in the development of air pollution policies. The French case indicates that also cultural and societal links may be important; thus, Skea and du Monteuil found that the German concern about damage to forests struck a more responsive chord in the neighbouring region of Alsace than elsewhere.

Second, Boehmer-Christiansen's study of the UK points to yet another variable: the extent to which abatement or damage costs hit the main constituencies of the party or parties in power. She attributes the low political saliency of damage to surface waters and forests in Britain in part to the fact that the areas most affected were found in Scotland and Wales, which were not important constituencies for the Conservative Party.[16] From her description it is equally striking that also the costs of abatement (or, rather, of restructuring the energy sector, from which emission cuts followed as a by-product) were to a large extent concentrated in groups that had little rapport with the Conservatives. The National Union of Mineworkers, and its leader at the time, Arthur Scargill, were in fact seen as a political 'enemy'. This probably helps explain both the initial caution to avoid measures that could trigger clashes with the coal-miners and also the government's long-term strategy of industrial restructuring, which in the longer run served to undermine the position of the union.

Kind(s) of values affected In P_{III} we suggested that environmental damage affecting productive assets, human health, or natural treasures tends to generate stronger demand for preventive action than damage affecting recreation or (non-vital) consumption only.

The sectors of the economy most directly affected by acidification were forestry and those branches of tourism that relied on access to unspoiled forests and fresh-water fishing. In economic terms these sectors were most important in the Nordic countries. Our case studies provide several indications that also damage to human health or natural treasures were matters of serious concern. Even in countries where general public concern with acidification has been relatively weak (such as Hungary and Spain) we can find instances of strong (local) concern over potential threats to human health in particular cities or industrial areas. Boehmer-Christiansen and Skea (1991, chapter 4) point to the difference between Germany and the UK in the emotional attachment to forests as one important factor in explaining why 'Waldsterben' generated so much more widespread and serious concern in Germany. Similarly, the widespread concern over the decline and even extinction of fish populations in Scandinavian lakes and rivers cannot be seen merely as a concern over damage to productive assets; it was perceived by many also as a threat to a cherished life-style and to an important element of Scandinavian nature. The overall picture seems consistent with P_{III}, but we would like to emphasise that environmental concern is by no means limited to instances of damage to productive assets, human health or particular national treasures. A considerable proportion of the societal demand and support for emission control that we find in our case studies seems to be best understood as embedded in a more general concern with the relationship between man and nature, and as reflecting basic values and beliefs rather than specific (material) interests.

P_{III} refers specifically to damage costs, but one of the conclusions that emerges from our case-studies is that type of value can be an important aspect to consider also with regard to abatement costs. The general lesson seems to be that the deeper a regulation penetrates into the behaviour of private citizens, and the more central to one's welfare or established life-style the behaviour affected, the more resistance a regulation tends to generate among the general public. Thus, measures requiring significant changes in the use of private cars seems to be a case in point, as the Swiss case well illustrates. It seems a reasonable prediction that in most Western societies a move to lower highway speed limits or restrict the use of private cares would tend to generate more resistance than a move to spend public funds on environmental protection, even if the *monetary* costs of the two measures to the average citizen were equal. Moreover, measures involving specific demands for behavioural change or costs that can easily be traced back to deliberate decisions on the part of specific actors are likely to generate more resistance than those that are indistinguishably incorporated into general taxes or prices, or are attributed to the operation of 'anonymous' market mechanisms (cf. Lane, 1986).

Values, beliefs and attitudes of the (attentive) public Over the years a large number of public opinion surveys have included questions about threats to the environment and attitudes towards environmental protection. In a very brief summary we cannot do justice to the richness of findings reported from these surveys, but some of the main conclusions may be summarised as follows.

In the hierarchy of basic values, environmental quality seems to rank higher in the more wealthy countries of Northern Europe than in the Mediterranean countries. Among the EU countries included in our sample, the Netherlands and Germany seem in a general assessment to be the 'greenest' while Spain seems to be least supportive of environmental values (see e.g. Hofrichter and Reif, 1990). The scores of the Nordic countries are generally above the European average. The gap does, however, seem to be diminishing as the support for environmental values has grown in Southern Europe. Moreover, in all European countries public concern with the environment waxes and wanes – with a peak in the late 1980s.

When questions are asked about *immediate* threats to their *local* environments, a somewhat different picture emerges. Although much concerned with environmental values in general, respondents in the Nordic countries tend to view their own countries as relatively clean and the overall state of their own local environment as good (Aardal 1993, p. 433). Respondents in Spain, Italy and Germany tend to perceive the immediate threats to local environmental values as somewhat more severe than respondents in the UK, Netherlands and France (Hofrichter and Reif 1991, p.134).

When asked about the salience of *particular* environmental problems, Eurobarometer data from 1986 and 1988 indicate that among the EU countries included in our study, concern with acidification was highest in the Netherlands and lowest in Spain and Italy, in absolute terms as well as relative to other environmental problems (Sprinz, 1992, pp.154–155). Although not directly comparable, other studies indicate that acidification was seen at the time as one of the most important environmental problems also in the Northern countries.

Many of the differences that we can find among countries in public opinion surveys may be due to culturally embedded attitudes towards uncertainty and risk. Thus, Boehmer-Christiansen and Skea (1991, p.61f;

280) suggest that German society 'has a greater predisposition towards pessimism and anxiety' (than the British). Although very hard to measure empirically, such differences could conceivably contribute to generating stronger demand for action in the more 'anxiety-ridden' societies. For example, a 'predisposition towards pessimism' would seem to favour the adoption of a precautionary rather than a more traditional wait-and-see approach to environmental protection.

Presence and strength of intermediate agents   As expected, significant differences exist in terms of the strength and involvement of intermediate actors articulating and aggregating environmental concerns, such as NGOs and 'green' political parties. We also find significant differences in the strength and involvement of 'counter-forces', notably polluting industries. As the data available to us are only in part directly comparable, the role played by intermediate agents in the various countries is hard to compare in a precise and reliable way. Table 13.4, in which we have tried to summarise the general picture as it emerges from our case studies and other, secondary sources, should therefore be read as a very crude approximation only.[17] Moreover, we should like to point out that some of our variables – in particular that of 'strength of green parties' – are problematic indicators of the articulation of environmental concerns. For one thing, other political parties may as actively promote environmental protection. In all countries the overall strength with which environmental concerns are articulated through parliaments depends more on what other parties do than on the strength of any 'green' party or parties. Second, as the case of Great Britain amply demonstrates, the strength of a political party is not simply a function of the proportion of votes it polls in elections; more important is the extent to which it emerges as a *pivotal* force in domestic politics.[18] Finally, their involvement in and influence upon acid rain policies depend as much on the relative priority given to that particular issue as on the strength of the party itself. While the validity and reliability of our indicators may be questioned, we nonetheless think that the overall assessment correctly identifies real and significant differences in the strength of intermediate actors articulating and shaping societal demand and support for environmental policies.

Table 13.4 Strength and involvement of intermediate actors articulating environmental concerns, in relative terms, as of the late 1980s

Country	Strength of environmental NGOs	Focus on acidification	Access to Government	Strength of 'Green' party/parties	Industrial Interests in abatement	Overall strength of intermediate proponents
FRG	Strong	High	Good/Intermediate	Strong	Strong	Strong
Netherlands	Strong	High	Good	Intermediate/strong	Intermediate	Strong/intermediate
Sweden	Intermediate/strong	High	Good	Intermediate	Intermediate/weak	Strong/intermediate
Switzerland	Intermediate/strong	High/intermediate	Good	Intermediate/strong	Intermediate	Strong/intermediate
Finland	Intermediate/strong	High	Good	Intermediate	Weak	Intermediate/intermediate
Norway	Intermediate/strong	High	Good	Weak	Weak	Intermediate/strong
France	Intermediate/weak	Low	Poor/Intermediate	Weak	Intermediate	Weak/intermediate
UK	Strong/intermediate	Low	Admin.: good Pol.: poor	Weak	Weak	Weak/intermediate
Italy	Intermediate +	Low	Poor/Intermediate	Intermediate	Weak/Intermediate	Weak
Spain	Weak +	Low	Poor	Weak	Weak	Weak

Legend:
+ = increasing.
Note that 'industrial interests in abatement' refers to positive interests in abatement activities or technology switches themselves rather than in the reduction of environmental damage.

Counter-forces related to 'dirty' energy or other polluting industries seem by and large to have been stronger in countries such as Spain, UK and Germany, all relying heavily on domestically produced coal, than in countries like Switzerland and Norway. There are important nuances to this picture, however. For example, German industries seem to have been more easily 'converted' to reliance on new technology than their British and Spanish counterparts. This seems to have softened their resistance to abatement measures significantly. In the case of Norway, its growing offshore petroleum industry has gradually served to turn an initial champion of international emission control into something of a mixed-motive actor, increasingly reluctant to commit itself to emission reductions that could interfere substantially with its oil and gas production. This can be seen clearly by comparing its role in pushing for the first sulphur protocol with its less heroic role in the cases of NO_x and greenhouse gas emissions

Conclusions: patterns of societal demand/support   Summing up, our assessment of net societal demand/support for emission control measures is given in table 13.5. Two of the key variables, pertaining to the domestic distribution of damage and abatement costs, have not been included in this table since the patterns emerging were either quite similar (centre–periphery variable) or too complex (concentration–diffusion variable) to be captured adequately by aggregate scores in a simple matrix. Again, we must emphasise that what we present is a very crude picture, based in major part on qualitative judgement rather than systematic and inter-calibrated quantitative measurement. We nonetheless think that the table below correctly reflects the general *pattern* within our sample of countries.

Table 13.5 Overall pattern of societal demand/support

Country	Kind(s) of values affected by damage to the environment	Public concern with support for environmental values*	Intermediate actors articulating demand / support	Overall assessment of societal demand / support
FRG (after 1982/3)	T, R (p) (h)	High-high	Strong	Strong
Netherlands	R (p) (h)	High-high/interm.	Strong/intermediate	Strong/intermediate
Sweden	P, R, (t)	High-high	Strong/intermediate	Strong
Norway	R, P (t)	High-high	Strong/intermediate	Strong
Finland	P, R, T	High-high	Strong/intermediate	Strong
Switzerland	R, (p) (h)	High-high/interm.	Strong/intermediate	Strong/intermediate
France	(h) (r)	Interm.-low	Weak	Weak
UK	(h) (r)	High/interm.-low	Intermediate/weak	Weak/intermediate
Italy	(h) (r)	interm.-low	Weak	Weak
Spain	(h)	low-low	Weak	Weak

Legend:

* = the first score (to the left of the -) refers to environmental values in general, the other to the problem of acidification specifically.

H = human health

P = productive assets (mainly those of forestry and tourism, to some extent agriculture)

R = recreation

T = «national treasure»

( ) = only to a minor extent or in limited geographical areas

Looking beyond such aggregate assessments of societal demand/ support we could ask whether certain *configurations* of demand and support are politically more effective than others. Our case studies seem to suggest that societal demand/support is particularly potent when it comes from *multiple* constituencies; in particular where there is a *confluence* of public concern about damage to environmental values and specific economic interests in particular pollution control measures. Germany is a good illustration. The emergence of widespread public concern with 'Waldsterben' and other environmental problems *and* the presence of industrial interests in new and environmentally more benign technologies and products seem to have created a potent and synergistic configuration of societal forces.

We can now see that the task of determining the *impact* of variance in societal demand upon national positions and implementation and compliance records is made difficult by the fact that the pattern described in table 13.5 is quite similar to the configuration of national damage and abatement costs. However, before trying to separate the impact of societal demand from that of other factors we must examine also the supply side.

Governmental Supply

In the domestic politics model we conceived of governmental supply as a function of four major determinants:

- The ideological profile of the government (cabinet) in power;
- The relative strength of the environmental branch of government;
- The extent to which government controls state policy; and
- The extent to which government controls society (relevant particularly for implementation).

Let us now briefly summarise our observations with regard to each of these factors.

The ideological profile of government (cabinet)   In chapter 3 we suggested that

> H_{II5}: Governments subscribing to a strong free-market ideology and cabinets of the extreme left (including communist parties) tend to give

lower priority to environmental protection than governments formed by moderate conservatives, liberal or social-democratic parties.

We do not really have a sufficiently wide range of variance in our sample to test this hypothesis as it stands. It is certainly true that neither of our extreme cases – the Thatcher government in the UK and the Communist regime in Hungary – will be remembered primarily for promoting environmental protection. More generally, we think there are sound theoretical arguments for expecting governments subscribing to laissez-faire principles or to traditional communist notions of development to be less ambitious in their environmental policies than governments close to the political centre. After all, pollution control most often takes the form of governmental intervention into economic activities. And the communist notion of development gives little weight to environmental values – at least in so far as they are not (linked to) productive assets.

Within what can be considered the normal range of variance for Western European governments, however, the overall impression is that a government's score on the conventional left–right dimension does not in itself tell us much about its concern with environmental values. This is consistent with findings from several other studies (see e.g. Jänicke, 1997). Electoral and political culture studies also point to environmentalism and post-materialism as dimensions that cut across most traditional political cleavages. The initial drive to establish an international regime for transboundary air pollution was led by a social-democratic government in Sweden and a conservative–centre government in Norway, and in neither case did their efforts generate significant domestic political controversy. In some countries (at least Switzerland, France and Finland) it seems that liberal and social-democratic parties have pushed for more ambitious air pollution policies than their conservative counterparts, but when we control for government/opposition status no fully consistent pattern emerges. Leaving aside the atypical case of Hungary, none of our case-study authors see the changes of governments that occurred during the period as leading to radical changes in environmental policies.

This is not to say that a government's ideological profile has no bearing whatsoever on its environmental policies. There may be important *indirect* effects. Thus, a government's ideological profile can have important implications for its views on the economy, and economic and industrial policies may have important side-effects for the environment. This is clearly seen in, for example, the fact that much of the reduction in

SO_2 emissions in the UK has come as a side-effect of privatisation (of gas and electricity supply, the automobile industry and the coal industry) and stronger reliance on market mechanisms. The transition from communism towards a more market-oriented economy in Hungary is also having multiple and diverse ramifications for the environment. What we see in these examples is not primarily the expression of ideologically based evaluations of environmental quality per se, but rather results of the fact that the pursuit of *other* goals – such as enhancing the overall efficiency of the economy – can have important consequences for the environment as well. Our study strongly suggests that in tracing the impact of a government's ideological profile on its environmental policies, we would be well advised to examine also such *indirect* causal paths.

Another observation is that governments tend to be guided not only by their own ideologies; they also tend to be particularly responsive to what they perceive to be important values and interests of their principal or critical *constituencies*. For any specific issue, the general rule-of-thumb seems to be that the more important source of (electoral) support the social groups involved in a particular set of polluting activities are, the less ambitious a government's pollution control measures tend to be. Conversely, the more important the victims or the producers of environmentally more benign substitutes, the more progressive environmental policies tend to be.[19] It is hard to imagine a Labour government administering such a harsh cure to the coal industry and the coal-miners as the Thatcher government did.[20] One reason why we find so little systematic difference in most of our countries between conservative and social-democratic parties in terms of acidification policies is probably that both sides get much of their electoral support from the same sectors of the economy (although from different social groups or classes). In Sweden, for example, their links to industry have earned both the conservatives and the social-democrats the unflattering nickname of 'concrete parties'.

Finally, we may point out that our case studies also provide several indications that policy style can make a difference. For example, O'Toole (1998) sees the heavy reliance on traditional legalistic instruments supported by modest fines rather than on precautionary measures as an important source of ineffectiveness in the Hungarian regulatory system. Boehmer-Christiansen points to the difference in policy approach between Germany and the UK – the former proactive and technology-oriented, the other pragmatic and relying more on basic natural science – as important in understanding differences in negotiating positions as well as actual achievements. The relevance of policy styles or political culture can be illustrated also by contrasting Lewanski et al's picture of Italian politicians as inclined to give priority to the expressive and symbolic aspects of policy with Boehmer-Christiansen's picture of the British government and civil service as pragmatic and self-confident – determined not to sign international agreements that they could not implement and whose rationale they had serious questions about.

Relative strength of the environmental branch of government In chapter 3 we hypothesised

> $H_{II6:}$ The greater the relative institutional capacity and involvement of the environmental branch of government, and the greater the 'political clout' of its leadership, the greater tends to be its influence over policy formation and implementation, and the more progressive will be national negotiating positions and the better the record of implementation and compliance.

The causal links between 'progressiveness' of environmental policies and the strength of the environmental branch of government are supposed to work both ways. Governments presumably build institutional capacity in order to achieve a substantive goal that they have already set for themselves. As pointed out by Weale (1992, p.14), creating or strengthening the environmental branch of government was typically the first major step that West European governments took to enhance their role in protecting the environment. In most European countries a significant build-up of institutional capacity for environmental management has taken place over the past two or three decades (see Jänicke and Weidner, 1997). These institutions are not only executive instruments for pursuing established policy goals, however; once in place, they can substantially *affect* the goals, priorities and modalities of public policies.

We can see some evidence of that happening in our case studies, The overall impression is, though, that the relative strength of the environmental branch of government can account for only a small proportion of the variance observed in national positions and implementation records. In a *cross*-national comparison we find that the pushers by and large were ahead of the draggers in building up their environmental ministries and agencies. For example, the Scandinavian countries and the UK established separate environmental ministries around

1970, while Italy did so only in 1986, and Spain followed in 1996 (Switzerland has a federal agency as its functional equivalent). There are also marked differences in terms of budget allocations; in the 1980s the Swedish ministry of the environment received between 1.4 and 2.2 per cent of the total state budget, compared to 0.7–1.4 per cent in Norway and Switzerland, and about 0.1 per cent in France and the UK.

There are some indications that institutional capacity is most important for exercising *leadership* in international environmental diplomacy. Leadership is not a role to which countries with weak environmental institutions can aspire. Yet, there is no evidence to suggest that relative strength of environmental ministries and agencies was the decisive factor in distinguishing pushers from draggers. *Intra-national* comparisons show beyond reasonable doubt that other variables must have been more important. It is, for example, hard to see that variance in institutional capacity can at all help us explain the changes that occurred in German, British and Norwegian policies over time.

Government (cabinet) control over state policy ('strength' of government)
We saw this dimension as a function of four more specific variables:

- the internal unity of the government (cabinet) itself (H_{II7});
- government (cabinet) control over the legislature (H_{II8});
- policy distance to major opposition party/parties (H_{II9}); and
- personal political authority of the head of government (H_{II10}).

In this section we cannot deal with each of these variables specifically. Our case studies shed most light on the significance of the former two. We will also look very briefly at one aspect of policy distance. By and large, however, acidification policies have not been subject to substantial domestic political controversy, so the range of variance within our sample is quite limited. Personal political authority is an elusive variable which we have not been able to measure systematically across cases.

Let us therefore briefly point out that the 'strength' of government seems to have had two opposite effects that are of interest in this context. First, it raises the threshold of access by societal actors to which the government is not already favourably inclined. The UK is a case in point. A one-party government, with a comfortable majority in the House of

Commons, and a prime minister with strong personal authority and involvement could easily afford to stay its course against criticism from the 'Greens' and environmental groups. Moreover, as pointed out by Boehmer-Christiansen, the British principle of collective responsibility served to prevent open public advocacy in favour of more (or, for that matter, less) ambitious policies from individual members of Cabinet. By contrast, the Swiss system – with its pluralism and provisions for direct democracy – seems to have a much lower threshold of access and probably also influence for societal groups (cf. Vogel, 1993, p.269).[21]

Second, we expected strength of government to facilitate implementation of and compliance with whatever agreements a government decides to sign. This hypothesis cannot really be put to test here. The reason is simply that no LRTAP agreement was concluded against strong resistance from opposition parties. We also hypothesised that initial agreement on negotiating positions would facilitate implementation (H_{II9b}). There is some evidence of that happening – 'positively' in, inter alia, the Netherlands and Switzerland, 'negatively' in Italy, where a closed policy circuit and last-minute changes of positions seem to have left the political system as well as society with no firm consensual basis for implementation of the NO_X and VOC protocols. We have to conclude, however, that neither strength of government nor the degree of prior consensus about negotiation positions are variables that can account for a substantial proportion of the variance observed in implementation and compliance records. This is consistent with the conclusion reached by Vogel (1993).

Governmental control over society   To reiterate, we depicted this dimension as a function of three main determinants:

- the degree of centralisation of formal authority at the national (federal) level (H_{III1});
- the extractive capacity of the state, measured as central government income in per cent of GDP (H_{III2}); and
- the extent to which targets of regulation belong to the public sector or are financially dependent upon funding from central government sources (H_{III3}).

The overall impression is one of moderate support for the hypotheses that centralisation of authority and high extractive capacity tend to facilitate implementation of international agreements, while evidence with regard to the third dimension is mixed and inconclusive. Support for the proposition that centralisation of authority matters can be found in cross-national as well as intra-national comparisons. Although strict control for other factors has not been possible, it seems that, other things being equal, governments of unitary nation-states (such as Norway and Sweden) by and large have had an easier task translating international environmental agreements into domestic policies than governments of federal (FRG/Germany, Switzerland) or highly decentralised systems (Spain).[22] This may be one of the reasons why we see a general trend in some countries (including the Netherlands and Switzerland) towards centralisation of responsibility for environmental management.

Looking at the causal mechanisms involved, there seems to be at least four main reasons why decentralised systems tend to face greater difficulties. One is simply that where the formal authority to set standards or grant emission permits is decentralised to the regional or local level, the authorities in charge will tend to be concerned primarily with *regional* or *local* rather than national, let alone international, impacts. This may certainly work both ways; thus, as Kux and Schenkel point out, some Swiss cantons acted as pace-setters for national environmental policy. However, the more emissions cause damage beyond the jurisdiction of the regional or local authority in charge, the more the regional or local cost-benefit calculus will differ from that of the nation as a whole. The greater the negative externalities caused by emissions from local sources, the more the level of emissions accepted by local government is likely to exceed the national optimum.

Second, in at least some countries (including Italy and Switzerland) decentralisation seems to have led to an obfuscation of the division of responsibility between or among the various levels of government. Moreover, it seems to be a widespread complaint on the part of authorities at the regional or municipal level that resources do not match responsibilities; more precisely, that the central government is more willing to delegate responsibilities than to provide the resources needed to do the job. Third, the technical expertise available to the regional and local level tend to be inferior to that available to the central government. Fourth, other things being equal, it seems that the further responsibilities are delegated, the less effective tend to be the lines of communication from central to local authorities. This is perhaps most striking in the case of Italy, where Lewanski et al. (1996) found that in many cases relevant authorities at the regional and local level were simply not aware of international commitments made by the Italian government, nor did they know what was expected of them in terms of follow-up activities.[23]

Comparing the various protocols (more precisely, those for SO_2 against those for NO_X and VOCs) it seems that the more responsibility for regulating the polluting activities is decentralised or divided *horizontally* among multiple agencies, the less powerful and consistent tends to be the drive for implementation. Interestingly, both the Scandinavian case studies point to the more complex division of responsibilities in the cases of NO_X and VOCs as important factors explaining why implementation records are significantly less impressive than for SO_2.

Summing up, it seems that decentralisation of management authority tends to lead to lower effectiveness under at least four circumstances:

- where pollution causes significant externalities beyond the geographical jurisdiction of regions or municipalities;
- where pollution stems from *mobile* sources;
- where regional and local governments lack the resources (expertise, manpower, finances) required to act; and
- where vertical (or horizontal) networks and procedures of communication are poorly developed.

We said that the evidence regarding public sector targets and targets financially dependent upon financial support from the government was mixed. In brief, ownership generally gives the government greater formal authority over a certain target, but this effect is at least to some extent balanced by the fact that governments-as-owners to some extent acquire the *interests* of the target. This is perhaps most clearly seen in the case of the Thatcher government, where the concern to make industries or companies that the government wanted to privatise attractive to private investors served as an argument against imposing costly requirements for emission control measures.

One of the observations made in several case studies is that a government's capacity to implement international environmental agreements cannot be understood simply as a function of structural properties such as the distribution of formal authority and budget allocations. More elusive qualitative aspects – including what might be called 'procedural streamlining' and the actual ability to penetrate society effectively – may be at least equally important. The case of Italy may be particularly instructive in this regard. Lewanski rates the implementation capacity of the Italian government as weak even though it would qualify for an intermediate score on several of the formal criteria suggested, including legal authority, budget and manpower.

Table 13.6 Patterns of governmental supply of environmental policy (within the domain of LRTAP) early 1980s–early 1990s

Country	Ideological profile	Strength of environmental branch	Strength of government (cabinet)	State control over society	Attention to acidification problem	Actual govern-mental supply
Sweden	Mostly soc. dem.	Strong	Strong	Strong	High	Strong
Netherlands	Various Coalitions	Intermediate/ strong	Varying, mostly intermediate	Strong	High/ intermediate	Strong/ intermediate
Norway	1981–86: conservative–centre, else mostly soc. dem.	Strong	Varying, mostly intermediate	Strong	High	SO₂: strong NOₓ, VOCs: intermediate weakening
FRG	1980–82: soc. dem–lib. 1982–present: cons.–liberal	Intermediate	Mostly strong	Intermediate	Up to 82: intermediate thereafter high, peaking mid 1980s	Before 1982: weak. After 1982–83: strong
Switzerland	Comprehens. Coalition	Intermediate	Intermediate	Weak/ intermediate	High	Strong/ intermediate
Finland	1982–87: soc.–agrar. 1987–91: cons.–soc.	Intermediate	Mostly intermediate	Strong	High	SO₂: strong NOₓ, VOCs: intermediate
France	1981–86: socialist 1986–93: soc. Pres. + non-soc. Parliament	Intermediate/ weak	1981–86: strong 1986–93: intermediate	Strong	Low (but high on energy)	Weak (but strong drive on energy switch)
UK	Conservative	Intermediate/ weak	Very strong, but declining from 1990	Strong (but declining)	Low	Weak, but increasing
Italy	Various coalitions	Intermediate/ weak	Varying, most often weak	Intermediate/ weak	Low	Positions: intermediate impl.: weak
Spain	1979–82:mod. cons. 1982–96:soc. dem.	Weak	Strong	Weak/ intermediate	Low	Weak, but increasing

Overall assessment of governmental supply A brief summary of scores and an aggregate assessment is provided in table 13.6 (below). Again, scores are relative to averages within our sample of cases and based on crude qualitative judgement rather than on systematic quantitative measurement. The three structural variables – strength of the environmental branch of government, strength of government itself, and state control over society – are all elements of institutional capacity, determining the *ability* to develop and implement environmental policies. Ideological profile and attention to the acidification problem can be seen as determinants of the political *will* to act. Capacity and will are both *necessary* conditions for governmental supply of environmental policies. Neither constitutes a sufficient condition; only in the presence of *both* will a government pursue an ambitious environmental policy. The overall score on governmental supply must therefore be conceptualised not merely as the *sum* of scores on the five dimensions included as determinants, but rather as the *product* of capacity and will.

Societies do not respond only through their governments Before we conclude this evaluation of the domestic politics model, we should add that targets of international environmental regulations will sometimes respond by positive adaptation themselves, even in the absence of effective implementation efforts by governments. To be sure, in environmental politics large industrial companies often serve as powerful actors lobbying successfully *against* the adoption or implementation of regulations that would increase their costs of operation. But it has become increasingly clear that there is a flip side of the coin (Jänicke and Weidner, 1997). For example, Lewanski et al. (1996) point out that in the case of Italy several large companies to some extent even made up for inefficient implementation on the part of the government. They did so by undertaking unilateral action that helped reduce emissions. More generally, there seems to be at least four mechanisms that provide large companies with incentives and/or capabilities to move on their own in the direction required by international environmental regulations.

One is simply the fact that large companies are often producers or at least users of cutting-edge technologies. In some cases they may actually reap substantial benefits from environmental regulations prohibiting or curbing the use of less advanced products. The support for air pollution control by major German companies (including car manufacturers) can probably at least in part be attributed to the fact that certain types of emission control measures would serve to strengthen their competitive edge in domestic as well as foreign markets.[24] Also large companies who are not themselves producers of environmentally more benign products will often, because of their advanced position in their field, be among the first to *apply* cutting-edge technologies. Second, their size and high profile make large companies highly visible targets for environmental NGOs and the media. A good environmental record is often seen as an important element of corporate images. This is one reason why many large companies have established their own environmental offices or sections. Third, companies with sufficient resources can keep informed about the preparations of national as well as international regulations in their field. Lewanski et al. (1996, p.120) point to the Italian Enichem as a case in point. The company has established its own office in Brussels, feeding back information about Community legislation in progress. This helps Enichem anticipate future legislation, and gives the company more lead time to adjust. It is no surprise that large companies act in advance of actual regulations more often than small firms do. Finally, MNCs operating production plants in several countries will often have to adjust to the strictest emissions or operation standards encountered. Once adjustments have been made in one country, the company will sometimes find that it might as well apply the new system(s) to all or most of its operations. The same applies to product standards. In this way MNCs may serve as important vehicles for diffusion of environmentally benign practices, transferring stricter environmental standards from one country to another (notably from forerunners to laggards).

Conclusions, Model II

In our presentation of the domestic politics model we described negotiating positions and implementation records as a function of societal demand/support for and governmental supply of environmental policies. Neither of these components was seen as inherently more important than the other, but we expected authoritarian and totalitarian systems to be less responsive to societal demands and hence more driven by 'supply-side' mechanisms than democratic systems.

Conventional wisdom suggests that policies designed to protect the environment most often develop as a more or less reluctant response to societal demand, articulated by environmental NGOs and 'green' parties. The picture emerging from our study is a more complex and balanced one. Governments not only respond, they sometimes lead the way and actively shape public opinion. Environmental groups are sometimes cast in roles of supporting followers rather than champions of governmental action. The drive for international control of acidifying emissions very much originated within bureaucratic-scientific networks rather than with environmental groups. In the early stages, the Swedish and Norwegian governments both encouraged and supported NGO involvement in the issue, and even tried to use them as agents for mobilising their British counterparts to put pressure on the British government. Also in other countries, such as Italy and Hungary, scientific-bureaucratic networks were important initiators.[25]

To be sure, there are cases in which societal demand played an important role. Most notably, the turn-around of FRG policy in the early 1980s owed much to public concern triggered by strong media attention. In some of the countries where policy development was essentially driven by scientific-bureaucratic networks in the early years, environmental NGOs today are active players in urging their governments to move on or at least deliver on earlier commitments. The point to be made here is exactly that we find *both* these configurations in our case studies. Moreover, there seems to be a fairly clear pattern. Our data seem to suggest that *leadership* in regime formation can come *only* from countries whose *governments* are themselves strongly committed to the cause and prepared to spend a considerable amount of resources to promote it. In international politics there is no such thing as leadership by reluctant response to demands from domestic pressure groups. More precisely, we would hypothesise that leadership in international environmental diplomacy requires the presence of at least two of three critical 'supply-side' factors: domestic scientific capacity, a fairly strong environmental ministry or agency, and a government favourably inclined towards environmental values. In the initial stage, domestic scientific capacity may be a *sine qua non*. At least in this particular case, societal demand seems to be more important in softening the resistance of laggards and nudging intermediates and bystanders to join.

In none of our cases did we find a severe and persistent incongruity between societal demand and governmental supply. True enough, there are instances where one side is ahead of the other. The overall impression, however, is that when significant gaps occur, they tend to be closed one way or the other, with moderate time-lags. Moreover, there are several indications that the two sometimes interact to reinforce one another. Faced with strong public concern, political parties sometimes engage in a competitive race 'bidding up' environmental policy to the point where supply may well exceed demand.

The close correspondence between demand and supply may be seen as an indication that democratic governance is indeed working. However, an equally plausible interpretation would be that supply and demand tend to move in tandem because they are both heavily influenced by a common background variable – notably the balance between national abatement and damage costs. The latter interpretation is consistent with the fact that the aggregate patterns of demand and supply correspond quite well to the picture derived from national cost/benefit calculations, and that demand as well as supply both seem to have varied (although by no means perfectly) with economic growth rates.

The fact that aggregate patterns of demand and supply correspond fairly closely to the abatement/damage cost balance implies that we are not in a position to distinguish with confidence the relative explanatory power of model II from that of model I. We may nevertheless point out that our case studies seem to confirm beyond reasonable doubt that many of the elements included in model II are indeed relevant and even important to the understanding of national positions as well as implementation and compliance records. For example, the domestic distribution of abatement and damage costs clearly makes a significant difference. So does the strength of environmental groups, occasional waves of media attention, and the level of governmental control over society. For other key variables we found at best weak or indirect effects. To the extent that the ideological profile of governments is a relevant dimension it seems to be so primarily through its *indirect* impact on environmental policies. 'Strength of government' can work both ways – to resist demands as well as to enhance supply. Vogel (1993, p.265) may well be right in stating that 'The content and intensity of public opinion have been more important than the structure of political institutions in determining cross-national and intertemporal differences in environmental policy'.

In an overall evaluation, it seems fair to conclude that model II enables us to penetrate *in greater depth* the processes of policy formulation and implementation. As a heuristic device it points us towards many of those factors that a 'political entrepreneur' might hope to be able to manipulate. It seems to give us a better grasp of the *multiple* mechanisms at work in this policy field. Even though we cannot in this case claim that it yields more accurate predictions at the rather general level of basic policy profiles, it seems that – everything else being constant – the more specific the aspect of behaviour that we want to predict or explain, the greater the marginal utility of moving beyond the narrow confines of the unitary, rational actor model (cf. Allison, 1971).

Model III (Social Learning and Policy Diffusion)

To recapitulate, our third model focuses on what might be called the *ideational* foundations of public policies. It depicts decision-makers as *information-seekers* and *persuaders*, and sees learning and policy diffusion as important mechanisms for the development of international regimes as well as domestic environmental policies. Moreover, it sees behaviour as at least to some extent guided by norms rather than self-interest. Transnational networks and international negotiations and organisations are seen as major arenas for the exchange of information and ideas.

We have dealt extensively with the informational functions of international organisations in a follow-up project (Hanf, Sprinz et al., 1997). Suffice it here to make a few important observations.

First, it is abundantly clear that knowledge and ideas played very important roles in the development of the LRTAP regime. Research and dissemination of more or less conclusive evidence of ecological damage were key elements in the Scandinavian campaign for international regulation of acid emissions. New scientific evidence – and even more so the *public discovery* of conclusions from scientific research – about damage to forests was one important factor behind the turn-around of FRG policy in the early 1980s. Less dramatically, we see evidence of research findings influencing beliefs and positions of governments as well as the general public in most of the other countries as well. Even the most outspoken and articulate sceptic, the Thatcher government, proved not to be entirely immune to the increasing convergence within the scientific community during the 1980s on new and more sombre estimates of the ecological impact of acidification.

Second, although information and ideas flow in multiple directions, we can see a fairly clear leader–follower pattern in the early and formative stages. The dominant pattern was one where others gradually adopted diagnostic conclusions and to some extent also policy ideas from Scandinavian and German sources. In particular, the publication of new German data about forest damage seems to had a significant impact in neighbouring countries. Our case studies provide strong evidence not only of learning (i.e. genuine adoption of new information and ideas) but also of what might be called policy 'imitation'. Thus, Lewanski et al. (1996) suggest that a wish to be seen as equally sophisticated and progressive as the presumably most 'advanced' countries (in particular, Germany and Sweden) was an important motive behind the decision of the Italian government to sign on to LRTAP protocols and declarations.

Third, transnational networks of experts, under the auspices of ECE, IIASA and others, served as important arenas for the development of consensual knowledge and the exchange of information and ideas. To some extent we can see what Haas (1990,1992) has called an 'epistemic community' developing, with ECE and IIASA as its organisational bases. There also seems to be a striking difference between the roles played by the two intergovernmental organisations involved. ECE served essentially as an *arena* for negotiations, exchange of information etc. In that capacity it played a very important role, particularly in the agenda-setting and negotiation stages. The EC entered the stage also as an *actor*, providing its own distinctive inputs into the internal decision-making processes and adding its institutional weight to the decisions produced. Several of the country studies report that EC directives and regulations have had considerable more 'teeth' and therefore a greater impact on domestic policies than ECE protocols. The combination of its greater institutional weight and its ability to reward and support as well as to punish gave the EC a major advantage in terms of inducing real changes in the domestic policies and behaviour of member countries – particularly the laggards. One important implication is that much of the compliance with LRTAP regulations that we observe can in fact be attributed primarily to *EC* directives and regulations.[26]

Dynamics and Context

Two additional questions remain to be addressed. One pertains to the internal *dynamics* of regime formation and implementation processes. In chapter 3 we formulated two competing propositions (P_{IVa} and P_{IVb}). One was based on the neo-functionalist notion that processes of international co-operation tend to gather their own momentum, leading participants to go for increasingly ambitious projects. The other was inspired by the economists' 'law of diminishing returns', suggesting that after the first and 'easy' steps have been taken only less profitable projects will remain. As the cost/benefit calculus for further steps becomes less and less positive, the drive for new projects will weaken.

Interestingly, we find some support for both these propositions. In the regime *formation* stage we can see fairly strong process-generated mechanisms of positive feedback at work. Levy's (1993) notion of 'tote-board-diplomacy' nicely captures important aspects of these internal dynamics. These mechanisms seem to have led some governments (at least Norway in the case of NO_X and Italy in the cases of NO_X and VOCs) to sign on to more ambitious emission reduction schemes than available calculations of damage and abatement costs would suggest. The bandwagon effect seems to have been felt even by the Thatcher government; the somewhat malicious and not entirely justified label 'dirty man of Europe' stuck. As we move into the stage of *implementation*, however, abatement costs become increasingly visible and painful, increasing also the domestic political costs of achieving reduction targets. If we are right that these opposite internal mechanisms work at different stages of the process, a plausible hypothesis will be that in international regimes *second-generation regulations tend to be particularly vulnerable to what we have called 'vertical disintegration'* (i.e. discontinuity between policy goals and principles on the one hand and the thrust of actual 'micro-deeds' on the other, cf. Underdal, 1979). Second-generation regulations are those that are most likely to be formulated at a time when the process-generated mechanisms of positive momentum are peaking, and before negative feedback from rising compliance costs and implementation failures can make a real impact.

The other observation pertains to the importance of policy *context* and exogenous events. Perhaps the most striking conclusion of all is that most of the emissions reductions that we have observed for pollutants regulated by LRTAP protocols are in fact due to other causes – the most important of which are economic recession (primarily in the former communist countries), energy switches (induced by, inter alia, changing price differentials and by economic and industrial policies, particularly in France and the UK), and technological developments. This applies even if we look at the principal pushers. Thus a Swedish study concludes that only about 24 per cent of the reductions achieved in domestic SO_2 emissions can be attributed to deliberate pollution control policies. As we have seen in chapter 2, the entire LRTAP regime in fact owes its very existence to a 'fortunate' coincidence of Scandinavian concern with transfrontier air pollution and an initiative taken by the Soviet Union to develop East–West detente. And Lewanski tells us that the decisions by the Italian government to sign on to protocols it would face a hard time implementing were to a large extent based on broader foreign policy considerations and a wish to be counted among the sophisticated and advanced countries of Europe.

This is not to say that the LRTAP regime did not make a real difference. In all countries included in our study, with the possible exceptions of France and Spain, some behavioural changes can be attributed to the regime. The international negotiation processes about emission control measures left some imprint also in at least some of the countries that did not sign one or more of the protocols, including the UK. The main point we want to make is, however, that environmental regimes and policies can hardly be understood adequately as isolated and self-contained policy segments. They are often heavily influenced by other policy concerns and by exogenous events, and they sometimes interact or even fuse with other policies. A quote from du Monteuil and Skea's (1996, p.38) study of the French case makes this point very clear: 'The changes in policy that did occur were due to the state accepting environmental goals *as integral to France's modernisation programme*' (italics mine). More generally, we can conclude that it is precisely when we have a *confluence* of different concerns – generating *positive policy synergy* – or when fortunate exogenous events interfere, that the most remarkable achievements seem to be made (cf. Gehring, 1994). This is probably true for most policy areas, but the fact that environmental damage typically occurs as a side-effect of other human activities implies that environmental policy in particular tends to be interwoven – in complex and sometimes problematic ways – with policies designed to manage these activities. Studying acidification or climate change policies in isolation from energy and industrial policies does not make much sense.

This may be bad news for the analyst (particularly for those involved in modelling political processes) as well as for the practitioner – but it certainly contributes to making environmental politics a fascinating field.

Acknowledgements

I gratefully acknowledge useful comments to an earlier draft from other members of the project team, in particular Sonja Boehmer-Christiansen, Kenneth Hanf, Rodolpho Lewanski, Larry O'Toole and Detlef Sprinz. I am equally grateful to Lynn P. Nygaard for her assistance in editing this final version.

Notes

1. The Netherlands is included in the comparative analysis as a tenth case.
2. A more detailed formulation of these propositions can be found in chapter 3, p.49f.
3. In table 13.1 we have included only the most relevant 'prototypical' profiles. Moreover, we have attached equal weight to all three dimensions. In some of our case studies we find interesting observations suggesting that in a more sophisticated analysis it might have been worthwhile to try to differentiate their impact. Thus, Hiienkoski suggests that the net import of pollutants was the principal determinant of Finnish activism in favour of *international* emission control regulations, while the damage cost variable seems to explain most of the variance observed in the level of ambitions for *domestic* environmental policy. For an analysis focusing on the pollution exchange balance, see Gehring (1994, pp. 327-334).
4. For a technically more sophisticated analysis examining the impact of single components, see Sprinz (1992).
5. Both are taken from Sprinz (1992, pp.153–154). Both indicators have limited validity, however. The former implicitly assumes that all geographical areas are equally important to a nation's welfare and that environmental policy will be based essentially on concern for the most damaged areas. The other focuses on only two subsets of damage costs, and although these are both important we know that local impact on human health was a more prominent concern in some countries (e.g. Hungary and Spain). In practical terms, the major discrepancies occur with respect to Finland (low on excess of critical loads, high on damage to forests and waterways) and the Netherlands (high on the former, low on the latter indicator).
6. Countries are ranked in *a*scending order with regard to abatement costs, and in *de*scending order with regard to damage costs and import–export of pollutants. The lower the aggregate score, the more 'progressive' should be negotiating positions and – with one reservation to be explored below – the better compliance records.
7. In principle, it is conceivable that *all* stand to gain from a particular regulation, or that *all* will lose. In a situation characterised by substantial asymmetries in abatement and damage costs as well as in pollution exchange balances, it is, however, most unlikely that all states will win or that all will lose from a regulation prescribing a standard reduction in emissions.
8. The location of 'break-even points' may be highly sensitive not only to assumptions about costs and damages but also to the choice of baselines and deadlines. Moreover, there are good reasons to believe that governments rarely define exact 'resistance points'; what they typically have are more or less elastic zones of acceptance,

indifference, and refusal (see Underdal, 1992, pp.236f).

9. Compared to its own ambitious national target, however, even Switzerland seems to fail.

10. Also Italy and, for the NO_x protocol, Spain as well, would appear as likely defectors, but in their cases genuine over-commitment would appear to be a sufficient cause. (Moreover, Lewanski et al. point out that one of the behavioural changes that reduced SO_2 emissions in Italy (increasing use of methane) in fact had the side-effect of increasing the emissions of NO_x and VOCs.)

11. When damage costs are measured in terms of damage to forests and waterways the Spearman rank correlation drops from .87 to .33. When the critical loads indicator is used the correlation drops from .72 to .09.

12. In terms of model I this observation would be interpreted to suggest that incentives to cheat have been effectively curbed, not that such incentives were not inherent in the structure of the problem.

13. Lewanski attributes this in part to the influence over negotiating positions wielded by the Ministry of Foreign Affairs, pursuing general foreign policy objectives, in part to a predisposition on the part of Italian politicians to be focus more on the expressive and symbolic aspects of politics than on the more mundane day-to-day business of implementation.

14. Lewanski concludes that such calculations played a minor role in the Italian case.

15. In a follow-up project, Helm and Sprinz (1995) have introduced the concept of 'political damage costs' as an analytical tool designed to capture the calculations that a government would make of the (domestic) political costs involved in choosing between alternative policies.

16. 'A vigorous British response to acid rain might have been conceivable if cherished environmental assets, particularly those located in Southern England, had been threatened' (Boehmer-Christiansen and Skea, 1991, p.277).

17. Note that the scores used compare each case to what seems to be the *average* of our set of cases. Labels like 'high' or 'low' should not be read as referring to absolute or cardinal scale values. Note that in some cases the difference between two or more countries is so small that differentiation of scores cannot be made with confidence. In such cases a column may tilt towards a score that deviates from the arithmetic average.

18. This is why we have rated the British Green party as *weak* even though its electoral support – it polled 15 per cent of the vote (but won no seats) at the 1989 elections to the European Parliament – could qualify for a higher rating.

19. This applies irrespective of whether the victims are found in particular sectors of the economy, geographical regions or social groups. Since the general concern with environmental values seems to be strongest among well-educated, (upper) middle-class people working in the tertiary sector, one may hypothesise that the most 'progressive' environmental policies are likely to be pursued by political parties relying heavily on the support of this particular social group.

20. Thus, emission reductions under a Labour government may, at least in the short run, have been less than under the Tory government, but such a difference could not have been attributed to differences in the overall concern with environmental values.

21. Note, though, that access and influence are not synonymous concepts.

22. In Spain, less than one fifth of total public expenditures on environmental measures were controlled by the national government, while nearly 60 per cent were in the hands of local government.
23. There is some evidence of this in Switzerland, too. Interestingly, in at least the Italian and Spanish cases, there is a marked difference between EU directives and regulations on the one hand and ordinary intergovernmental agreements on the other. It seems that the former enter the system through more well-established channels and trigger standard operating procedures familiar at the regional and local as well as at the national level. In addition, EU directives and regulations have a different legal status and some of them are supported by financial incentives as well.
24. The German support for modernisation of Hungarian plants seems at least to some extent to combine public virtue with private profit, more specifically with the cultivation of potentially important markets for their own products (O'Toole et al., 1998). In political terms this configuration seems particularly potent. For a more general examination of industrial interests in abatement, see Sprinz (1992, chapter 6).
25. Hungary under communist rule is somewhat of a deviant case in that societal demand could make itself heard only within very narrow confines. Environmental policy was therefore very much a question of governmental priorities. Moreover, one might even ask to what extent Hungary under communism can be seen as an autonomous actor. As specified in chapter 3, the domestic politics model seems generally less applicable and relevant for authoritarian and totalitarian political systems.
26. The latter were, though, to some extent influenced by the former.

References

In some cases, this chapter refers to unpublished and more extensive earlier drafts of the case-studies included in this volume. These earlier drafts are not listed below.

Aardal, B. (1993), *Energi og miljø. Nye stridsspørsmål i møte med gamle strukturer*, Oslo: Institutt for samfunnsforskning.
Allison, G.T. (1971), *Essence of Decision*, New York: Little, Brown & Co.
Amann, M. (1989), *Potential and Costs for Control of NO_X Emissions in Europe*, Laxenburg: IIASA.
Amann, M. and Kornai, G. (1987), *Cost Functions for Controlling SO_2 Emissions in Europe*, Laxenburg: IIASA.
Boehmer-Christiansen, S. and Skea, J. (1991) *Acid Politics*, London: Belhaven Press.
Gehring, T. (1994), *Dynamic International Regimes*, Frankfurt Am Main: Peter Lang.
Haas, P.M. (1990), *Saving the Mediterranean*, New York: Columbia University Press.
Haas, P.M. (1992), 'Introduction: Epistemic Communities and International Policy Coordination', *International Organization*, vol. 46, pp.1-35.
Hanf, K., Sprinz, D. et al., (1997), *International Governmental Organizations and National*

Participation in Environmental Regimes: The Organizational Components of the Acidification Regime, Rotterdam: Erasmus University, Final Report to the European Commission.

Helm, C. and Sprinz, D. (1995), *The Effectiveness of International Environmental Regimes: Disentangling Domestic From International Factors*, Unpublished paper, Potsdam Institute for Research on Climate Change Impact.

Hofrichter, J. and Reif, K. (1990), 'Evolution of Environmental Attitudes in the European Community', *Scandinavian Political Studies*, vol. 13, pp.119-146.

Jänicke. M. (1997), 'The Political System's Capacity for Environmental Policy', in M. Jänicke and H. Weidner (eds.), *National Environmental Policies: A Comparative Study of Capacity-Building*, Berlin: Springer.

Jänicke, M. and Weidner, H. (1997), 'Germany', in M. Jänicke and H. Weidner (eds.), *National Environmental Policies: A Comparative Study of Capacity-Building*, Berlin: Springer.

Levy, M.A. (1993), 'European Acid Rain: The Power of Tote-Board Diplomacy', in P.M. Haas, R.O. Keohane and M.A. Levy (eds.), *Institutions for the Earth*, Cambridge, MA; the MIT Press.

Mäler, K.G. (1989), *The Acid Rain Game*, Unpublished paper prepared for the ESF Workshop 'Economic Analysis and Environmental Toxicology', Amsterdam, May 1989.

Moravcsik, A. (1993), 'Introduction: Integrating International and Domestic Theories of International Bargaining', in P.B. Evans, H.K. Jacobson and R.D. Putnam (eds.), *Double-Edged Diplomacy*, Berkeley, CA: University of California Press.

O'Toole. L.J. jr. (1998), *Institutions, Policy and Outputs for Acidification: The Case of Hungary*, Aldershot: Ashgate.

Schelling, T.C. (1967), 'What is Game Theory?', in James C. Charlesworth (ed.), *Contemporary Political Analysis*, New York: The Free Press.

Sprinz, D. (1992). *Why Countries Support International Environmental Agreements: The Regulation of Acid Rain in Europe*, Ann Arbor: University of Michigan, unpublished Ph.D.-dissertation.

Underdal, A. (1979), 'Issues Determine Politics Determine Policies', *Cooperation and Conflict*, vol. 14, pp. 1-9.

Underdal, A. (1992), 'Designing Politically Feasible Solutions', in Raino Malnes and Arild Underdal (eds.), *Rationality and Institutions*, Oslo: Scandinavian University Press.

Vogel, D. (1993), 'Representing Diffuse Interests in Environmental Policymaking', in R.K. Weaver and B.A. Rockman (eds.), *Do Institutions Matter?*, Washington, D.C.: The Brookings Institution.

Weale, A. (1992), *The New Politics of Pollution*, Manchester: Manchester University Press.